普.通.高.等.学.校
计算机教育"十二五"规划教材

C# 面向对象
程序设计
（第 2 版）

OOP WITH C# (2nd edition)

郑宇军 ◆ 编著

人民邮电出版社
北京

图书在版编目（CIP）数据

C#面向对象程序设计 / 郑宇军编著. -- 2版. -- 北京：人民邮电出版社，2013.7（2022.7重印）
普通高等学校计算机教育"十二五"规划教材
ISBN 978-7-115-29761-7

Ⅰ. ①C… Ⅱ. ①郑… Ⅲ. ①C语言－程序设计－高等学校－教材 Ⅳ. ①TP312

中国版本图书馆CIP数据核字(2012)第284588号

内 容 提 要

本书以面向对象的软件工程思想为主线，细致深入地讲解了 C#语言面向对象程序设计的方法和技巧，内容涵盖面向对象的基本概念、基于接口的设计、泛型程序设计方法、Windows 和 WPF 窗体界面、文件和数据库访问，以及 ASP.NET 和 Silverlight 网站设计，并通过一个贯穿全书的"旅行社管理系统"案例展现了如何运用面向对象技术和 C#语言来进行实际软件系统开发。全书提供了丰富的示例代码和课后习题。

本书适合作为高等院校计算机及相关专业的教材，也可供专业开发人员自学参考。示例源代码和教学课件可在人民邮电出版社教学服务与资源网（http://www.ptpedu.com.cn）上下载。

◆ 编　著　郑宇军
　责任编辑　刘　博
　责任印制　彭志环　杨林杰

◆ 人民邮电出版社出版发行　北京市丰台区成寿寺路 11 号
　邮编　100164　电子邮件　315@ptpress.com.cn
　网址　http://www.ptpress.com.cn
　北京七彩京通数码快印有限公司印刷

◆ 开本：787×1092　1/16
　印张：25.25　　　　　2013 年 7 月第 2 版
　字数：697 千字　　　2022 年 7 月北京第 17 次印刷

定价：49.80 元

读者服务热线：(010)81055256　印装质量热线：(010)81055316
反盗版热线：(010)81055315
广告经营许可证：京东市监广登字20170147号

第 2 版前言

自 2009 年 6 月本书第 1 版出版以来，颇受广大读者的欢迎，也有幸被许多老师选为程序设计专业教材。经过 3 年多的时间，编者结合教学实践和软件开发中的经验体会、C#语言的最新升级以及许多读者热情的反馈建议，对原书进行了系统的修订。

这次修订保留了原教材特点，坚持以面向对象的软件工程思想为主线、紧密贴合实际应用需求；同时增加了一些新的知识点，加强了教材内容的新颖性和趣味性，以便进一步提升教学效果。修改的主要方面包括：

（1）以 C# 4.0/4.5 版本为主进行讲解（但语言的核心要素并没有根本变化，新版本的优势主要体现在应用的多样性和开发的方便性上）。

（2）在第 6 章中增加了对 Lambda 表达式的介绍。

（3）将原来的第 7 章和第 13 章压缩合并在一起，在新的第 13 章中对 WPF（Windows Presentation Foundation）进行了较为细致的讲解。

（4）在第 15 章中新增了"LINQ 对象数据查询"一节，使读者能够对 LINQ 数据访问技术有一个初步的了解。

（5）新增了第 16 章"Silverlight 客户端应用程序"，对 Silverlight 富客户端开发技术进行了深入讲解。

本书配套的教学课件和案例程序代码也进行了相应的升级，有关内容可在人民邮电出版社网站上进行下载。

本书第 1 章由王侃执笔，第 2 章和第 15 章由杨军伟执笔，第 3 章～第 14 章和第 16 章由郑宇军执笔，吴晓蓓、凌海风、江勋林、郑艳华、宋琴等也参与了本书的部分文字编写工作。全书由郑宇军统稿。

此次编写工作得到了我校王卫红老师的悉心指导和帮助。在 C#程序设计课程教学实践中，与简珍峰老师和王松老师的经验交流使编者受益良多。同时，感谢我校计算机学院 09 级的徐静、张璐滢、何俊丽、王岳春、顾唯超、诸伊娜等同学，他们在学习开发过程中的创造性给了编者很多启发。

尽管我们做了最大的努力，书中的不足与疏漏之处仍在所难免，恳请广大读者批评指正。我们的 E-mail 地址是：bookzheng@yeah.net（邮件主题请注明"CSharp 程序设计"）。

编者
2012 年 8 月
于浙江工业大学　郁文楼

目 录

第1章 面向对象程序设计概述 ... 1
1.1 计算机程序设计语言 ... 1
1.2 面向对象的基本概念 ... 2
1.2.1 对象 ... 2
1.2.2 类 ... 2
1.2.3 消息和通信 ... 2
1.2.4 关系 ... 3
1.2.5 继承 ... 3
1.2.6 多态性 ... 4
1.2.7 接口和组件 ... 4
1.3 面向对象的开发方法 ... 5
1.3.1 面向对象的分析 ... 5
1.3.2 面向对象的设计 ... 5
1.4 案例研究——旅行社管理系统的分析与设计 ... 6
1.5 小结 ... 8
1.6 习题 ... 8

第2章 C#和 Visual Studio 开发环境基础 ... 9
2.1 C#语言和.NET 技术简介 ... 9
2.2 C#程序的基本结构 ... 10
2.2.1 注释 ... 10
2.2.2 命名空间 ... 11
2.2.3 类型及其成员 ... 11
2.2.4 程序主方法 ... 12
2.2.5 程序集 ... 12
2.3 Visual Studio 开发环境 ... 13
2.3.1 集成开发环境概述 ... 13
2.3.2 创建控制台应用程序 ... 14
2.3.3 创建和使用动态链接库程序 ... 15
2.3.4 创建 Windows 应用程序 ... 15
2.3.5 创建 ASP .NET 应用程序 ... 16
2.4 小结 ... 18
2.5 习题 ... 18

第3章 C#语法基础 ... 19
3.1 数据类型 ... 19
3.1.1 简单值类型 ... 19
3.1.2 复合值类型 ... 21
3.1.3 类 ... 23
3.1.4 数组 ... 25
3.1.5 类型转换 ... 27
3.2 操作符和表达式 ... 30
3.2.1 算术操作符 ... 30
3.2.2 自增和自减操作符 ... 31
3.2.3 位操作符 ... 31
3.2.4 赋值操作符 ... 32
3.2.5 关系操作符 ... 33
3.2.6 逻辑操作符 ... 33
3.2.7 条件操作符 ... 34
3.3 控制结构 ... 35
3.3.1 选择结构 ... 35
3.3.2 循环结构 ... 38
3.3.3 跳转结构 ... 42
3.4 案例研究——旅行社管理系统中结构和枚举 ... 44
3.5 小结 ... 46
3.6 习题 ... 46

第4章 类和对象 ... 47
4.1 成员概述 ... 47
4.1.1 成员种类 ... 47
4.1.2 成员访问限制 ... 48
4.1.3 静态成员和非静态成员 ... 49
4.1.4 常量字段和只读字段 ... 50
4.2 方法 ... 51
4.2.1 方法的返回值 ... 52
4.2.2 参数类型 ... 52

4.2.3	方法的重载	55
4.3	类的特殊方法	56
4.3.1	构造函数和析构函数	56
4.3.2	属性	59
4.3.3	索引函数	61
4.3.4	操作符重载	62
4.4	this 对象引用	65
4.5	常用类型	65
4.5.1	Object 类	65
4.5.2	String 类	66
4.5.3	StringBuilder 类	72
4.5.4	Math 类	72
4.5.5	DateTime 结构	73
4.6	案例研究——旅行社业务类的实现	74
4.6.1	省份、城市和景点类	74
4.6.2	旅游线路和方案类	76
4.6.3	旅行团和游客类	78
4.7	小结	81
4.8	习题	81

第 5 章　继承和多态 …… 82

5.1	继承	82
5.1.1	基类和派生类	82
5.1.2	隐藏基类成员	84
5.1.3	base 关键字	86
5.1.4	对象的生命周期	87
5.2	多态性	89
5.2.1	虚拟方法和重载方法	89
5.2.2	抽象类和抽象方法	92
5.2.3	密封类和密封方法	94
5.3	案例研究——旅行社业务类的实现和精化	97
5.3.1	会员类	97
5.3.2	职员类	98
5.4	小结	103
5.5	习题	103

第 6 章　委托和事件 …… 105

6.1	委托和方法	105
6.1.1	通过委托来封装方法	105
6.1.2	委托的加减运算	107
6.1.3	传递委托对象	107
6.1.4	Delegate 类型成员	109
6.2	匿名方法和 Lambda 表达式	110
6.2.1	匿名方法	110
6.2.2	Lambda 表达式	111
6.2.3	外部变量	111
6.3	事件处理	112
6.3.1	委托发布和订阅	112
6.3.2	事件发布和订阅	114
6.3.3	使用 EventHandler 类	117
6.3.4	在事件中使用匿名方法	118
6.4	Windows 控件事件概述	120
6.5	案例研究——旅行团基本事件处理	122
6.5.1	旅行团事件发布	122
6.5.2	旅行团事件处理	123
6.6	小结	126
6.7	习题	126

第 7 章　Windows Form 应用程序设计 …… 127

7.1	图形用户界面概述	127
7.2	位置、坐标、颜色和字体	128
7.2.1	Size 和 SizeF 结构	128
7.2.2	Point 和 PointF 结构	128
7.2.3	Color 结构	129
7.2.4	Font 和 FontFamily 类	129
7.3	窗体、消息框和对话框	130
7.3.1	窗体	130
7.3.2	消息框	132
7.3.3	对话框	134
7.4	常用 Windows 控件	135
7.4.1	Control 类	135
7.4.2	标签、文本框和数值框	137
7.4.3	按钮、复选框和单选框	139
7.4.4	组合框和列表框	141
7.4.5	日历控件	143
7.4.6	滑块、进度条和滚动条	144
7.4.7	图片框控件	145
7.4.8	容器控件	146

7.4.9	列表视图和树型视图	147
7.5	菜单栏、工具栏和状态栏	151
	7.5.1 菜单栏	151
	7.5.2 工具栏	152
	7.5.3 状态栏	153
7.6	案例研究——旅行社信息窗体和登录窗体	154
	7.6.1 旅行社对象及其信息窗体	154
	7.6.2 系统用户及登录窗体	156
7.7	小结	158
7.8	习题	158

第8章 对象持久性——文件管理 … 159

- 8.1 文件和流 … 159
  - 8.1.1 File 类 … 159
  - 8.1.2 使用文件流 … 161
  - 8.1.3 FileInfo 类 … 163
- 8.2 流的读写器 … 164
  - 8.2.1 二进制读写器 … 164
  - 8.2.2 文本读写器 … 165
- 8.3 文件对话框 … 168
- 8.4 基于文件的对象持久性 … 170
  - 8.4.1 实现对象持久性 … 170
  - 8.4.2 .NET 中的自动持久性支持 … 172
- 8.5 案例研究——旅行社信息和系统用户的持久性 … 177
  - 8.5.1 旅行社对象的持久性 … 177
  - 8.5.2 系统用户对象的持久性 … 177
- 8.6 小结 … 180
- 8.7 习题 … 180

第9章 异常处理 … 181

- 9.1 异常的基本概念 … 181
- 9.2 异常处理结构 … 183
  - 9.2.1 try-catch 结构 … 183
  - 9.2.2 try-catch-finally 结构 … 184
  - 9.2.3 try-finally 结构 … 186
- 9.3 异常的捕获和传播 … 187
  - 9.3.1 传播过程 … 187
  - 9.3.2 Exception 和异常信息 … 188
  - 9.3.3 异常层次结构 … 190
- 9.4 自定义异常 … 192
  - 9.4.1 主动引发异常 … 192
  - 9.4.2 自定义异常类型 … 193
- 9.5 使用异常的指导原则 … 196
- 9.6 案例研究——旅行社管理系统中的异常处理 … 197
  - 9.6.1 文件 I/O 异常处理 … 198
  - 9.6.2 旅行社业务异常 … 199
- 9.7 小结 … 201
- 9.8 习题 … 201

第10章 基于接口的程序设计 … 202

- 10.1 接口的定义和使用 … 202
  - 10.1.1 接口的定义 … 202
  - 10.1.2 接口的实现 … 203
- 10.2 接口与多态 … 204
  - 10.2.1 通过接口实现多态性 … 204
  - 10.2.2 区分接口方法和对象方法 … 206
- 10.3 接口和多继承 … 208
  - 10.3.1 多继承概述 … 208
  - 10.3.2 基于接口的多继承 … 209
  - 10.3.3 解决二义性 … 213
- 10.4 接口与集合 … 216
  - 10.4.1 集合型接口及其实现 … 216
  - 10.4.2 列表、队列和堆栈 … 217
  - 10.4.3 自定义集合类型 … 219
- 10.5 案例研究——旅行社管理系统中的集合类型 … 221
  - 10.5.1 职员列表与数据绑定 … 221
  - 10.5.2 使用自定义集合 … 224
- 10.6 小结 … 229
- 10.7 习题 … 230

第11章 泛型程序设计 … 231

- 11.1 为什么要使用泛型 … 231
- 11.2 泛型类 … 232
  - 11.2.1 泛型类的定义和使用 … 232
  - 11.2.2 使用"抽象型"变量 … 234
  - 11.2.3 使用多个类型参数 … 235

11.2.4 类型参数与标识 ················235
11.2.5 泛型的静态成员 ··············237
11.3 类型限制 ································239
11.3.1 主要限制 ························239
11.3.2 次要限制 ························239
11.3.3 构造函数限制 ··················240
11.4 泛型继承 ································240
11.5 泛型接口 ································243
11.5.1 泛型接口的定义 ··············243
11.5.2 泛型接口的实现 ··············244
11.5.3 避免二义性 ······················247
11.5.4 泛型接口与泛型集合 ······248
11.6 泛型方法 ································252
11.6.1 泛型方法的定义和使用 ··252
11.6.2 泛型方法的重载 ··············254
11.6.3 泛型方法与委托 ··············254
11.7 案例研究——旅行社管理
系统中的泛型集合 ··················256
11.7.1 使用泛型列表 List<T> ····256
11.7.2 泛型优先级队列 ··············258
11.8 小结 ··259
11.9 习题 ··260

第 12 章　C#中的泛型模式：
　　　　　可空类型和迭代器 ···········261

12.1 可空类型 ································261
12.1.1 可空类型：值类型+null ···261
12.1.2 可空类型转换 ··················266
12.1.3 操作符提升 ······················266
12.2 遍历和迭代 ····························267
12.2.1 可遍历类型和接口 ··········267
12.2.2 迭代器 ······························270
12.2.3 迭代器代码 ······················273
12.2.4 使用多个迭代器 ··············274
12.2.5 自我迭代 ··························276
12.3 案例研究——旅行社管理系统中的
可空值与迭代器 ······················279
12.3.1 旅行社业务对象中的可空值 ···279
12.3.2 遍历游客集合 ··················280
12.4 小结 ··281

12.5 习题 ··281

第 13 章　WPF 应用程序设计 ········282

13.1 WPF 窗体和控件 ····················282
13.1.1 创建一个 WPF 程序 ········282
13.1.2 窗体和布局 ······················284
13.1.3 控件内容模型 ··················286
13.1.4 文本框控件 ······················290
13.1.5 范围控件 ··························291
13.2 使用 XAML 设计界面 ············292
13.2.1 XAML 文档和元素 ··········292
13.2.2 元素属性和事件 ··············293
13.2.3 资源和样式 ······················295
13.3 绘制图形 ································298
13.3.1 画刷 ··································298
13.3.2 形状 ··································300
13.3.3 图形变换 ··························303
13.3.4 打印输出 ··························304
13.4 动画和多媒体 ························305
13.4.1 基于属性的动画 ··············305
13.4.2 故事板和事件触发器 ······307
13.4.3 基于路径的动画 ··············309
13.4.4 播放多媒体文件 ··············310
13.5 案例研究——旅行社管理
系统的 WPF 界面 ····················312
13.5.1 构建系统主界面 ··············312
13.5.2 新建、修改和删除业务对象 ···314
13.5.3 信息打印输出 ··················316
13.5.4 Windows Form 集成 ········317
13.6 小结 ··318
13.7 习题 ··318

第 14 章　C# Web 应用程序设计 ····319

14.1 ASP .NET 技术概述 ···············319
14.2 ASP .NET Web 窗体和基本对象 ···320
14.2.1 Web 窗体 ··························320
14.2.2 请求和响应 ······················321
14.2.3 服务器对象 ······················324
14.2.4 应用程序、会话、视图和缓存 ···325
14.3 HTML 控件 ·····························327

14.3.1 从HTML元素到HTML控件 ……327
14.3.2 HtmlControl类型 ……328
14.3.3 HtmlAnchor、HtmlTextArea和HtmlSelect控件 ……329
14.3.4 HtmlTable控件 ……331
14.3.5 HtmlInputControl控件 ……333
14.4 Web服务器控件 ……335
14.4.1 标准窗体控件 ……335
14.4.2 验证控件 ……340
14.5 案例研究——旅游信息查询网站 ……341
14.5.1 网站母版页 ……341
14.5.2 网站首页与线路浏览 ……343
14.5.3 旅行团方案页面 ……346
14.5.4 景点信息页面 ……347
14.6 小结 ……348
14.7 习题 ……348

第15章 对象持久性——数据库存取和LINQ查询 ……349

15.1 关系数据库概述 ……349
15.1.1 关系表和对象 ……349
15.1.2 关系数据库语言SQL ……351
15.2 ADO .NET数据访问模型 ……354
15.2.1 非连接类型 ……354
15.2.2 连接类型 ……358
15.3 LINQ对象数据查询 ……362
15.4 案例研究——旅行社管理系统的数据库解决方案 ……366
15.4.1 数据表格设计 ……366
15.4.2 数据库连接管理 ……367
15.4.3 实现业务对象的数据库存取 ……368
15.4.4 终端数据访问 ……372
15.5 小结 ……374
15.6 习题 ……374

第16章 Silverlight客户端应用程序 ……375

16.1 Silverlight应用开发基础 ……375
16.2 Silverlight程序架构 ……376
16.3 处理键盘和鼠标事件 ……379
16.3.1 处理键盘事件 ……379
16.3.2 处理鼠标事件 ……380
16.4 模板和自定义控件 ……381
16.4.1 使用控件模板 ……381
16.4.2 创建自定义控件 ……383
16.5 案例研究——使用必应地图服务 ……386
16.5.1 开发前的准备工作 ……386
16.5.2 创建程序并添加必应地图控件 ……387
16.5.3 地图、图层与图片系统 ……388
16.5.4 旅游景点地图导航 ……390
16.6 小结 ……394
16.7 习题 ……394

第1章
面向对象程序设计概述

本章简要地回顾了计算机程序设计语言和软件开发方法的发展历程，并由此引出了面向对象的基本思想、概念和方法。掌握这些基本知识能为深入学习 C#面向对象程序设计打下良好的基础。

1.1 计算机程序设计语言

人类使用自然语言，而计算机执行的是机器指令；程序设计语言是人与计算机之间进行交流的工具，它定义了一套代码规则，程序设计人员遵循这些规则所编写出来的程序可被翻译成计算机能够"理解"的形式。

程序设计语言可以分为低级语言和高级语言。低级语言包括机器语言和汇编语言，使用它们进行编程需要对机器结构有深入的了解，而且代码晦涩难懂、不利于人们的理解和交流。高级语言则更加接近自然语言，比较符合人们的思维方式，因此大大提高了程序设计的效率，并使得人们通过"阅读程序文本"来理解"计算过程"成为可能。高级语言程序在计算机上有两种处理方式：一是由专门的解释程序来直接解释执行高级语言代码，二是由专门的编译程序将其翻译为低级语言代码而后执行。目前在程序设计的各个主要领域，高级语言已基本上取代了低级语言。

Fortran 语言是第一个被广泛使用的高级语言，其程序由一个主程序和若干个子程序组成，通过将不同的功能分配到独立的子程序中，能够有效地实现程序的模块化。20 世纪七八十年代非常流行的 Pascal 语言提供了丰富的数据类型和强有力的控制结构，使用它能够方便地编写结构化的应用程序，其程序结构中的一个模块就是一个过程，因此也被称为面向过程的语言。当然，最为流行的结构化程序设计语言莫过于 C 语言，它兼顾了诸多高级语言的特点，同时还提供了指针和地址等低级操作的能力，因此既适合于开发应用程序，又适合于开发系统程序，此外它还有良好的可移植性，成为程序设计语言诞生以来最为成功的范例之一。

简而言之，结构化程序设计采用自顶向下、分而治之的方法，对目标系统进行功能抽象和逐步分解，直至每个功能模块都能以一个过程或函数来实现为止。这样就将复杂系统划分为一系列易于控制和处理的软件模块，其特点是结构良好，条理清晰，功能明确。对于需求稳定、算法密集型的领域（如科学计算领域），上述方法是有效并适用的。

随着信息技术的飞速发展，计算机软件从单纯的科学和工程计算渗透到社会生活的方方面面，软件的规模也越来越大，复杂性提高，此时结构化方法逐步暴露出诸多问题和缺陷，这主要体现在：

- 功能与数据相分离，使用的建模概念不能直接映射到问题领域中的对象，不符合人们对现实世界的认识和思维方式。
- 自顶向下的设计方法，限制了软件模块的可复用性，降低了开发效率。
- 当系统需求发生变化时，系统结构往往需要大幅调整，特别是对于较复杂的系统，维护和扩展都变得非常困难。

为了解决上述问题，一种全新的、强有力的软件开发方法——面向对象（Object Oriented，OO）的方法应运而生。它是在结构化等传统方法的基础上发展起来的，也使用抽象和模块化等概念；但和传统方法相比，其根本性的变化在于：不再将软件系统看成是工作在数据上的一系列过程或函数的集合，而是一系列相互协作而又彼此独立的对象的集合。这不仅更为符合人们的思维习惯，有助于保持问题空间和解空间在结构上的一致性，同时能够有效控制程序的复杂性，提高软件的生产效率。近二十年来，面向对象的方法学得到了迅速发展和广泛应用，Java、C++、C#等面向对象的程序设计语言也成为了当今世界上计算机软件的主流开发语言。

1.2 面向对象的基本概念

本节介绍面向对象技术中一些最为基本的概念。

1.2.1 对象

客观世界中的事物都是对象（Object），这既包括有形的物理对象（如一个人、一只狗、一本书等），也包括抽象的逻辑对象（如一个几何图形、一项商业计划等）。对象一般都有自己的属性（Attribute），而且能够执行特定的操作（Operation）。例如，一个人可以描述为"姓名张三，身高170，体重65"，这里的"姓名"、"身高"、"体重"就是对象的属性，而"张三"、"170"、"65"则是对应的属性值；该对象还可以执行"走路"、"看书"等操作。属性用于描述对象的静态特征，而操作用于描述对象的动态特征。

在面向对象的模型中，软件对象就是对客观世界中对象的抽象描述，是构成软件系统的基本单位。但软件对象不应也不可能描述现实对象的全部信息，而只应包含那些与问题域有关的属性和操作。例如，在一个学籍管理系统中，通常会关心每个"学生"对象的"姓名"、"学号"、"专业"等属性信息，而他们的"发型"、"鞋号"等信息则不属考虑范围。

1.2.2 类

类（Class）是指具有相同属性和操作的一组对象的集合，它描述的不是单个对象而是"一类"对象的共同特征。例如在学籍管理系统中可以定义"学生"类，而"张三"、"李明"、"王娟"这些学生就是属于该类的对象，或者叫做类的实例（Instance）；它们都具有该类的属性和操作，但每个对象的属性值可以各不相同。

类是面向对象技术中最重要的结构，它支持信息隐藏和封装，进而支持对抽象数据类型（Abstract Data Type，ADT）的实现。信息隐藏是指对象的私有信息不能由外界直接访问，而只能通过该对象公开的操作来间接访问，这有助于提高程序的可靠性和安全性。例如"学生"类可以有"生日"和"年龄"这两个属性，那么可以把它们都定义为私有的，不允许直接修改；再定义一个根据生日计算年龄的私有操作，以及一个修改生日的公共操作，这样用户就只能通过对象的生日来间接修改其年龄，从而保证其年龄和生日的合法性。

类将数据和数据上的操作封装为一个有机的整体，类的用户只关心其提供的服务，而不必了解其内部实现细节。例如对于"借书证"类的"刷卡"操作，用户可以只关注该操作返回的借书人信息（如借书权限、是否欠费等），而不去管磁条中是怎样存储借书人的有关信息的。

1.2.3 消息和通信

对象具有自治性和独立性，它们之间通过消息（Message）进行通信，这也是对客观世界的形象模拟。发送消息的对象叫做客户（Client），而接收消息的对象叫做服务器（Server）。按照封装原则，对象总是通过公开其某些操作来向外界提供服务；如果某客户要请求其服务，那么就需要向服务器对象发送消息，而且消息的格式必须符合约定要求。消息中至少应指定要请求的服务（操作）名，必要时还应提供输入参数和输出参数。例如，某"学生"对象需要办理借书证，那么就要请求"图书馆"对象的"办理图书证"服务，并在消息中提供自己的姓名、学号等消息；"图书馆"对象检查这些消息合格后，创建一个新的"借书证"对象并返回给该学生。

1.2.4 关系

在很多情况下，单个对象是没有作用的。例如"轮子"、"车厢"、"发动机"等对象单独放在那里都没有什么意义，而只有将它们组成一个"汽车"对象才能发挥各自的作用。再如"学生"对象也不能是孤立的，而是要和"班级"、"老师"、"考试"等对象进行交互才能有意义。对象之间的关系可在类级别上进行概括描述，典型的有以下几种。

- 聚合（Aggregation）：一个对象是另一个对象的组成部分，也叫部分—整体关系，如"轮子"与"汽车"的关系，"学生"和"班级"的关系等。
- 依赖（Dependency）：一个对象对另一个对象存在依赖关系，且对后者的改变可能会影响到前者，如"借书证"对象依赖于某个"学生"对象，当"学生"对象不存在了（如该学生毕业或退学），相应的"借书证"对象也应被销毁。
- 泛化（Generalization）：一个对象的类型是另一个对象类型的特例，也叫特殊——一般关系，其中特殊类表示对一般类内涵的进一步细化，如"学生"类可进一步细化为特殊的"本科生"和"研究生"类。从泛化关系可以引出面向对象方法中另一个重要概念——继承。
- 一般关联（Association）：对象之间在物理或逻辑上更为一般的关联关系，主要是指一个对象使用另一个对象的服务，如"老师"和"学生"之间的教学关系。根据语义还可将关联关系分为多元关联和二元关联，二元关联还可进一步细分为一对一关联、一对多关联以及多对多关联等。聚合和依赖有时也被视为特殊的关联关系。

上述 4 种关系的简单示例如图 1-1 所示。

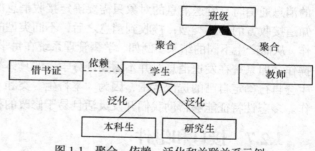

图 1-1　聚合、依赖、泛化和关联关系示例

1.2.5 继承

在泛化关系中，特殊类可自动具有一般类的属性和操作，这叫做继承（Inheritance）；而特殊类还可以定义自己的属性和操作，从而对一般类的功能进行扩充。例如"学生"类可以从"人"这个类中继承，这样就继承了"人"的"姓名"、"身高"等属性，而"学号"、"专业"等则是"学生"类自己特有的属性。在类的继承结构中，一般类也叫作基类或父类，特殊类也叫作派生类或子类。

继承的概念是从生物学中借鉴而来的，它可以具有多层结构。例如，动物可分为脊椎动物和无脊椎动物，脊椎动物又可分为哺乳动物、鱼类、鸟类、爬行动物、两栖动物等，这种划分可以持续很多层。在分类过程中，低级别的类型通常继承了高级别类型的基本特征。图 1-2 所示为这一简单的动物继承关系。不过，在实际软件系统的建模过程中，继承的层次结构不宜过细过深，否则会增加理解和修改的难度。

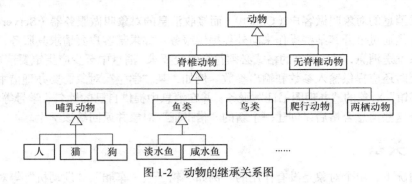

图1-2 动物的继承关系图

继承具有可传递性。例如"学生"类从"人"类继承,"研究生"类再从"学生"类继承,那么"本科生"和"研究生"类也就自动继承了"人"的"姓名"、"身高"等属性。这样派生类就能够享受其各级基类所提供的服务,从而实现高度的可复用性;当基类的某项功能发生变化时,对其的修改会自动反映到各个派生类中,这也提高了软件的可维护性。

自然界中还存在一种多继承的形式,例如,鸭嘴兽既有鸟类的特征又有哺乳动物的特征,那么可以把它看成是鸟类和哺乳动物共同的派生类;再如一名在职研究生可能同时具有老师和学生的身份。在面向对象的软件开发中,多继承具有较大的灵活性,但同时也会带来语义冲突、程序结构混乱等问题。目前,C++和Eiffel等语言支持多继承,而Java和C#则不支持,但它们可通过接口等技术来间接地实现多继承的功能。

1.2.6 多态性

多态性(Polymorphism)是指同一事物在不同的条件下可以表现出不同的形态。在面向对象的消息通信时,发送消息的对象只需要确定接收消息的对象能够执行指定的操作,而并不一定要知道接收方的具体类型;接收到消息之后,不同类型的对象可以作出不同的解释,执行不同的操作,从而产生不同的结果。例如,学籍管理系统在每学期开始时会要求每个学生对象执行"选课"操作,但系统在发送消息时并不需要区分学生的具体类型是本科生还是研究生,而不同类型的学生会自行确定自己的选课范围,因为"本科生"类和"研究生"类会各自定义不同的"选课"操作。多态性特征能够帮助我们开发灵活且易于修改的程序。

1.2.7 接口和组件

随着软件规模和复杂度的不断增长,现代软件开发越来越强调接口(Interface)和组件(Component)技术,而接口也已成为面向对象不可或缺的重要元素。组件是指可以单独开发、测试和部署的软件模块,接口则是指对组件服务的抽象描述。一个组件中可以只有一个类,也可以包含多个类。

接口是一种抽象数据类型,它所描述的是功能的"契约",而不考虑与实现有关的任何因素。例如,可以定义一个名为"图书借阅"接口,并规定其中包括"图书目录查询"、"借书"和"还书"这三项功能。一个类如果声明支持某接口,它就必须支持接口契约中规定的全部功能。例如"图书馆"类要声明支持"图书借阅"接口,它就至少要为"图书目录查询"、"借书"和"还书"这三项操作提供具体的实现机制。

接口一旦发布就不应再作修改,否则就会导致所有支持该接口的类型都变得无效。而组件一经发布,也不应取消它已声明支持的接口,而是只能增加新的接口。例如"图书借阅"接口发布后,我们不能简单地试图为该接口增加一项"图书复印"功能,否则很多已支持该接口的类型就会出错；更合理的方式是定义一个新的"图书复印"接口，那些具有复印能力的类型可以声明支持这个新接口，而不具备复印功能的类型则保持不变。

对于服务的使用方（客户）而言，它既不关心服务提供者的实际类型，也不关心实现服务的具体细节，而只需要根据接口去查询和使用服务即可。例如读者可以向任何一个声明支持"图书借阅"接口的对象发送图书查询请求，并在查到自己所需图书后发送借阅请求，而不必考虑服务的提供者究竟是图书馆、书店还是别的什么机构。接口将功能契约从实现中完全抽象出来，能够有效地实现系统职责分离，同时弥补继承和多态性的功能不足，进而实现良好的系统设计。

1.3 面向对象的开发方法

从20世纪80年代开始，面向对象技术得到了飞速的发展，业界也涌现了一系列面向对象的软件开发方法学，其中代表性的有Booch方法、Wirfs-Brock的责任驱动设计方法、Rambaugh的对象模型技术、Coad/Yourdon分析和设计方法、Jacobson的面向对象软件工程方法等。为了避免不同符号所引起的混乱，Booch、Rambaugh和Jacobson等共同参与制定了统一建模语言UML（Unified Modeling Language），采用统一的概念和符号来描述对象模型，支持软件开发的全过程。目前，UML已成为世界通用的面向对象的建模语言，适用于各种开发方法。

1.3.1 面向对象的分析

面向对象的分析（Object Oriented Analysis，OOA）就是运用面向对象的方法对目标系统进行分析和理解，找出描述问题域和系统责任所需要的对象，定义对象的基本框架（包括对象的属性、操作以及它们之间的关系），最后得到能够满足用户需求的系统分析模型。OOA主要有以下5项任务。

（1）识别问题域中的对象和类。通过对问题域和系统责任的深入分析，尽可能地找出与应用有关的对象和类，并从中筛选出真正有用的对象和类。

（2）确定结构。找出对象和类中存在的各种整体—部分结构和一般—特殊结构，并进一步确定这些结构组合而成的多重结构。

（3）确定主题。如果系统包含大量的对象和类，那么可划分出不同的应用主题域，并按照主题域对分析模型进行分解。

（4）定义属性。识别各个对象的属性，确定其名称、类型和限制，并在此基础上找出对象之间的实例连接。

（5）定义服务。识别各个对象所提供的服务，确定其名称、功能和使用约定，并在此基础上找出对象之间的消息连接。

OOA的结果是系统分析说明书，其中包括使用类图和对象图等描述的系统静态模型，使用例图、活动图和交互图等描述的系统动态模型，以及对象和类的规约描述。模型应尽量与问题域保持一致，而不考虑与目标系统实现有关的因素（如使用的编程语言、数据库平台和操作系统等）。

1.3.2 面向对象的设计

面向对象的设计（Object Oriented Design，OOD）是以系统分析模型为基础，运用面向对象的方法进行系统设计，解决与系统实现有关的一系列问题，最后得到符合具体实现条件的系统设计模型。OOD主要有以下4项任务。

（1）问题域设计。对问题域中的分析结果作进一步的细化、改进和增补，包括对模型中的对象和类、结构、属性、操作等进行组合和分解，并根据面向对象的设计原则增加必要的新元素类、属性和关系。这部分主要包括以下设计内容：
- 复用设计，即寻找可复用的类和设计模式，提高开发效率和质量。
- 考虑对象和类的共同特征，增加一般类以建立共同协议。
- 调整继承结构，如减少继承层次、将多继承转换为单继承、调整多态性等。
- 改进性能，如分解复杂类、合并通信频繁的类、减少并发类等。
- 调整关联关系，如将多元关联转换为二元关联、将多对多关联转换为一对多关联等。
- 调整和完善属性，包括确定属性的类型、初始值和可访问性等。
- 构造和优化算法。

（2）用户界面设计。对软件系统的用户进行分析，对用户界面的表现形式和交互方式进行设计。这部分主要包括以下设计内容：
- 用户分类。
- 人机交互场景描述。
- 系统命令的组织。
- 详细的输入和输出设计。
- 在需要时可增加用于人机交互的对象和类，并使用面向对象的方法对其进行设计。

（3）任务管理设计。当系统中存在多任务（进程）并发行为时，需要定义、选择和调整这些任务，从而简化系统的控制结构。这部分主要包括以下设计内容：
- 识别系统任务，包括事件驱动任务、时钟驱动任务、优先任务和关键任务等。
- 确定任务之间的通信机制。
- 任务的协调和控制。

（4）数据管理设计。识别系统需要存储的数据内容和结构，确定对这些数据的访问和管理方法。这部分主要包括以下设计内容：
- 数据的存储方式设计，目前主要有文件系统和数据库系统两种方式。
- 永久性类（Persistent Class）的存储设计，包括其用于存储管理的属性和操作设计。
- 永久性类之间关系的存储设计。

OOA 和 OOD 之间不强调严格的阶段划分，设计模型是对分析模型的逐步细化，主要是在问题域和系统责任的分析基础上解决各种与实现有关的问题。OOA 阶段一些不能确定的问题可以遗留到 OOD 阶段解决，开发过程中也允许存在反复和迭代。

1.4 案例研究——旅行社管理系统的分析与设计

旅行社管理系统是为旅行社提供业务支持的信息系统。旅行社设计旅游线路，并推出各种类型的旅行团。游客可查询旅游信息并报名参加旅行团，旅行社根据报名情况组织发团。通过简要的分析，可知系统的基本功能需求包括：

（1）旅行社设计旅游线路，每条旅游线路包含若干个旅游景点。
（2）旅行社基于旅游线路定制旅行团。
（3）游客上网查询旅游景点、线路和旅行团信息。
（4）游客报名参加旅行团，旅行社业务员负责接受或拒绝报名申请。
（5）满足条件后，旅行社为旅行团安排导游并发团。

在上述分析的基础上，可以找到一系列可能的对象，并将其抽象到不同的类。最为基本的类有以下几个：
- 景点类 Scene，具有景点名称、位置、种类、票价等属性。
- 旅游线路类 Line，具有线路名称、景点集合、用时等属性。
- 旅行团类 Tour，具有旅游线路、时间、等级、人数、价格、导游、游客等属性。
- 游客类 Customer：具有游客姓名、身份证号、性别、生日等属性，还可以执行报名和取消报名等操作。

景点所在的位置应通过所属的省市来描述，那么需要定义 Province 和 City 类，它们和 Scene 类之间是聚合的关系。为了加强客户关系管理，旅行社还积极发展游客成为注册会员。为此可定义 Customer 的派生类 Member，它在 Customer 的基础上增加了用户名、密码、会员积分等属性。

一个成熟的旅行社通常会有多个固定的组团方案，如每天一次的北京一日游、每周一次的四川三日游等。那么可以把这些方案抽象为一个 Package 类，由该类维护旅游线路、等级、价格等信息，这样基于同一方案的多个旅行团就可以共享这些信息，而每个旅行团对象只需要关心自己的时间、导游、游客等独立信息。图 1-3 所示为这些类之间的基本关联关系。

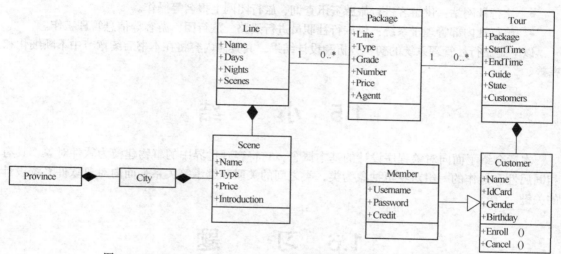

图 1-3　Scene、Line、Package、Tour、Customer 等主要业务类的关系图

系统还需要提供旅行社的内部管理功能，那么可定义旅行社类 TravelAgency，它具有名称、总经理、法人代表、通讯地址、电话等属性。为了对旅行社的职员进行管理，可将旅行社的所有职员抽象为一个 Staff 类，它具有姓名、身份证号、性别、年龄、学历等属性，不同类型的职员可定义为 Staff 的不同派生类，主要包括：
- 导游类 Guide，具有导游证号、导游证有效期等属性。
- 业务员类 Agent，负责处理游客报名、组团管理等操作。
- 主管类 Director，除处理旅行社业务外，还负责管理一般职员。
- 经理类 Manager，基本属性和 Director 相同，还可执行雇佣和解雇其他职员等操作。

另一些岗位（如前台、保安等）在系统中没有专门的功能，那么可作为一般 Staff 对待。系统所关心的职员类型的基本继承关系如图 1-4 所示。

上述类型都是旅行社管理系统的业务类，它们都不可避免地要使用到大量的通用数据类型，如整数、字符串、日期时间等，这大都可以在程序设计语言的类型系统中找到。系统所使用的用户界面中也会用到大量的控件类，如窗体、按钮、网页、图片等，这也主要由商业开发工具来提供。当然，随着开发过程的深入，还可能发现更多需要的类型，如酒店、机场、车站等，这里可将它们先记录为候选类。

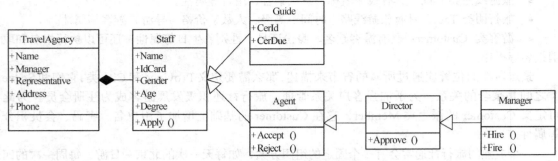

图 1-4　旅行社及其职员类的关系图

从设计角度出发，整个旅行社管理系统可划分为以下 3 部分。
- 业务类库：管理和维护上述基本业务类。
- 旅行社网站：供游客进行旅游资讯查询、旅行团网上报名等操作。
- 旅行社内部管理子系统：供旅行社职员进行线路、旅行团、游客等信息管理工作。

以上是旅行社管理系统的初步分析和设计结果，其具体内容将在本书后续章节中不断细化和完善。

1.5　小　　结

本章介绍了面向对象程序设计的基本概念，它将客观世界中的事物建模为软件对象，具有相同属性和操作的一组对象可抽象为类，类之间的关联和继承等关系是面向对象软件系统设计的关键。

1.6　习　　题

1. 简述对象和类的概念，并说说类在软件系统设计中的重要性。
2. 面向对象技术中的继承是指什么？试举例说明你在日常生活中看到的继承例子。
3. 将图 1-3 和图 1-4 合并成为一个更为完整的关系图。
4. 查阅有关统一建模语言 UML 的资料，初步了解 UML 与面向对象技术的关系。

第 2 章
C#和 Visual Studio 开发环境基础

本章从 C#语言和.NET 技术的概貌入手，依次介绍了 C#程序的组成结构和 Visual Studio 开发环境，读者将从这里开始自己的 C#程序开发之路。

2.1 C#语言和.NET 技术简介

C 语言曾经是最为流行的一门结构化程序设计语言，C++则在 C 语言的基础上增加了对面向对象的支持。但严格来说，C++并不是完全面向对象的程序设计语言；为了和 C 语言相兼容，C++保留了许多低级特性，因此具有较大的灵活性和较强的底层控制能力。但它也导致了 C++学习困难、程序过于复杂、安全性难以保证等问题。

正如低级语言被高级语言逐步取代一样，程序设计语言的发展就是不断增强抽象描述能力、屏蔽底层实现细节、提高软件生产率的过程。随着面向对象的优越性被广泛接受，人们需要更加符合现代软件开发要求的面向对象程序设计语言，C#因此应运而生。它汲取了 C++、Java、Delphi 等多种语言的精华，具有语法简洁、类型安全、完全面向对象等特点，自 2000 年一经推出便取得了巨大的成功。

C#语言简单易学，它将内存管理、设备驱动、控制优化等底层操作交由.NET Framework 实现，这样开发人员就能够把注意力集中在问题域模型和程序逻辑上，而不必去关注过多的底层细节。概括地说，.NET 是一个建立在开放网络协议和标准之上的计算平台，.NET Framework 则是.NET 平台上的基础编程框架，它由以下两部分组成。

- 公共语言运行时（Common Language Runtime，CLR）。它提供了.NET 应用程序的运行时环境，负责管理代码的执行、提供元数据类型支持和各种系统服务。
- .NET 类库。它定义了功能丰富的类型集合，能够为应用程序提供基本类型、通用数据结构、Windows 和 Web 界面设计、数据库访问、XML Web Service、异常处理等各种组件服务。

.NET Framework 支持 C#、Visual Basic .NET、Visual C++ .NET 等多种语言，这些高级语言代码会被编译为一种通用中间语言 IL（Intermediate Language）代码，该语言类似于低级语言，但其代码与具体的硬件平台无关；之后 CLR 再针对特定的平台将 IL 程序翻译为机器指令，加载所需的资源并管理程序的执行。这也使得.NET 平台上不同语言程序能够方便地进行通信，解决了困扰人们已久的多语言集成的难题。

随着语言的流行和用户的增加，C#也得到了不断的改进和完善。2005 年 C#升级到了 2.0 版本，其最大特点是增加了对泛型程序设计（Generic Programming）的支持。2008 年 C#推出 3.5 版，其中增加了 Lambda 表达式、隐式类型、扩展方法等特性，并支持一种新的面向对象的数据访问模型——LINQ（Language Integrated Query）模型。2010 年 C#推出 4.0 版，新增特性包括可选参数和命名参数、元组（Tuple）类型、动态绑定、逆变和协变等。目前 C#语言的最新版本是 4.5，其提供了增强的异步操作和国际化支持等功能。.NET Framework 也随之进行了相应的版本升级，为.NET 应用开发提供了更为有效的支持。

2.2　C#程序的基本结构

下面先看一个非常简单的 C#程序，它用于在屏幕输出一行文本"欢迎光临！"。

```
//程序 P2_1
using System;
namespace P2_1
{
    public class Program
    {
        public static void Main()
        {
            Console.WriteLine("欢迎光临！");
        }
    }
}
```

可在任何一种文本编辑器（如 Windows 记事本或 Microsoft Word）中输入上述代码，将文件存为"P2_1.cs"（后缀名.cs 表示 C#源程序文件）；而后打开命令行窗口，使用 C#编译器 csc.exe 编译程序（必要时指定源文件所在的目录）就能生成的可执行文件 P2_1.exe。程序的编译和运行结果如图 2-1 所示。

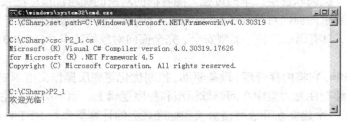

图 2-1　编译和执行 C#程序 P2_1

　　　　C#编译器 csc.exe 位于 Windows 目录下的\Microsoft.NET\Framework\v*.*子目录中，其中 v*.*表示.NET Framework 的版本号，如"C:\Windows\Microsoft.NET\Framework\v4.0.30319"。使用命令行编译时，可将该目录加入系统 Path 环境中。通过命令"csc/?"可以查看编译器使用选项的帮助信息，有关编译器更为详细的内容可参看 MSDN 帮助。

接下来分析一下该程序的基本结构。

2.2.1　注释

程序 P2_1 的第 1 行以连续两个反斜杠"//"开头，表示程序的注释，那么在它同行右侧的内容会被编译器忽略，不对程序的运行产生任何影响。如果要写多行注释，可以每一行都以"//"开头，或是将所有注释内容都放在一对标记"/*"和"*/"之间，例如：

```
/* 程序 P2_1
该程序用于在屏幕输出一行文本"欢迎光临！" */
```

但多行注释标签不可以嵌套。例如对于下面的代码，编译器会将第一行开头的"/*"到第二行末尾的"*/"之间的内容视为注释，第三行的标记"*/"视为非法代码：

```
/* 程序 P2_1
/* 该程序用于在屏幕输出一行文本"欢迎光临！" */
*/
```

注释可以出现在程序代码的任何位置，主要用于对代码的功能和用途等进行描述，从而提高程序的可读性，便于理解和修改程序。优秀程序员都应当养成注释代码的良好习惯。

2.2.2 命名空间

程序中常常需要定义很多的类型，为了便于类型的组织和管理，C#引入了命名空间的概念。一组类型可以属于一个命名空间，而一个命名空间也可以嵌套在另一个命名空间中，从而形成一个逻辑层次结构，这就好比目录式的文件系统组织方式。

程序 P2_1 的第 2 行通过关键字"using"引用了一个.NET 类库中的命名空间"System"，之后程序就可以自由使用该命名空间下定义的各种类型。程序的第 3 行则通过关键字"namespace"定义了一个新的命名空间"P2_1"，在其后的一对大括号"{ }"中定义的所有类型都属于该命名空间。

C#语言是大小写敏感的，比如关键字"using"不能写成"USING"，"namespace"不能写成"Namespace"。

命名空间的使用还有利于避免命名冲突。不同的开发人员可能会使用同一个名称来定义不同的类型，这在程序相互调用时就会产生混淆，而将这些类型放在不同的命名空间中就可以解决此问题。

2.2.3 类型及其成员

在 C#语言中，类是最为基本的一种数据类型，类的属性叫做"字段"（Field），类的操作叫做"方法"（Method）。类使用关键字"class"来定义，程序 P2_1 就定义了一个名为"Program"的类，并为其定义了一个方法"Main"，在其中执行文本输出的功能：

```
public static void Main()
{
    Console.WriteLine("欢迎光临！");
}
```

这里 Main 方法的功能是通过调用 Console 类的 WirteLine 方法来完成的，方法的输入参数是用一对双引号括起来的字符串，表示要输出的文本内容。如果要显式定义字符串对象，那么 Main 方法中的代码可改写为如下内容，其中 string 表示字符串类型，而 s 是该类型的一个对象（也叫变量）：

```
string s = "欢迎光临！";
Console.WriteLine(s);
```

Console 类是.NET 类库的 System 命名空间下定义的一个类，表示对控制台窗口的抽象。由于程序已引用了该命名空间，因此在 Main 方法的代码中可以直接使用该类。如果删除程序第二行的 using 引用代码，那么在使用 Console 类时还需要指定该类所属的命名空间：

```
System.Console.WriteLine("欢迎光临！");
```

Console 类是控制台应用程序与用户交互的基础，表 2-1 列出了其常用的一些输入/输出方法。

表 2-1　　　　　　　　　　　　　Console 类的常用成员方法

方　　法	输入参数	返回值	作　　用
Read	无	整数	读入下一个字符
ReadKey	无	ConsoleKeyInfo 对象	读入一个字符
ReadLine	无	字符串	读入一行文本，至换行符结束
Write	任意对象	无	输出一行文本
WriteLine	任意对象	无	输出一行文本，并在结尾处自动换行

同理，在其他程序中使用程序 P2_1 中的 Program 类也可以采用这两种方式：一是直接使用全称"P2_1.Program"；二是先在其他程序中引用命名空间 P2_1，而后使用简称"Program"。不过，如果存在命名冲突（如其他程序中也定义了名为"Program"的类），那么就必须使用全称来加以区分。

程序 P2_1 虽然简单，但我们从中看到了 C#应用程序的基本结构：命名空间下包含类，类可以包含成员字段和成员方法，方法中又包含执行代码。这种包含关系都是通过一对大括号"{}"来表示的。

2.2.4　程序主方法

程序的功能是通过执行方法代码来实现的，每个方法都是从其第一行代码开始执行，直至最后一行代码结束，期间可以通过代码来调用其他方法，从而完成各式各样的操作。应用程序的执行必须要有一个起点。C#程序的起点就是由 Main 定义的，程序总是从 Main 方法的第一行代码开始执行，在 Main 方法结束时停止运行。

因此，对于 C#可执行程序，其中必须有一个类定义了 Main 方法，那么编译器就会确定该方法作为程序的入口。如果有多个类都定义了 Main 方法，那么还需要通过编译选项 main 明确指定主方法所属的类。例如下面的编译命令就指定 Program 类中的 Main 方法作为程序主方法：

```
csc /main:Program P2_1.cs
```

2.2.5　程序集

人们使用代码编写的是源程序文件，它必须通过编译后才可执行，而编译生成的程序模块叫做程序集（Assembly）。程序集是.NET 应用程序的基本单元，一个软件系统可以是一个程序集，但更多时候是多个相互调用的程序集组成的集合。

.NET Framework 中提供了 C#的编译器和运行环境。因此只要安装了该框架（可从 Microsoft 网站免费下载），那么开发人员就可以使用各种文本编辑器来编写 C#程序代码，而后编译和执行程序。

程序集可以是 exe 可执行文件格式，也可以是 dll 动态链接库文件；后者主要是为其他程序提供各种类型和服务，本身并不能直接启动，因此程序中可以不包含 Main 方法。例如，下面的编译命令使用了 target 选项来将程序 P2_1 编译为动态链接库文件 P2_1.dll：

```
csc /target:dll P2_1.cs
```

下面的程序 P2_2 则调用了程序 P2_1 中的 Program.Main 方法：

```
//程序 P2_2
using System;
namespace P2_2
{
```

```
    public class Program
    {
        Console.WriteLine("请输入姓名:");
        string name = Console.ReadLine();
        Console.Write(name);
        Console.Write(",");
        P2_1.Program.Main();
    }
}
```

那么，在编译程序 P2_2 时，就需要添加对 P2_1.dll（或 P2_1.exe）的引用，相应的编译选项为 reference（可简写为 r）：

```
csc /reference:P2_1.dll P2_2.cs
```

程序 P2_2 的运行结果如图 2-2 所示。

图 2-2　程序 P2_2 的输出结果示例

2.3　Visual Studio 开发环境

2.3.1　集成开发环境概述

对于大型软件系统开发，仅仅使用编译器命令是远远不够的。集成开发环境（Integrated Development Environment，IDE）将代码编辑器、编译器、调试器、图形界面设计器等工具和服务集成在一个环境中，能够有效提高软件开发的效率。

Visual Studio.NET 是最流行的.NET 应用程序集成开发环境，开发的每个程序集对应一个 Visual Studio 项目（Project），而多个相关的项目又可组成一个 Visual Studio 解决方案（Solution）。

提示

使用 Visual Studio 标准版、专业版或企业版，可使用 Visual C#、Visual Basic .NET、Visual C++ .NET 等不同的语言来创建程序项目，并可以将这些项目包含在同一个解决方案中。

启动 Visual Studio 开发环境，可以看到如图 2-3 所示的主界面，主要包括以下几部分。
- 菜单栏：位于标题栏的下方，其中包含了用于开发、维护、编译、运行和调试程序以及配置开发环境的各项命令。
- 工具栏：位于菜单栏的下方，提供了常用命令的快捷操作方式。
- 代码编辑区：位于开发环境中央，是编辑代码或设计程序的主区域。
- 输出窗口：位于代码编辑区的下方，用于输出当前操作得到的结果。
- 解决方案资源管理器：位于开发环境的右侧，它通过树型视图对当前解决方案进行管理，解决方案是树的根节点，解决方案中的每个项目都是根节点的一个子节点，项目节点下则列出了该项目中使用的各种文件、引用和其他资源。
- 服务器资源管理器：位于开发环境的左侧，用于快速访问本地或网络上的各项服务器资源。
- 属性窗口：位于解决方案资源管理器的下方，用于查看或编辑当前所选元素（如项目、文件、控件等）的具体信息。

图 2-3　Visual Studio 开发环境

- 状态栏：位于开发环境的底部，用于对光标位置、编辑方式等当前状态给出提示。

图 2-3 所示为 Visual Studio 各窗口的默认位置，而这些窗口还可根据用户需要来移动、调整、打开或关闭，或是通过"视图"菜单来控制它们的显示；其中大部分窗口还可以通过选项卡的方式来进行切换，如代码编辑区可一次打开多个源文件，这就能最大程度地利用有限的屏幕空间。其他常用的窗口还有管理程序中的类及其关系的类视图（可与解决方案资源管理器进行切换显示）、作为控件集合的工具箱（可与服务器资源管理器进行切换显示）等。

Visual Studio 解决方案将被保存为 .sln 文件，而单个 C#程序项目则会被保存为 .csproj 文件。接下来介绍使用 Visual Studio 创建 4 种基本的 C#应用程序项目的步骤。

2.3.2　创建控制台应用程序

如果要在 Visual Studio 开发环境中创建程序 P2_1，那么可通过菜单命令"文件→新建→项目"打开如图 2-4 所示的对话框，在左侧的项目类型视图中选择"Visual C#"，在右侧的模板视图中选择"控制台应用程序"，输入项目名称，必要时可指定项目存放的位置及所属的解决方案（在对话框左上角的下拉框中还可选择目标程序所依赖的 .NET Framework 版本），而后按下"确定"按钮。此时 Visual Studio 就会帮助我们自动完成下列工作。

图 2-4　"新建项目"对话框

（1）将.NET 类库中的基本程序集添加到项目引用中。
（2）生成 C#源文件 Program.cs（用户可在随后修改其文件名），其中包含对常用命名空间的引用，以及程序命名空间、主程序类和 Main 方法的基本框架。
（3）生成项目配置文件，在其中保存项目的基本信息。

之后开发人员就可以编辑源文件中的程序代码，必要时还可增加新的源文件、程序集引用和其他资源，并通过菜单命令编译和运行程序。此外，使用快捷键 F6 可以直接编译程序，使用快捷键 F5 可以直接调试运行程序，或是按 Ctrl+F5 组合键不调试而直接运行程序。

2.3.3　创建和使用动态链接库程序

如果要创建动态链接库程序，那么可在图 2-4 所示的"新建项目"对话框中选择"类库"模板，之后同样可以编写代码来定义各种类型及其成员。不过此类项目只能编译成 dll 动态链接库文件，而不能直接运行。例如，旅行社管理系统中的业务类库就可以创建为一个动态链接库程序项目，其输出可供旅行社内部管理子系统和旅行社网站使用。

在解决方案资源管理器的项目节点下，展开"引用"子节点就可以看到当前项目中包含的对其他程序集（包括动态链接库和可执行文件）的引用，如图 2-5 所示。如果要添加新的引用，那么可通过菜单命令"项目→添加引用"打开如图 2-6 所示的对话框，在其中选择指定的程序集并按下"确定"按钮。其中".NET"选项卡下列出的是.NET 类库中的程序集；在"浏览"选项卡下指定程序集文件的路径名，则可引用自定义的程序集；对于组织在同一个解决方案中项目，相互之间可通过"项目"选项卡来进行引用。在程序的编译和部署过程中，Visual Studio 会自动处理当前程序与其引用程序集之间的关联。

图 2-5　在解决方案中管理程序集引用

图 2-6　"添加引用"对话框

2.3.4　创建 Windows 应用程序

如果创建带图形界面的 Windows 应用程序，那么可在图 2-4 所示的"新建项目"对话框中选择"Windows 应用程序"模板。这时 Visual Studio 会为项目自动生成两个 C#源程序文件，一是 Form1.cs，在解决方案资源管理器中双击该文件即可打开窗体的设计视图（见图 2-7），此时在 Visual Studio 工具箱中可看到一系列可用的 Windows 窗体控件，其中"公共控件"选项卡下包含了按钮、文本框、单选框、复选框等常用的 Windows 控件，通过鼠标拖放就能把这些控件添加到窗体中。

同样，在 Visual Studio 开发环境中按 F5 功能键可以调试运行程序，在程序启动时将看到一个和设计视图内容基本一致的 Windows 窗体。而如果在解决方案资源管理器中选中 Form1.cs，单击鼠标右键并选择"查看代码"，就可以查看其源代码，其中有如下的类定义：

```
public partial class Form1 : Form
```

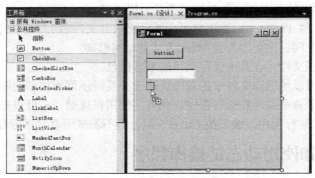

图 2-7 工具箱与窗体设计视图

Form1 就是程序定义的 Windows 窗体类，它继承自从 .NET 类库中的 Form 类，而 Form 类提供了 Windows 窗体的基本框架以及与用户交互的基本功能。在解决方案资源管理器中选择当前项目，通过菜单命令"项目→添加 Windows 窗体"还可以向项目中加入新的窗体。

另一个程序文件是 Program.cs，它定义了包含 Main 方法的主程序类 Program，其基本代码如下：

```
//程序 P2_3(Program.cs)
using System;
usingSystem.Windows.Forms;
namespace P2_3
{
    static class Program
    {
        static void Main()
        {
            Application.EnableVisualStyles();
            Application.SetCompatibleTextRenderingDefault(false);
            Application.Run(new Form1());
        }
    }
}
```

代码中的 Application 类表示当前的 Windows 应用程序，其 Run 方法就用于在程序启动时打开主窗体 Form1，它和 Form 类都在 System.Windows.Forms 命名空间下定义。在 Windows 应用程序项目中一般不用修改 Program.cs 中的代码。

Visual Studio 中创建的 C# Windows 应用程序还包含一个隐含源文件 Form1.Designer.cs，通过设计视图对窗体进行设置所生成的代码都将保存在该文件中。通常开发人员也无须手动修改该文件中的代码。

例如，在旅行社管理系统中，若是旅行社内部管理子系统只在单位办公计算机上使用，那么可将其创建为一个 Windows 应用程序项目。

2.3.5 创建 ASP .NET 应用程序

在 Visual Studio 中还可以方便地创建 ASP .NET Web 应用程序。在如图 2-4 所示的"新建项目"对话框中选择"ASP .NET Web 应用程序"模板，Visual Studio 就会创建一个基本的 Web 网站框架，其中包括默认主页 Default.aspx、"关于"页面 About.aspx，以及其他相关程序和配置文件。选中当前项目，通过菜单命令"添加→新建项"，在打开的对话框中选择"Web 窗体"，就可以向项目中加入新的 aspx 网页。

如果不希望使用任何自动生成的页面,那么可在"新建项目"对话框中选中左侧"Visual C#"节点下的"Web"节点,再在右侧选择"ASP .NET 空 Web 应用程序"模板,这样就可以从零开始创建一个 ASP .NET 程序。

提示

在 Visual Studio 中还可以通过菜单命令"文件→新建→网站"来创建 ASP .NET 网站,这样创建的程序项目与使用的"ASP .NET Web 应用程序"模板得到的结果基本相同。二者的区别在于,前者的项目是配置在当前服务器的文件系统上,编译后就可通过网络进行访问;后者则主要用于本地开发和调试,之后需要部署到服务器上才能进行远程访问。

确切地说,ASP .NET 程序项目的每一个网页都包含两个文件:一个设计文件(后缀名为.aspx)和一个 C#源代码文件(后缀名为.aspx.cs)。打开 aspx 设计文件,在 Visual Studio 工具箱中可看到一系列可用的 Web 窗体控件,如按钮、文本框、单选框、复选框等,其中许多控件的外观或功能都和 Windows 控件类似,通过鼠标拖放同样可把这些控件添加到网页中。此外,网页的设计视图左下方还有"设计"、"拆分"、"源"3 个小按钮,单击"设计"按钮可看到图形化的网页界面,单击"源"按钮则切换到网页的 HTML 视图,单击"拆分"按钮则可以同时显示 HTML 视图和图形界面,此时对 HTML 内容的修改可直接反映到界面上,如图 2-8 所示。

图 2-8 网页设计视图

运行 ASP .NET 应用程序,系统会自动启动浏览器,并在其中显示相关的网页内容,如图 2-9 所示。

图 2-9 ASP .NET 应用程序运行示例

2.4 小　　结

　　C#是.NET Framework 上的核心编程语言，具有语法简洁、类型安全和完全面向对象的特点，能够有效提高软件开发的生产率。
　　类是 C#应用程序的核心要素，类中包含成员字段和成员方法，程序功能主要通过方法代码来实现。C#使用命名空间来对类进行组织和管理。
　　Visual Studio 是.NET 应用程序的集成开发环境，在其中可以方便地创建和维护 C#控制台应用程序、动态链接库、Windows 应用程序、ASP .NET Web 应用程序等各种程序项目。

2.5 习　　题

1. C#程序的执行过程是怎样的？哪些方法会在程序中执行？
2. Visual Studio 开发环境包含哪些基本菜单？它们主要用于完成哪些工作？
3. 在 Visual Studio 中创建控制台应用程序和 Windows 应用程序时，默认引用了哪些系统程序集文件？
4. 简述从 C#源代码到可执行程序指令的转换过程。
5. 创建一个 C#类库动态链接库程序，在其中定义一个方法来计算两个数的和；再创建一个 C#控制台应用程序，在其中调用动态链接库程序的求和方法。

第 3 章
C#语法基础

在深入探讨 C#面向对象程序设计之前,首先需要掌握 C#语言的基本语法结构。本章将讲解了 C#语言中的数据类型、操作符和表达式、程序控制结构这几项基本要素。

3.1 数据类型

数据类型是对客观数据对象的抽象,它将数据和对数据的操作封装为一个整体。C#语言中的数据类型分为两大类:值类型和引用类型。值类型包括整数、字符、实数、布尔数等简单值类型,以及结构(Struct)和枚举(Enum)这两种复合值类型;引用类型则包括类、接口(Interface)、委托(Delegate)和数组。这些类型在本质上都是面向对象的。委托和接口将分别在第 6 章和第 10 章中介绍。

3.1.1 简单值类型

1. 整数类型

整数类型是对数学中的整数的抽象,但由于受到计算机存储限制,程序设计语言中的值类型总是要设置取值范围限制。C#定义了以下 8 种整数类型。

- int:32 位整数,取值范围为-2147483648(-2^{31})~2147483647($2^{31}-1$)。
- uint:32 位无符号整数(即正整数),取值范围为 0~4294967295($2^{32}-1$)。
- long:64 位长整数,取值范围为-9223372036854775808(-2^{63})~9223372036854775807($2^{63}-1$)。
- ulong:64 位无符号整数,取值范围为 0~18446744073709551615($2^{64}-1$)。
- short:16 位短整数,取值范围为-32768(-2^{15})~32767($2^{15}-1$)。
- ushort:16 位无符号短整数,取值范围为 0~65535($2^{16}-1$)。
- sbyte:8 位字节型整数,取值范围为-128(-2^{7})~127($2^{7}-1$)。
- byte:8 位无符号字节型整数,取值范围为 0~255($2^{8}-1$)。

例如,下面的代码就先定义了一个整型变量 x,而后将其赋值为 80:
```
int x;
x = 80;
```
上述两行代码还可以合并为:
```
int x = 80;
```
如果是同类型的多个变量,那么 C#允许将它们的声明语句简写到一行代码中,例如:
```
int x1 = 10, x2, x3 = 20; //x2 还未被赋值
```
而下面这些语句则是错误的:

```
uint x = -10;  //错误：无符号整数不能取负数
byte y = 500;  // 错误：超出了取值范围
long z;
Console.WriteLine(z);  //错误：z 在被赋值前不能使用
```

定义变量时应尽量选择最为适合的类型：过短的类型可能不足以表达变量的变化范围，过长的类型则会造成资源浪费。例如要表示人的年龄 byte 类型就足够了，要表示人口的数量则 uint 类型会比较适合。

2. 字符类型

C#中使用 char 来表示字符类型。由于使用了 16 位 Unicode 字符集，char 类型不仅包含基本的 ASCII 字符，还能够表示汉字等各国语言符号。注意单个字符值要用一对单引号括起来；如果使用了双引号，那么它表示的就是只有一个字符的字符串。例如：

```
char a = 's';
char b = '人';
string s = "人";
```

对于像单引号、回车符这样的特殊字符，C#中使用加斜杠 "\" 的转义符来表示。例如：

```
char a1 = '\'';  //a1 表示单引号
char a2 = '\\';  //a2 表示单斜杠
char a3 = '\r';  //a3 表示回车符
```

表 3-1 列出了 C#中常见的一些转义符格式。

表 3-1　　　　　　　　　　　　　　C#常用转义符

转义符	含　义	转义符	含　义
\'	单引号	\n	换行
\"	双引号	\r	回车
\\	单斜杠	\f	换页
\a	警报（Alert）	\t	水平 tab
\b	退格（Backspace）	\v	垂直 tab
\e	取消（Esc）	\0	空字符

char 类型的变量实际上仍以整数方式进行存储，其取值范围与 ushort 相同，因此一个字符会使用 16 位字节的内存空间。

3. 实数类型

C#定义了以下 3 种实数类型。

- float：32 位单精度浮点数类型，取值范围为 $\pm 1.5 \times 10^{-45} \sim \pm 3.4 \times 10^{38}$。
- double：64 位双精度浮点数类型，取值范围为 $\pm 5.0 \times 10^{-324} \sim \pm 1.7 \times 10^{308}$。
- decimal：128 位十进制小数类型，取值范围为 $\pm 1.0 \times 10^{-28} \sim \pm 7.9 \times 10^{28}$。

除了 double 类型之外，float 和 decimal 类型的变量值应在小数后分别加上后缀 F 和 M，以便让编译器知道以何种类型处理这些数值，例如：

```
double x = 1.2;
float y = -0.5F;
decimal z = 3.2M;
```

实数类型使用的位数较多，相应的计算也会消耗更多的资源，因此在程序中应尽量使用低精度的类型。有时还可以使用整数来替代实数类型，例如，在货币计算中如果只需要精确到小数点后两位，那么使用"分"而不是"元"来做单位，就可以将实数运算转换为整数运算。此外，计算机中的实数运算也受到精度和范围的限制，有时候并不能得出数学上完全正确的结果。

4. 布尔类型

bool 来表示布尔类型，其可能取值只有两个：true（真）或 false（假）。例如：

```
bool b1 = true;
bool b2 = flase;
```

布尔类型在计算机内部实际上是使用二进制 1 和 0 表示的，其运算效率在各种数据类型之间也是最高的。

3.1.2 复合值类型

1. 结构

像 int、double 这些简单值类型都是在.NET 类库中预定义的。很多情况下，人们需要将不同的简单值类型组合起来使用，这时可使用结构类型。考虑"复数"这个概念，.NET 类库没有其类型定义，那么可在 double 类型的基础上定义该类型，它由两个 double 类型的字段组成，其中 a 表示复数的实部，b 表示虚部：

```
struct ComplexNumber //定义结构类型
{
    public double a;
    public double b;
}
```

之后就可以创建该类型的变量，并通过圆点连接符"."来访问其公有成员，例如：

```
ComplexNumber c1; //创建结构类型的变量
c1.a = 2.5;
c1.b = 5;
```

结构类型支持信息隐藏和封装，也可以包含成员方法，支持对象创建和消息通信；不过它不支持继承和多态，因此是一种"部分面向对象"的类型。在下面的示例程序中，ComplexNumber 结构还包含了一个成员方法 Write，程序主方法就调用该方法在控制台输出了两个复数：

```
//程序 P3_1
using System;
namespace P3_1
{
    class Program
    {
        static void Main()
        {
            ComplexNumber c1; //定义结构变量 c1
            c1.a = 1.5;
            c1.b = 3;
            ComplexNumber c2 = c1; //定义结构变量 c2
            c2.a = c2.b;
            Console.Write("c1 = ");
            c1.Write();
            Console.Write("c2 = ");
            c2.Write();
        }
    }

    struct ComplexNumber
    {
```

```
        public double a;
        public double b;

        public void Write()
        {
            Console.WriteLine("{0}+{1}i", a, b);
        }
    }
}
```

上面最后一行代码使用了 Console 类的 WriteLine 方法的格式化输出功能，它表示用方法的第 2 个参数值替换第 1 个字符串参数中的 {0} 标记、用第 3 个参数值替换第 1 个参数中的 {1} 标记，……依次类推。那么程序 P3_1 的输出内容如下：

```
c1 = 1.5+3i
c2 = 3+3i
```

实际上，前一小节中介绍的简单值类型在.NET 类库内部都是以结构方式来定义的，这些结构不仅包含了变量所存储的值，还提供了类型本身的有关消息。例如它们都提供了 MinValue 和 MaxValue 字段来表示类型的最小值和最大值，实数类型还提供了 NegativeInfinity 和 PositiveInfinity 字段来表示正无穷大和负无穷大，此外它们还能够通过成员方法 ToString 来得到数值的字符串表示，通过方法 Parse 来将字符串转换为数值。下面给出了这样一些代码示例：

```
short a = short.MaxValue; //x = 32767
short b = short.MinValue; //x = -32768
ushort c = ushort.MinValue; //y = 0
double x = double.PositiveInfinity;
double y = -x; //y = double.NegativeInfinity;
string s = a.ToString(); //s = "32767"
int i = int.Parse(s); //i = 32767
```

2. 枚举

还有一种情况是要对变量的取值范围作特殊的限定，那么可将其定义为枚举类型，在其中"枚举"出所有可能的取值项。考虑"星期"这个概念，其取值可以是星期一到星期日，那么可以用如下方式定义枚举类型 Weekday：

```
enum Weekday
{
Monday, Tuesday, Wednesday, Thursday, Friday, Saturday, Sunday
}
```

枚举类型的变量值是通过"类型名"加连接符"."再加"枚举项"来使用的，例如星期一就是"Weekday.Monday"：

```
Weekday w1 = Weekday.Monday;
```

枚举类型的成员项都是使用文字来进行说明的，这有助于对程序的理解，但它们的实际存储类型都是整数。例如对于枚举 Weekday，其第一个值 Monday 对应整数 0，Tuesday 对应整数 1，……，最后一个 Sunday 对应整数 6。而如果将 Sunday 移到最前面，那么该枚举值就成为 0，Monday 变为 1，……，依次类推。

C#还允许在程序中明确指定枚举值对应的整数值。例如对于下面的定义，Monday～Sunday 所对应的整数值将分别为 1～7：

```
enum Weekday
{
    Monday = 1, Tuesday, Wednesday, Thursday, Friday, Saturday, Sunday
}
```

提示
　　程序中不应滥用枚举类型。事实上大多数值都是有范围的；除了程序中常用的一些特殊概念之外，大部分还是应直接使用整数类型，并在程序逻辑中控制范围，如"月份"在1~12之间，"年级"在1~6之间，"鞋码"在10~30之间等，而不是将它们都定义为枚举。

3.1.3 类

如果要把程序 P3_1 中 ComplexNumber 的类型由结构换成类，只需要将其定义的关键字由 struct 改为 class 即可：

```
class ComplexNumber
{
    public double a;
    public double b;

    public void Write()
    {
        Console.WriteLine("{0}+{1}i", a, b);
    }
}
```

但此时编译程序，Main 方法的第 2 行代码将会出错，提示"使用了未赋值的局部变量 c1"。这是因为结构是值类型，声明变量时就会分配内存；而类是引用类型，如果只声明变量，得到的就是一个什么都没有的空对象（null），当然也就不能访问其成员。纠错办法是将 Main 方法的第 1 行代码改为如下内容：

```
ComplexNumber c1 = new ComplexNumber(); //创建 c1 对象
```

C#中的 new 关键字用于显式创建类的变量，这样才得到了在程序中真实存在的内存对象。那么修改后得到的完整程序代码如下：

```
//程序 P3_2
using System;
namespace P3_2
{
    class Program
    {
        static void Main()
        {
            ComplexNumber c1 = new ComplexNumber(); //创建 c1 对象
            c1.a = 1.5;
            c1.b = 3;
            ComplexNumber c2 = c1; //创建 c2 对象
            c2.a = c2.b;
            Console.Write("c1 = ");
            c1.Write();
            Console.Write("c2 = ");
            c2.Write();
        }
    }

    class ComplexNumber
    {
        public double a;
```

```
        public double b;

        public void Write()
        {
            Console.WriteLine("{0}+{1}i", a, b);
        }
    }
}
```

但编译运行该程序，会看到控制台的输出结果发生了变化：

```
c1 = 3+3i
c2 = 3+3i
```

这就说明了值类型和引用类型的第 2 个区别：将值类型的一个变量赋值给另一个变量，那么原始变量的数据会被复制给新变量，之后两个变量是相互独立的；而将引用类型的一个变量赋值给另一个变量，实际上新变量只是包含了指向原始对象的指针，对其中任何一个变量的修改都会影响到另一个变量。在程序 P3_2 中使用引用类型的类时，对 c2 对象的修改也就同时改变了 c1 对象。图 3-1 所示为这种值类型和引用类型变量在存储上的区别。

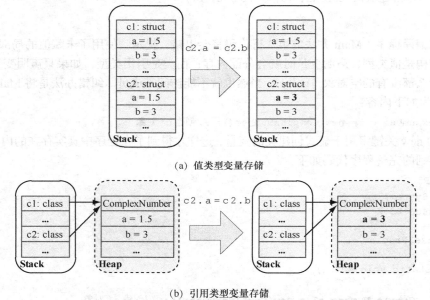

(a) 值类型变量存储

(b) 引用类型变量存储

图 3-1　值类型和引用类型的 ComplexNumber 变量存储示意图

和结构相比，类是一种完全面向对象的类型，使用它能够充分享受继承和多态性等面向对象的优越性。定义一个新类时，如果要使其从另一个类中继承，那么应在新类的名称后接冒号再接基类的名称。

下面的代码先定义了一个 Student 类，而后定义了其派生类 Graduate，这样 Graduate 就继承了 Student 的 3 个字段和 1 个方法，而 supervisor 字段则是 Graduate 类自己独有的。

```
class Student //基类
{
    Public string name;
    public int age;
    public int grade;
    public void Register() { }
```

```
}
class Graduate : Student  //派生类
{
    public string supervisor;
```

前面出现过的 string 类型也是.NET 类库中定义的一个类，但 C#语言对字符串有着特殊的处理方式：对于两个 string 变量 s1 和 s2，语句"s1 = s2"并不会使它们指向同一个对象，因此不存在修改一个字符串而影响到另一个字符串的情况。但所有的 string 对象仍是引用类型；和其他对象类型一样，可将空值 null 赋给 string 变量，而这对于任何值类型的变量都是不允许的：

```
string s1 = null;  //正确
int i1 = null;  //错误
```

3.1.4　数组

数组是一种聚合类型，它表示具有相同类型的一组对象的集合，这些对象叫做数组元素。数组元素的类型可以是值类型，也可以是引用类型。

1．一维数组

一维数组的声明是在数组元素的类型声明后加上一对中括号，例如，声明一个整型数组：

```
int[] nums;
```

作为引用类型，数组对象同样需要初始化之后才能使用，而且 new 关键字后面的类型声明的中括号里需要指定数组的长度，例如：

```
nums = new int[3];
```

而要访问数组元素，那么需要在数组对象后的中括号里指定数组元素的下标位置，其中第 1 个数组元素的下标为 0，第 2 个的下标为 1，……依次类推，例如：

```
nums[0] = 3;  // 数组的第一个元素赋值为 3
nums[1] = 6;  // 数组的第二个元素赋值为 6
nums[2] = 9;  // 数组的第三个元素赋值为 9
Console.WriteLine(nums[1]);  //输出数组的第二个元素值
```

使用数组要防止"越界"，即要访问的数组元素的下标不能超过数组的长度范围，如对上面的数组不能访问 nums[3]。但这种错误在编译时不会被发现，只有在运行到该行代码时才会报错，所以在编程时应格外注意。

为了简化代码，C#还允许在初始化数组对象的同时为其中的元素赋值，例如：

```
int[] xs = new int[3] { 10, 20, 30};
double[] ys = new double[4] {1.25, 3.5, 3.75, 5};
```

上面代码还可以进一步简写为：

```
int[] xs = { 10, 20, 30};
double[] ys = {1.25, 3.5, 3.75, 5};
```

对于一个数组变量，通过其 Length 属性可以判断数组的长度，例如：

```
int[] nums = {3, 6 ,9};
Console.WriteLine(nums.Length);  //输出 3
```

当数组元素的类型为值类型时，数据直接存放在数组中；而当数组元素的类型为引用类型时，数组中存放的只是各个引用对象的地址，这时不排除多个数组元素指向同一个对象的可能。看下面的程序示例：

```
//程序 P3_3
using System;
```

```csharp
namespace P3_3
{
    class Program
    {
        static void Main()
        {
            ComplexNumber c1 = new ComplexNumber();
            c1.a = 1.5;
            c1.b = 3;
            ComplexNumber[] cs = new ComplexNumber[4];
            cs[0] = c1;
            cs[1] = new ComplexNumber();
            cs[1].a = 4;
            cs[1].b = 5;
            cs[2] = c1;
            cs[3] = cs[1];
            cs[0].a = 8;
            cs[3].b = 9;
            Console.Write("c1 = ");
            c1.Write();
            Console.Write("cs[1] = ");
            cs[1].Write();
            Console.Write("cs[2] = ");
            cs[2].Write();
        }
    }

    class ComplexNumber
    {
        public double a;
        public double b;
        public void Write()
        {
            Console.WriteLine("{0}+{1}i", a, b);
        }
    }
}
```

由于使用的 ComplexNumber 是引用类型，c1、cs[0]、cs[2] 同时指向一个对象，而 cs[1] 和 cs[3] 同时指向另一个对象。程序 P3_3 的输出内容如下：

```
c1 = 8+3i
cs[1] = 4+9i
cs[2] = 8+3i
```

2. 多维数组

多维数组的声明是在元素类型后的中括号里增加逗号，并在初始化时指定每一维的长度。例如，下面的语句创建了一个三行两列的二维整型数组：

```csharp
int[,] nums = new int[3,2];
```

访问数组元素时同样需要指定每一维的下标，例如：

```csharp
nums[0, 0] = 1;
nums[0, 1] = 2;
nums[1, 0] = 3;
nums[1, 1] = 5;
nums[1, 2] = 8;//错误：数组越界！
```

对于多维数组变量，其 Length 属性返回的是整个数组的总长度，通过其 GetLength 方法还可以获取每一维的长度。下面的程序就输出了 3 个数组的总长度和各维长度，其中创建数组时还使用了不同简写方式，书写数组元素值时要注意不同层次上大括号对的匹配性（C#编译器可自动推断出数组每一维的长度）：

```
//程序 P3_4
using System;
namespace P3_4
{
  class Program
  {
    static void Main()
    {
      int[,] x = new int[2, 3] { {1, 2, 3}, {3, 5, 8} };
      int[,] y = new int[,] { {10, 50}, {25, 75}, {50, 150}, {100, 80} };
      int[, ,] z = { { {1,2}, {3,5}, {8,13} }, { {1,2}, {3,5}, {8,13} } };
      Console.Write("数组 x 长度为: ");
      Console.WriteLine(x.Length);
      Console.WriteLine("各维长度: {0} * {1}", x.GetLength(0), x.GetLength(1));
      Console.Write("数组 y 长度为: ");
      Console.WriteLine(y.Length);
      Console.WriteLine("各维长度: {0} * {1}", y.GetLength(0), y.GetLength(1));
      Console.Write("数组 z 长度为: ");
      Console.WriteLine(z.Length);
      Console.WriteLine("各维长度: {0} * {1} * {2}", z.GetLength(0), z.GetLength(1), z.GetLength(2));
    }
  }
}
```

程序 P3_4 的输出内容如下：

```
数组 x 长度为: 6
各维长度: 2 * 3
数组 y 长度为: 8
各维长度: 4 * 2
数组 z 长度为: 12
各维长度: 2 * 3 * 2
```

为了提高程序的可读性，创建数组时还是应尽量写出数组各维的长度。此外，数组的维数过高不仅会大量占用系统资源，还会增加程序的复杂度以及数组越界的危险性，因此要控制使用三维以上的多维数组。

3.1.5 类型转换

相容的数据类型的变量之间可以进行类型转换。有的转换是系统默认的，这叫做隐式（Implicit）转换；有的转换则需要明确指定转换的类型，这叫做显式（Explicit）转换。显式转换不能保证成功，还有可能发生信息丢失，使用时要加以注意。

1. 值类型之间的转换

从低精度的简单值类型到高精度的简单值类型可以进行隐式转换，这包括以下 3 种情况：
- 如果一种整数类型的取值范围被另一种整数类型的取值范围所涵盖，那么从前者到后者可进行隐式转换。
- 从整数类型到实数类型可进行隐式转换。
- 从 float 类型到 double 类型可进行隐式转换。
- 从 char 类型到 ushort、uint、int、ulong、long、float、double、decimal 这些类型可进行隐式转换（这和 ushort 类型到其他类型的隐式转换是等价的）。

下面的代码给出了从低精度整数类型到高精度整数类型的转换示例：

```
sbyte x1 = 100;
short x2 = x1;
int x3 = x2;
long x4 = x3;
```

不满足上述隐式转换条件的简单值类型之间只能进行显式转换，转换的办法是在要转换的值之前加上一对圆括号，并在括号中写上要转换到的目标类型。例如：

```
double y1 = 13.56;
int y2 = (int)y1;
```

显式转换时要注意值的范围。例如，将实数转换为整数后，实数原来的小数部分就会丢失。再看下面的代码示例：

```
long z1 = 100;
int z2 = (int)z1;
short z3 = (short)z2;
sbyte z4 = (sbyte)z3;
```

执行完上述转换后，z1~z4 的值都还是 100。但如果将 z1 值改成 1000，z2 和 z3 的值仍然不变，但 z4 的值就会变成-24，这是因为：整数 1000 所对应的 16 位二进制码为 0000001111101000，但转换到 8 位整数类型 sbyte 时就只剩下了后 8 位 11101000，对应十进制整数就是-24。这时程序不会报错，有经验的开发人员也能推断出预期结果，但在程序中应避免这种情况的出现，而决不能将其作为一种非常规的"技巧"来使用。

整数类型和 char 类型之间的转换所表达的是字符的整数编码，此时要注意数字字符和数字本身的区别。C#还规定从各种整数类型到 char 类型都只能进行显式转换。看下面的代码示例：

```
int x = 6;
char a = (char)x; //显式转换
Console.WriteLine(a);
char b = '6';
int y = b; //隐式转换
Console.WriteLine(b);
Console.WriteLine(y);
```

整数编码 6 对应的字符是黑桃符号"♠"，因此第 3 行代码将输出该符号；而字符"6"对应的编码是 54，因此第 6 行和第 7 行代码将分别输出 6 和 54（Console 类会自动根据参数的类型来选择输出的格式）。

自定义的枚举类型在本质上也是整数，但枚举类型和整数类型之间必须使用显式转换，唯一的例外是常数 0 可以直接赋值给枚举变量。看下面的程序示例：

```
//程序 P3_5
using System;
namespace P3_5
{
    class Program
    {
        static void Main()
        {
```

```
            Weekday w1 = 0;
            Weekday w2 = (Weekday)3; //显式转换
            Weekday w3 = (Weekday)100; //显式转换
            Console.WriteLine(w1);
            Console.WriteLine(w2);
            Console.WriteLine(w3);
            int x = (int)Weekday.Friday; //显式转换
            Console.WriteLine(x);
        }
    }

    enum Weekday
    {
        Monday, Tuesday, Wednesday, Thursday, Friday, Saturday, Sunday
    }
}
```

该程序前两行输出的 w1 和 w2 的值将分别为 "Monday" 和 "Thursday"；而 w3 被赋予的值已经超出了枚举的定义范围之外，因此程序第三行将直接输出整数 100；而将枚举值 "Friday" 转换为整数时，得到的值为 4。

此外，布尔类型与整数等其他类型之间不存在任何转换关系（这和 C/C++ 等语言中的布尔类型是不同的）。

2. 引用类型之间的转换

引用类型之间转换的基本原则是：从派生类的对象到基类的对象可以进行隐式转换；而从基类对象到派生类对象只能进行显式转换，且不一定成功。这一点很容易理解，因为特殊对象同时也是一般对象，而一般对象不能保证是特殊对象。转换之后，两个变量指向的仍是同一个对象。

例如在下面的程序中，Graduate 是 Student 的派生类，那么 Graduate 对象 g1 可以隐式转换为 Student 对象 s1；这样 s1 实际上仍是一个 Graduate 对象，因此可通过显式转换得到 g2；而 Student 对象 s2 并不是一个 Graduate 对象，试图将其显式转换为 Graduate 对象 g3 时就会失败（错误信息为 "无法将类型为 P3_6.Student 的对象强制转换为类型 P3_6.Graduate"）。

```
//程序 P3_6
using System;
namespace P3_6
{
    class Program
    {
        static void Main()
        {
            Graduate g1 = new Graduate();
            g1.name = "陈亮";
            Student s1 = g1;  //隐式转换
            Student s2 = new Student();
            s2.name = "宋燕燕";
            Graduate g2 = (Graduate)s1; //显式转换
            Console.WriteLine(g2.name);
            Graduate g3 = (Graduate)s2; //错误：转换失败！
        }
    }

    class Student
```

```
{
    public string name;
    public int age;
    public int grade;
    public void Register() { }
}

class Graduate : Student
{
    public string supervisor;
}
```

进一步推广到数组的情况，C#规定：如果两个数组的维数相同，而它们的数组元素之间可以进行引用转换，那么两个数组之间也可以进行引用转换，且转换方式与数组元素之间的转换方式相同。例如从一个 Graduate[]数组可隐式转换为一个 Student[]数组，反之则要进行显式转换。例如：

```
Graduate[] gs1 = new Graduate[] { g1, g2 };
Student[] ss1 = gs1;
Graduate[] gs2 = (Graduate[])ss1;
```

但要记住一点：在元素为值类型的数组之间（如 int[]数组和 long[]数组之间）不能进行任何转换。

3. 值类型和引用类型之间的转换

在 C++等传统语言当中，整数、布尔等值类型属于语言的内置类型，而这些内置类型本身不是面向对象的。C#语言则采用了一种全新的观点：程序设计语言的整个类型系统是一个有机的整体，所有的变量都是对象，所有的类型都有一个共同的基类——object 类。

object 类本身是引用类型，那么其他引用类型都可以与它进行转换。object 同时又是所有值类型的基类，那么所有值类型的变量都可以隐式转换为 object 类型，这个过程叫做装箱（Boxing）；而 object 类型可以显式转换到值类型，这个过程叫做拆箱（Unboxing）。这样值类型和引用类型两部分就有机地联系在了一起。下面给出了装箱和拆箱的简单示例代码：

```
int x = 3;
object y = x; //装箱
int z = (int)y; //拆箱
```

3.2 操作符和表达式

和数学运算中的概念类似，表达式由操作数和操作符（也叫运算数和运算符）组成，其作用是将操作符施加于操作数以得到相应的计算结果，而结果的数据类型由操作数的数据类型决定。对于包含两个或两个以上操作符的复合表达式，那么操作符执行的顺序取决于它们的优先级（高优先级的操作符先执行）和结合性（同一优先级的操作符从左向右或是从右向左执行）。使用圆括号可以控制表达式的计算顺序。

3.2.1 算术操作符

C#中的算术操作符有加（+）、减（-）、乘（*）、除（/）和取模（%），它们可以作用于各种整数和实数类型，从而完成基本的算术运算。例如：

```
int x1 = 12, x2 = 9;
```

```
int y1 = x1 + x2; //y1 = 21
int y2 = x1 *x2; //y2 = 108
```
算术操作符都是按从左向右执行，且乘、除和取模操作的优先级要高于加减操作，例如：
```
int x = 2 * 4 - 1;//从左向右计算，x = 7
int y = 6 + 4 / 2; //先计算 4 / 2，y = 8
```
使用括号可以改变表达式的求值顺序，确保括号内的表达式总是被单独计算的，例如：
```
int x = 2 * (4 - 1); //x = 6
int y = (6 + 4) /2; //y = 5
```
除操作符用于求两个数的商，模操作符用于求两个数相除的余数。对于相同类型的操作数，表达式的计算结果总是和操作数的类型相同，因此整数相除和取模的结果仍为整数，而实数相除和取模的结果为实数，这一点和初等数学中的余数概念是有所区别的。例如：
```
int x1 = 10 / 3; //x1 = 3
int x2 = 10 % 3; //x2 = 1
double y1 = 5.4 / 1.5; //y1 = 3.6
double y2 = 5.4 % 1.5; //y2 = 0.9
```
枚举变量可与整数进行加减运算，这实际上是先将枚举转换为整数，运算之后再将结果重新转换为枚举。两个枚举变量相减将得到一个整数，但它们不能直接相加。例如：
```
Weekday w1 = Weekday.Friday + 1; //w1 = Weekday.Saturday
Weekday w2 = Weekday.Thursday - 2;//w2 = Weekday.Tuesday
int i1 = w1-w2; //i1 = 4
int i2 = w1+w2; //错误，枚举变量不能相加
```

3.2.2 自增和自减操作符

由于计算机程序中会经常用到加 1 和减 1 运算，C#中定义了自增操作符"++"和自减操作符"--"。和算术操作符不同，这两个操作符都只有一个操作数，而它们的优先级也高于算术操作符。

在表达式中，当操作符"++"放在变量之前时，那么变量值先被加 1，而后表达式再使用该变量；而当操作符"++"放在变量之后时，表达式先使用该变量，而后变量值被加 1。操作符"--"的情况也是类似的。例如在下面的代码中，6 次输出的内容将分别是 100、102、102、49、49、48。

```
int x = 100;
Console.WriteLine(x++); //先输出 100，而后计算 x = x + 1 = 101
Console.WriteLine(++x); //先计算 x = x + 1 = 102，而后输出 102
Console.WriteLine(x); //输出 102
int y = 50;
Console.WriteLine(--y); //先计算 y = y - 1 = 49，而后输出 49
Console.WriteLine(y--); //先输出 49，而后计算 y = y - 1 = 48
Console.WriteLine(y); //输出 48
```

3.2.3 位操作符

位操作符是对数据按二进制位进行运算的操作符，具体包括以下几种。
- 取反操作符"~"：作用于一个操作数，且必须放在操作数前，表示对操作数的各二进制位取反（0 变为 1，1 变为 0）。
- 左移位操作符"<<"：作用于两个操作数，表示将左操作数的各二进制位依次左移，左边的高位被舍弃，右边的低位顺序补 0，移动的位数由右操作数指定。
- 右移位操作符">>"：作用于两个操作数，表示将左操作数的各二进制位依次右移，右

边的低位被舍弃，左边的高位则对正数补 0、负数补 1，移动的位数由右操作数指定。
- 与操作符 "&"：作用于两个操作数，表示对两个操作数的对应二进制位依次进行与运算。
- 或操作符 "|"：作用于两个操作数，表示对两个操作数的对应二进制位依次进行或运算。
- 异或操作符 "^"：作用于两个操作数，表示对两个操作数的对应二进制位依次进行异或运算。

取反操作符的优先级高于算术操作符，其他的则低于算术操作符。与、或、异或这三个操作符可以作用于整数类型和布尔类型，它们的运算规则如表 3-2 所示；其他的位操作符则只能作用于整数类型（或是能够转换为整数的其他类型）。

表 3-2　　　　　　　　　　　　与、或、异或运算规则

左操作数	右操作数	与	或	异或
0	0	0	0	0
0	1	0	1	1
1	0	0	1	1
1	1	1	1	0
false	false	false	false	false
false	true	false	true	true
true	false	false	true	true
true	true	true	true	false

进行位运算时，要注意有符号整数的二进制首位表示符号位（0 表示正数，1 表示负数）。下面给出了一些简单的位运算代码示例，其各行输出内容分别是 –21、9、80、–2、20、–10、–30。

```
byte x = 20;     //二进制 00010100
sbyte y = -10;   //二进制 11110110
Console.WriteLine(~x); //11101011 (-21)
Console.WriteLine(~y); //00001001 (9)
Console.WriteLine(x << 2); //01010000 (80)
Console.WriteLine(y >> 3); //11111110 (-2)
Console.WriteLine(x & y); //00010100 (20)
Console.WriteLine(x | y); //11110110 (-10)
Console.WriteLine(x ^ y); //11100010 (-30)
```

位运算的效率一般会高于加减乘除等算术运算。以移位运算为例，在整数的有效范围内将其左移 n 位相当于乘以 2 的 n 次方，右移 n 位则相当于除以 2 的 n 次方。例如，下面的两行代码得到的 x 和 y 的值是相同的，但第二行代码的效率会大大高于第一行：

```
int x = 256 * 256; //x = 65536
int y = 256 << 8;  //y = 65536
```

3.2.4　赋值操作符

前面已经看到了简单赋值操作符 "=" 的用法，它表示将右操作数的值赋予左操作数，前提是右操作数的类型与左操作数相同，或是可隐式转换为左操作数的类型。

为了简化程序代码，C#还提供了 10 个复合赋值操作符 "+="、"-="、"*="、"/="、"%="、"<<="、">>="、"&="、"|="、"^="，它们实际上是将加、减、乘、除、取模、左移位、右移位、与、或、异或这些运算和简单赋值结合起来。例如 "x += y" 等价于 "x = x + y"，"x &= y" 等价于 "x = x & y"，等等。

和一般算术运算不同，赋值操作符属于右结合的操作符，即表达式会按照从右向左的顺序进行赋值运算。例如执行完下面的语句后，x 的值为 100，而 y 和 z 的值均为 1000：

```
int x = 10, y = 10;
int z = y *= x *= 10;//先计算 x *= 10, 再计算 y *= x, 最后执行 z = y
```
赋值操作符在所有操作符中的优先级最低,即表达式总是计算等号右边的部分,而再进行赋值运算。例如表达式"x += y + z"会先计算 y 与 z 的和,再将其加到 x 上。

3.2.5 关系操作符

关系操作符用于对指定的条件进行判断,并返回一个布尔值来表示判断结果。其中"==""!=""<""<="">"">="这 6 个操作符又叫比较操作符,它们分别用于判断左操作数是否等于、不等于、小于、小于等于、大于、大于等于右操作数。默认情况下,它们可用于值类型之间的大小比较,而"=="和"!="操作符还可用于比较两个引用类型的变量是否指向同一个对象。下面的示例代码说明了这一点:

```
int x1 = 5;
int x2 = x1;
Console.WriteLine(x1 == x2); //输出 true
object o1 = x1;
object o2 = x2;
object o3 = o1;
Console.WriteLine(o1 == o2); //o1 和 o2 指向不同的对象,输出 false
Console.WriteLine(o1 == o3); //o1 和 o3 指向同一个对象,输出 true
```

C#中将相等操作符"=="误写为赋值操作符"="是一个语法错误,例如对于整数 x 和 y,语句"bool b = (x = y)"不能通过编译,这有助于发现代码错误。此外,C#语言不支持连续的关系比较,如语句"bool b = (x == y == z)"和"bool b = (x > y > z)"都是错误的。

C#还有一个特殊的关系操作符"is",它用于判断左操作数是否属于右操作数所指定的类型,例如:

```
int x = 5;
object o = x;
Console.WriteLine(o is object); //输出 true
Console.WriteLine(o is int); //输出 true
Console.WriteLine(o is long); //输出 false
```

操作符"is"主要用于引用类型的判断,只要左操作数的类型为右操作数,或是右操作数的派生类型,表达式就返回 true。但如果左操作数的值为 null,那么返回值始终为 false。

3.2.6 逻辑操作符

C#中提供了 3 个逻辑操作符。
- 逻辑与操作符"&&":有两个操作数,用于判断它们的值是否都为 true。
- 逻辑或操作符"||":有两个操作数,用于判断它们的值是否至少有一个为 true。
- 逻辑非操作符"!":只有一个操作数,用于判断其值是否为 false。

以枚举 Weekday 为例,判断某个枚举变量 w1 是否为周末的表达式可以写成:
```
bool b1 = (w1 == Weekday.Saturday) || (w1 == Weekday.Sunday);
```
而判断 w1 是否为工作日的表达式可以写成以下两种形式:
```
bool b2 = !((w1 == Weekday.Saturday) || (w1 == Weekday.Sunday));
bool b3 = (w1 != Weekday.Saturday) && (w1 != Weekday.Sunday);
```

综合使用这些逻辑操作符，可以对各种复杂的逻辑组合条件进行判断。例如判断某一个整数变量year是否表示闰年的语句可以写成：

    boolbLeap = (year % 400 == 0) || ((year % 4 == 0) && (year % 100 != 0));

在逻辑操作符的求值过程中，有时不需要计算完整个表达式就可得到结果，这称之为逻辑表达式的"短路"效应。例如对于上面这行代码，当year的值为2000时，第一对括号中的条件子表达式的值为true，那么bLeap的值就直接确定为true，而不再需要计算操作符"||"右侧的部分。

再看一个考核程序示例，考核小组由5名考官（含1名主考、2名本单位考官和2名外单位考官）组成，考核通过的条件是：主考同意，且至少有1名本单位考官和1名外单位官同意。将每名考官的评分依次放在一个布尔数组x中，下面的程序P3_7就通过一个复合逻辑表达式来判断考核结果，程序同时还输出了表达式中进行关系判断的次数：

```
//程序 P3_7
using System;
namespace P3_7
{
    class Program
    {
        static void Main()
        {
            bool[] x = new bool[5];
            Console.WriteLine("请依次输入每名考官的评分,1通过,0不通过: ");
            x[0] = (Console.ReadLine() == "1");
            x[1] = (Console.ReadLine() == "1");
            x[2] = (Console.ReadLine() == "1");
            x[3] = (Console.ReadLine() == "1");
            x[4] = (Console.ReadLine() == "1");
            int i = 0;
            bool b = (x[i++]) && (x[i++] || x[i]) && (x[++i] || x[++i]);
            Console.WriteLine("投票结果为: {0}", b);
            Console.WriteLine("判断次数为: {0}", i);
        }
    }
}
```

程序P3_7的输出结果如图3-2所示：利用逻辑运算的短路效应，只要主考不同意考核就不通过，此时只需进行一次判断；要确定考核通过，至少需要进行3次判断。

3.2.7 条件操作符

C#中还提供了一个条件操作符"?:"，它的特殊性在于它有三个操作数，使用时符号"?"和":"分别位于三个操作数之间，其形式为：

图3-2 程序P3_7的输出结果示例

    b ? x : y

其中第一个操作数b为布尔类型，如果其值为true则返回x的值，否则返回y的值。例如，下面的语句可用于计算两个整数x和y的最小值：

    intz = (x <= y) ? x : y;

注意在任何情况下，条件表达式都不会对x和y同时进行求值，但它仍要求x和y的类型是一致的（相同或是能进行隐式转换）。例如，下面最后一行代码是错误的，因为整数i和字符串s属于不同的类型：

```
int i = 5;
string s = "abc";
Console.WriteLine((i % 2 == 0) ? i : s);  //错误: i和s类型不同!
```
而将该行语句改为如下内容就是正确的了:
```
Console.WriteLine((i % 2 == 0) ? i.ToString() : s);  //正确
```
条件操作符的优先级仅高于赋值操作符,但低于其他的任何操作符。如果一个表达式中使用了多个条件操作符,要注意它们是从右向左结合的,例如求三个整数的最大值的表达式可以写成:
```
int w = x >= y ? x >= z ? x : z : y <= z ? z : y
```
该语句相当于:
```
int w = (x >= y) ? ((x >= z) ? x : z) : ((y <= z) ? z : y)
```
不过为了提高程序的可读性,还是应尽量使用括号来明确条件表达式的求值顺序。

3.3 控制结构

方法中的各行代码通常是按照顺序依次执行的。若要改变代码的执行流程,就要使用到控制结构。结构化程序设计中最常见的三种控制结构分别是选择、循环和跳转。

3.3.1 选择结构

选择结构根据不同的条件来选择要执行的语句,C#中提供了 if 和 switch 两种选择结构。

1. if 选择结构

最简单的选择结构是 if 语句,其形式为:
```
if (b-exp)
{
    statement;
}
```
其中 b-exp 表示一个布尔类型的表达式,其值为 true 时才会执行其后的语句 statement(如果该语句只有一行代码,那么其外的一对大括号可以省略)。例如下面的代码判断整数 i 是否为一个偶数,如是则将 i 除以 2:
```
if (i % 2 == 0)
    i = i / 2;
```
稍微复杂一点的选择结构是 if-else 语句,其形式为:
```
if (b-exp)
{
    statement1;
}
else
{
    statement2;
}
```
那么,程序在 b-exp 的值为 true 时执行语句 statement1,否则执行语句 statement2。下面的代码就使用了 if-else 语句来求两个整数最小值(注意不要把变量 z 的声明放在 if-else 结构之内,否则之后的语句将不能使用该变量):
```
int z;
if (x <= y)
    z = x;
else
```

```
        z = y;
Console.Write(z);
```

如果要实现两条以上的选择路径,那么可在 if-else 语句中嵌入多个 else if 分支,形如:
```
if (b-exp1)
{
    statement_1;
}
else if (b-exp2)
{
   statement_2;
}
else if…
else
{
    statement_n;
}
```

程序首先检查布尔表达式 b-exp1 的值,为 true 时执行语句 statement_1;否则检查 b-exp2 的值,为 true 时执行语句 statement_2;…;依次类推,如果所有条件都不成立,则执行最后一个 else 后的语句 statement_n。

设整数 x 表示考试分数,1~5 分别对应不及格、及格、中、良、优,那么下面的代码可根据分数来输出成绩的等级:
```
if (x == 5)
    Console.WriteLine("优");
else if (x == 4)
    Console.WriteLine("良");
else if(x == 3)
    Console.WriteLine("中");
else if(x == 2)
    Console.WriteLine("及格");
else
    Console.WriteLine("不及格");
```

以上 3 种 if 语句的选择流程可用图 3-3 来说明。

if-else 语句中还可以嵌套 if-else 语句,此时每个 else 分支都与其前面最近的一条不带 else 分支的 if 语句组成一个 if-else 对。不过在这种情况下,还是应尽量使用大括号对来明确 if-else 语句的具体结构。程序 P3_8 就使用了嵌套 if-else 语句来计算三个整数的最大值:
```
//程序 P3_8
using System;
namespace P3_8
{
    class Program
    {
        static void Main()
        {
            Console.Write("请输入 x 的值:");
            int x = int.Parse(Console.ReadLine());
            Console.Write("请输入 y 的值:");
            int y = int.Parse(Console.ReadLine());
            Console.Write("请输入 z 的值:");
            int z = int.Parse(Console.ReadLine());
            Console.Write("最大值为:");
```

```
            if (x >= y)
            {
              if (x >= z)
                  Console.WriteLine(x);
               else
                  Console.WriteLine(z);
            }
            else
            {
              if (y >= z)
                  Console.WriteLine(y);
               else
                  Console.WriteLine(z);
            }
          }
        }
```

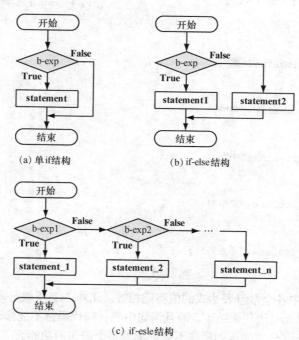

图 3-3 if 选择结构流程图

2. switch 选择结构

实现多条选择路径的另一种方式是使用 switch 语句，其形式为：

```
switch(exp)
{
  casec-exp1:
     statement_1;
     break;
  casec-exp2:
     statement_2;
     break;
  case …
  default:  //可选
     statement_n;
```

```
        break;
    }
```
其中，switch 标签后的 exp 为控制表达式，其类型可以是整数（包括字符和枚举）和字符串，而各个 case 标签后的常量表达式 c-exp 的类型必须与控制表达式的类型一致（相同或是能进行隐式转换）。程序执行时会将 exp 的值与表达式 c-exp1、c-exp2、……，依次进行比较，如果相等则执行对应 case 标签下的 statement 语句，然后执行 break 语句来结束 switch 语句；如果控制表达式的值与所有常量表达式的值均不匹配，则执行 default 标签后的语句，这里 default 部分是可选的。switch 选择结构执行的流程如图 3-4 所示。

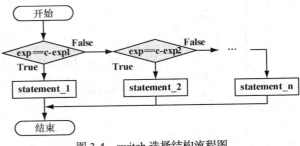

图 3-4　switch 选择结构流程图

以前面判断成绩等级代码为例，使用 switch 语句可以将其改写为如下代码：

```
switch (x)
{
    case 5:
        Console.WriteLine("优");
        break;
    case 4:
        Console.WriteLine("良");
        break;
    case 3:
        Console.WriteLine("中");
        break;
    case2:
        Console.WriteLine("及格");
        break;
    default:
        Console.WriteLine("不及格");
        break;
}
```

显然，switch 语句中各个常量表达式的值不能相等。此外，C#还规定每个 case 分支都必须以 break、return、goto 或 throw 语句来结束，而且语句中的任何代码都不能修改控制表达式的值，这就使得 switch 语句中 case 分支的排列顺序不会影响到整个语句的功能。

 对于经常使用的选择语句，如果能够判断出不同条件发生的概率，那么应尽量将概率较高的 if 或 case 分支放在选择结构中靠前的位置，这样能够减少程序中进行判断的次数，提高程序效率。例如考试成绩为良的学生数量最多，那么可以将相应的条件(x == 4)或(case 4)作为选择语句的第一个分支。

3.3.2　循环结构

循环结构用于代码段的重复执行。C#中提供了 while、do-while、for 和 foreach 这 4 种循环结构。

1. while 循环结构

while 循环结构指定一个条件，并在该条件为真时反复执行循环体。其形式为：

```
while(b-exp)
```

```
{
    statement;
}
```
设存款本金为 1000 元，年利率为 8%，用 while 循环来计算 5 年后的本息合计金额：
```
decimal x = 1000;
int i = 0;
while (i < 5)
{
    x = x * (1 + 0.08M);
    i++;
}
Console.WriteLine(x);
```
循环结构在控制台程序中的一个典型应用就是反复执行某项功能，直至用户按下特定的键后才退出程序。程序 P3_9 就给出了这样一个例子：
```
//程序 P3_9
using System;
namespace P3_9
{
    class Program
    {
        static void Main()
        {
            Console.WriteLine("按Q键退出，按其他键继续");
            while (Console.ReadKey().KeyChar != 'Q')
            {
                Console.WriteLine("欢迎光临");
                Console.WriteLine("按Q键退出，按其他键继续");
            }
        }
    }
}
```

2. do-while 循环结构

do-while 循环结构的形式为：
```
do
{
    statement;
}
while(b-exp);
```
也就是说，它首先执行 statement 语句，而后判断循环条件 b-exp 是否为 true，是则再次执行 statement 语句。以计算存款本息为例，使用 do-while 循环可将代码改写为：
```
decimal x = 1000;
int i = 0;
do
{
    x = x * (1 + 0.08M);
    i++;
}
while (i < 5)
```
和 while 循环不同的是，即使循环条件一开始就为 false，do-while 循环也能保证 statement 至少被执行一次。下面的程序是使用 do-while 循环来改写程序 P3_9，改写后控制台至少将输出一次"欢迎光临"：

```
//程序 P3_10
using System;
namespace P3_10
{
    class Program
    {
        static void Main()
        {
            do
            {
                Console.WriteLine("欢迎光临");
                Console.WriteLine("按Q键退出，按其他键继续");
            }
            while (Console.ReadKey().KeyChar != 'Q');
        }
    }
}
```

3. for 循环结构

for 循环是最为复杂也是最为灵活的一种循环结构，其形式为：

```
for(for-initializer;for-condition; for-iterator)
{
    statement;
}
```

其中 for-initializer、for-condition 和 for-iterator 分别表示循环初始化表达式、循环判断表达式和循环迭代表达式，它们都是完整的C#语句，而且都是可选的，之间用分号分隔。for 循环的执行顺序为：

（1）如存在循环初始化表达式 for-initializer，则执行它。

（2）如存在循环判断表达式 for-condition，则对它求值。

（3）若表达式 for-condition 为空，或其值为 true，则执行循环迭代表达式 for-iterator 和语句 statement，然后转第（2）步。

（4）若表达式 for-condition 的值为 false，则循环终止。

图 3-5 对 while、do-while 和 for 这 3 种循环结构的流程进行了比较。

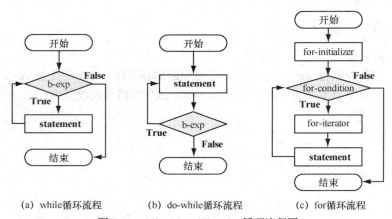

图 3-5　while、do-while、for 循环流程图

下面给出了使用 for 循环来计算存款本息的示例代码：

```
for (int i = 0; i< 5; i++)
```

```
{
    x = x * (1 + 0.08M);
}
```

也就是说,通常循环初始化表达式用于循环变量的初始化,循环判断表达式用于确定是否继续循环,而循环迭代表达式用于修改循环变量。再如程序 P3_7,可以改用 for 循环来依次输入每位考官的评分(注意循环变量 j 不能和之后的变量 i 重名):

```
static void Main()
{
    bool[] x = new bool[5];
    Console.WriteLine("请依次输入每名考官的评分,1 通过,0 不通过: ");
    for(int j=0; j<5; j++)
        x[j] = (Console.ReadLine() == "1");
    int i = 0;
    bool b = (x[i++]) && (x[i++] || x[i]) && (x[++i] || x[++i]);
    Console.WriteLine("投票结果为: {0}", b);
    Console.WriteLine("判断次数为: {0}", i);
}
```

复杂的循环程序常常需要用到多重循环结构。在下面的程序 P3_11 中,首先通过两重 for 循环对二维数组的每个元素进行赋值,而后在控制台依次输出这些元素(输出结果实际上就是一个如图 3-6 所示的九九乘法表):

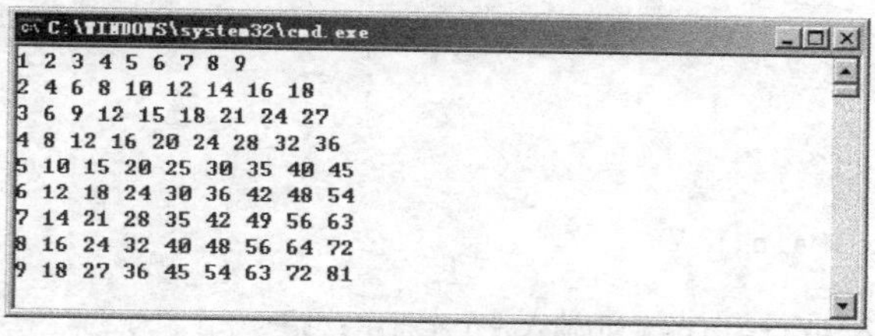

图 3-6 程序 P3_11 的输出结果

```
//程序 P3_11
using System;
namespace P3_11
{
    class Program
    {
        static void Main()
        {
            int[,] x = new int[9, 9];
            for (int i = 0; i < 9; i++)
                for (int j = 0; j < 9; j++)
                    x[i, j] = (i + 1) * (j + 1);
            for (int i = 0; i < 9; i++)
            {
                for (int j = 0; j < 9; j++)
                {
                    Console.Write(x[i, j]);
                    Console.Write(' ');
                }
                Console.WriteLine();
            }
        }
    }
}
```

4. foreach 循环结构

顾名思义，foreach 就是遍历集合中的元素，并对每个元素执行一次循环操作。例如下面的代码将在控制台依次输出数组 x 的各个元素：

```
int[] x = { 3, 7, 11 };
foreach(int i in x)
{
    Console.WriteLine(i);
}
```

上面的 foreach 语句指定了集合的元素类型（int）和标识（i），并在 in 关键字之后指定了集合对象 x，那么在循环过程中 x 的每一个元素会被依次赋值给 i，并执行循环体内所指定的操作。

foreach 循环语句有一个限制，即在循环体内不允许修改集合的元素。例如不能使用下面的代码将整数数组的每个元素加 1：

```
foreach(int i in x)
    i++;  //错误：不能修改集合元素 i 的值！
```

C#语言内部规定：如果一个对象支持"可枚举"的接口 IEnumerable 或 IEnumerable<T>，那么就可以将该对象作为 foreach 遍历的集合对象。C#的数组类型就默认支持该接口，因而支持 foreach 遍历。另一个支持"可枚举"接口的例子是字符串类型 string，通过 foreach 语句可以遍历其中的每个字符。例如下面的程序就用于统计字符串中字符"a"出现的次数：

```
//程序 P3_12
using System;
namespace P3_12
{
    class Program
    {
        static void Main()
        {
            Console.WriteLine("请输入一个字符串:");
            string s = Console.ReadLine();
            int i = 0;
            foreach (char ch in s)
            {
                if (ch == 'a')
                    i++;
            }
            Console.WriteLine("a 共出现{0}次", i);
        }
    }
}
```

3.3.3 跳转结构

如果循环条件一直为 true，那么循环语句就会无休止地执行下去，这叫做"死循环"。例如在前一小节计算存款本息的代码中，如果忘记在循环体中修改循环变量的语句"i++"，程序就会发生死循环，这属于应用程序的致命错误。

避免死循环有两种手段：一是改变循环条件，二是使用跳转语句强制"跳出"循环。

1. break 语句

在 switch 选择结构中我们已经看到过 break 语句，它表示"跳出"switch 语句。同样的，在任何一种循环结构中使用 break 语句都能够跳出当前循环，并继续执行循环语句之后的代码。例如程序 P3_9 可用 break 语句改写为程序 P3_13 的形式：

```
//程序 P3_13
using System;
namespace P3_13
{
    class Program
    {
        static void Main()
        {
            while (true)
            {
                Console.WriteLine("欢迎光临");
                Console.WriteLine("按 Q 键退出，按其他键继续");
                if (Console.ReadKey().KeyChar == 'Q')
                    break;  //跳出 while 循环
            }
        }
    }
}
```

在循环结构中，break 语句常常和选择结构配合使用，从而在程序满足特定条件时跳出循环。某些情况下程序还需要分辨出循环是正常结束的还是强制跳出的，下面的代码就演示了这样一种技术，它用于在整数数组 x 中找到并输出第一个大于 100 的数：

```
bool bf = false;
foreach(int i in x)
{
    if (i> 100)
    {
        Console.WriteLine(i);
        bf = true;
        break;  //跳出 foreach 循环
    }
}
if (!bf)
    Console.WriteLine(没有找到大于 100 的数);
```

在多重循环中，还要注意跳转语句只能跳出当前所在的循环层。

2. continue 语句

break 语句在跳出循环后将执行循环语句之后的代码，而 continue 语句用于在跳出循环后继续下一次循环，也就是说它只是跳过当前这次循环尚未执行的部分。例如下面的代码用于计算数列 1+2+4+5+7+8+ … +100 之和（不含 3 的倍数）：

```
int s = 0;
for (int i = 1; i <= 100; i++)
{
    if (i % 3 == 0)
        continue;  //跳过 3 的倍数
    s += i;
}
Console.WriteLine(s);
```

3. return 语句

return 语句用于整个方法的返回，此时程序的控制权转移至该方法的调用者；如果是在程序主方法 Main 中，return 语句就结束了整个程序。例如程序 P3_13 中的 break 语句可以替换为 return 语句，而程序的功能不会发生任何变化。

return 语句不一定要出现在循环结构中；但只要在循环结构中出现了该语句，循环（不论是单重还是多重循环）就会被立即终止。本书 4.2 节将对 return 语句作进一步说明。

4. goto 语句

goto 语句用于无条件跳转到指定的语句去执行。例如，在下面的程序中，Main 方法的最后一行代码前定义了一个"end"标签，而循环体中的 goto 语句中指定了该标签，那么条件满足时程序控制权就会无条件转移：

```csharp
//程序 P3_14
using System;
namespace P3_14
{
    class Program
    {
        static void Main()
        {
            int[] x = { 24, -316, 57, 200, 106, -10, 0, 1000, 99 };
            int j = 0;
            foreach (int i in x)
            {
                if (i > 100)
                {
                    j = i;
                    goto end;
                }
            }
            end: Console.WriteLine("x中第一个大于100的元素为{0}", j);
        }
    }
}
```

goto 语句是典型的非结构化控制语句，滥用 goto 语句会导致程序结构混乱，大大降低程序的可理解性和可维护性，因此在程序中应限制使用。事实上，结合循环结构以及 break 或 continue 语句，总是能够替换程序中的 goto 语句。例如，将程序 P3_14 中的 foreach 循环改写为如下内容，就可以消除之后的 goto 语句：

```csharp
foreach (int i in x)
{
    if (i > 100)
    {
        j = i;
        Console.WriteLine("x中第一个大于100的元素为{0}", j);
        break;
    }
}
```

3.4 案例研究——旅行社管理系统中结构和枚举

本节继续对旅行社管理系统作进一步分析，识别并定义所需的一些复合值类型。创建动态链接库程序项目 TravelLib，在其中定义有关旅行社业务的基础类型。考虑到游客种类包括普通游客、学生、儿童和老人，但不必要为每一种都定义一个类，而是可以通过一个枚举变量来标识，那么枚举类型的定义如下：

```csharp
enum CustomerType
{
    Common, Student, Child, Old
}
```

不是所有的游客都拥有或使用身份证,那么可定义下面的枚举类型来区分不同的证件类型(这里使用中文来命名枚举值):

```csharp
enum IDCardType
{
    身份证, 驾驶证, 护照, 户口簿, 军官证, 警官证, 离休证, 归侨证, 台胞证
}
```

 IDCardType 的定义中使用了中文来命名枚举值。事实上,C#中的类型、变量、枚举值等很多元素都支持中文名,不过除了英文单词难以表达或容易引起歧义的情况外,通常还是要避免使用中文。

在此基础上可定义结构类型 IDCard,其信息包括证件类型 Type、证件号 Id,以及证件的有效期 Due:

```csharp
struct IDCard
{
    public IDCardType Type;
    public string Id;
    public DateTime Due;
}
```

再考虑职员的学历以及当前状态,其取值范围可分别通过如下的枚举类型来表述:

```csharp
enum Degree
{
    高中, 中专, 大专, 本科, 硕士, 博士
}

enum StaffState
{
    在岗, 试用, 随团, 出差, 休假, 离职, 其他
}
```

很多对象都存在状态变迁的情况,而用枚举值来刻画对象所处的状态会很方便。注意 StaffState 定义的最后一个枚举值"其他"用于涵盖各种不确定或未知的情况。再如旅行团的状态可以定义为:

```csharp
enum TourState
{
    报名, 满员, 发团, 完成, 取消, 其他
}
```

系统还需要区分不同的旅游地区,假设只考虑国内旅游的情况,那么可定义如下的 Area 类型:

```csharp
enum Area
{
    华北, 东北, 华东, 中南, 西南, 西北, 其他
}
```

继续分析业务类的相关属性,可以发现很多内容都可以通过枚举来描述,例如:

- 景点类型 SceneType,取值包括自然山水、海滨沙滩、草原、沙漠、公园、博物馆、名胜古迹等。
- 组团类型 PackageType,取值包括每日发团、按周发团、按月发团、一次性团等。

- 组团等级 PackageGrade，取值包括经济、标准、豪华等。
- 酒店等级 HotelGrade，取值包括普通、两星、三星、四星、五星、农家等。
- 餐饮等级 DinnerGrade，取值包括简易、标准、高级、野餐等。

在处理景点门票、交通工具等其他信息时，还可能用到更多的结构和枚举类型，详细的设计和编码留给读者自己来完成。

3.5 小　　结

C#中的数据类型分为值类型和引用类型，其中值类型的变量直接包含数据，而引用类型的变量实际上只包含指向实际数据的指针。兼容的数据类型之间可以进行类型转换，例如低精度的值类型到高精度的值类型、派生类型到基础类型可进行隐式转换，而高精度的值类型到低精度的值类型、基础类型到派生类型可进行显式转换。

在数据类型上可以进行各种操作，C#定义的操作符主要包括算术操作符、自增减操作符、位操作符、赋值操作符、关系操作符、逻辑操作符和条件操作符。在包含多个操作符时，表达式的求值顺序由操作符的优先级决定；括号可被看做是优先级最高的操作符，因此总是可以使用括号来进行控制表达式的运算过程。

大多数程序不仅需要顺序执行，而且需要通过选择结构、循环结构和跳转语句来控制程序的流程。其中循环结构的灵活运用是实现许多高效程序的关键，而且要特别注意防止"死循环"的出现。

3.6 习　　题

1. 在 3.1.1 小节中介绍的简单值类型，其.NET 类库中的结构原型的名称分别是什么？
2. 说明值类型和引用类型最主要的三个不同之处。
3. 编写程序，验证 3.1.5 小节中介绍的显式值类型转换可能发生的数据丢失情况。
4. 对于两个整数 x 和 y，令 a = y / x，b = y % x，那么表达式 "y == (a * x + b)" 的结果是否总是为真？如果是两个实数呢？
5. 写出下面代码的输出结果：

```
int x = 0;
Console.WriteLine((x++) + (x++) + (x++));
Console.WriteLine((++x) + (++x) + (++x));
Console.WriteLine((--x) + (--x) + (--x));
Console.WriteLine((x--) + (x--) + (x--));
```

6. 位操作符和逻辑操作符有什么相同和不同之处？
7. 比较使用 "?:" 操作符的条件表达式和只有两个分支的 if-else 选择语句，二者是否总是可以互换？
8. 编写程序，由用户在控制台依次输入本金、利率和存款年数，程序输出本息合计的结果。要求能够反复执行计算，直至用户按下 Q 键退出程序。
9. 整数 n 的阶乘 $n! = n \times (n-1) \times (n-2) \times \cdots \times 1$。试分别使用 while、do-while 和 for 循环写出计算整数阶乘的代码。
10. continue 语句的功能总是可以用 break 语句 + 选择结构来实现，试举例说明其原因。
11. 编写程序，合并一个字符串中所有相邻的重复字符。
12. 编写程序，计算等差数列 $2+4+6+\cdots+100$ 之和，以及等比数列之和 $2+2^2+2^3+\cdots+2^{10}$。

第4章
类和对象

前面已经介绍了类的基本概念，本章将深入讲解 C#语言中类的用法，重点是如何使用类的成员方法来实现各种操作。在 4.5 节还介绍了 .NET 类库中的几个常用类型，读者可通过它们来加深对有关概念的理解。

4.1 成员概述

4.1.1 成员种类

类和结构都是复合类型，它们可以包含数据成员、函数成员和嵌套成员，其中数据成员指的就是字段。在类型定义时可以指定字段的初始值，例如：

```
class ComplexNumber
{
    public double a = 0;
    public double b = 0;
}
```

但此时字段并没有获得实际的存储空间，其值只有在创建对象时才会被分配：

```
ComplexNumber c1 = new ComplexNumber();
```

如果类型的定义中没有指定字段的初始值，使用对象时也没有给字段赋值，那么字段将会被赋予其类型的默认值：整数（包括字符和枚举）和实数类型的默认值为 0，布尔类型的默认值为 false，而所有引用类型的默认值为 null。例如对于下面的类 Student，在使用语句 Students1 = new Student ()创建对象后，其字段 name 和 age 的值将分别为 null 和 0：

```
class Student
{
    public string name;
    public int age;
    public Address address;

    public struct Address  //嵌套成员
    {
        public string Province;
        public string City;
        public string Detail;
    }
}
```

注意 Student 类中还包含了一个 Address 结构的定义，这种在其他类型中定义的类型就是嵌套成员。

类和结构的变量都可以称为对象,但使用时要牢记类是引用类型,而结构是值类型。类和结构都可以有函数成员——方法,包括一般方法和特殊方法,它们将分别在 4.2 节和 4.3 节中进行详细介绍。

4.1.2 成员访问限制

访问限制是实现信息隐藏和封装的重要手段,它是指类型及其成员可以定义不同的访问级别,其他对象必须符合权限才能进行访问。访问级别是通过访问限制修饰符(Modifier)来规定的,C#提供了以下 4 种访问限制修饰符。

- private(私有):用于成员访问限制,表示不允许外部对象访问该成员。
- public(公有):用于类型和成员访问限制,修饰成员时表示允许外部对象访问该成员,修饰类型时表示允许外部程序集使用该类型。
- protected(保护):用于成员访问限制,表示只允许当前类及其派生类的对象访问该成员,而不允许其他外部对象访问。
- internal(内部):用于类型和成员访问限制,表示不允许外部程序集使用该类型或访问该成员。

在设计一个类时,只有那些需要外部对象使用的成员才应被定义为公有;不希望外部对象访问、但希望派生类继承的成员应被定义为保护的;其他成员都应被定义为私有的。

例如对于 Student 类,如果希望公开学生姓名,但不公开其 E-mail 地址,那么可以将前者修饰为 public,后者修饰为 private;发送邮件的功能则可以封装在公有的 SendEmail 方法中,那么外部对象仍可以调用该方法来发送邮件,而不必去知道学生的 E-mail 地址。看下面的代码示例:

```
public class Student
{
    public string name;
    private string email;

    public void SendEmail(string title, string text)
    {
        MailMessage m1 = new MailMessage("admin@qh.edu", email, title, text);
        SmtpClient.Send(m1);
    }
}
```

而如果 Student 还有一个派生类 Graduate,该类的另一个方法 Register 中也要使用到 E-mail 地址,那就应当将 Student 中 email 字段的修饰符由 private 改为 protected,否则下面的代码就是不合法的:

```
public class Graduate : Student
{
    public void Register()
    {
        Console.WriteLine("姓名:{0}", name);
        Console.WriteLine("电子邮件:{0}", email);  //错误:不能访问私有的基类字段
    }
}
```

为了强调消息隐藏,C#中成员的默认访问限制修饰符为 private,因此没有指定访问限制的成员默认都是私有的。而如果采用简写方式一次声明多个成员字段,那么它们的访问限制也是相同的,如"public string name, email;"定义的两个字段都是公有的。

4.1.3 静态成员和非静态成员

通常所说的成员都是指非静态成员，或者叫做实例成员，即成员属于类型的实例（对象）所有。例如对于前面的 Student 类，每个学生对象都有自己的 name 和 email 字段。

另一种情况是静态成员，即成员属于类型本身所有，而不是随着具体对象的变化而变化。C# 中使用关键字 static 来定义静态成员，例如在 Student 类中可通过下面的代码来定义一个静态字段 School：

```
public static string School = "华中理工大学";
```

在使用时，静态成员是通过类名而不是对象名来访问的，例如：

```
Student s1 = new Student();
Console.WriteLine(s1.name);  //访问非静态成员
Console.WriteLine(Student.School);  //访问静态成员
```

对于类的静态字段，程序会在首次使用到该类时为其分配存储空间，而且该类的所有对象共享这一字段。类似的，可以定义类的静态方法，它也是通过类而非对象来进行调用的。下面的程序就为 Student 类定义了静态字段 School 和静态方法 WriteSchoolInfo，而另一个非静态方法 WritePersonalInfo 则同时使用到了类的静态字段和非静态字段：

```
//程序 P4_1
using System;
namespace P4_1
{
    class Program
    {
        static void Main()
        {
            Student.WriteSchoolInfo();
            Student s1 = new Student();
            s1.name = "王小红";
            Student s2 = new Student();
            s2.name = "周军";
            s2.Department = "数学系";
            Student.School = "华中科技大学";
            Student.WriteSchoolInfo();
            s1.WritePersonalInfo();
            s2.WritePersonalInfo();
        }
    }

    publicclass Student
    {
        public static string School = "华中理工大学";
        public string name, Department = "计算机系";
        publicint age;

        public void WritePersonalInfo()
        {
            Console.WriteLine("{0} {1} {2} ", School, Department, name);
        }

        public static void WriteSchoolInfo()
        {
            Console.WriteLine(School);
```

            }
        }
    }

在修改了类的静态字段后,所有 Student 对象(包括其派生对象)所访问的字段值都将发生变化。程序 P4_1 的输出如下:

> 华中理工大学
> 华中科技大学
> 华中科技大学计算机系王小红
> 华中科技大学数学系周军

在该程序中还应注意:WritePersonalInfo 方法中使用的 School 实际上是 Student.School 的简写。此外,静态方法中的代码只能使用类的静态成员,而不能直接使用非静态成员。这是因为静态方法属于类所有,在其中不能判断非静态成员究竟属于哪个对象。而实例方法既能够使用静态成员,也能够使用非静态成员。

对于一个类而言,如果它的所有成员都是静态的,那么还可以使用 static 关键字将其定义为静态类。例如,Console 类就是一个静态类,对它的所有成员访问都是通过类本身进行的。此外,对静态类不能使用 new 关键字来创建对象。

4.1.4 常量字段和只读字段

数学中有一些常量的值是不会发生变化的,比如圆周率π、万有引力常数 G 等。为了防止在程序中改变这些值,可以使用关键字 const 将其定义为类型的常量字段。例如下面的 Circle 类就定义了一个常量字段 PI,并在 GetArea 方法中使用了该字段:

```
public class Circle
{
    public const double PI = 3.1415927;
    public double r;

    public double GetArea()
    {
        return PI * r * r;
    }
}
```

常量字段必须在定义时就进行赋值,之后只能读取、而不允许修改其值。常量字段默认也是静态的(但无须再用关键字 static 来进行修饰),即字段属于类型本身所有。因此在下面计算圆周长的代码中,对 PI 的访问只能通过类型名 Circle 进行:

```
Circle c = new Circle();
c.r = 5;
double p = 2 * Circle.PI * c.r;
Circle.PI = 3;   //错误:不允许修改常量字段!
```

提示　　常量字段的类型一般都是值类型或字符串类型;对于 string 以外的引用类型,其常量字段的值必须是 null,这在实际应用中通常没有什么意义。

还有一类常量是针对单个对象的,如每个人的身份证号、每张银行卡的卡号都不应被改变,那么可使用关键字 readonly 将其定义为只读字段,它只能在对象创建时被赋值,而后不允许再修改。例如下面程序中的 BankCard 类就定义了一个只读字段 id,并在其静态字段 count 中记录已有的对象数量,从而使每个新创建的 BankCard 对象的 id 值自动递增:

```
//程序 P4_2
using System;
namespace P4_2
{
    class Program
    {
        static void Main()
        {
            for (int i = 0; i < 5; i++)
            {
                BankCard c = new BankCard();
                Console.WriteLine(c.id);
            }
        }
    }

    public class BankCard
    {
        public readonly int id;
        private static int count = 0;

        public BankCard()
        {
            id = ++count;
        }
    }
}
```
运行该程序将在控制台依次输出 1、2、3、4、5。其中的构造函数将在 4.3.1 小节中介绍。

4.2 方　　法

方法是最基本的函数成员，对象所能执行的操作都是通过方法来定义的。方法的声明应依次包括以下 4 部分。
- 返回类型：表示方法返回的计算结果的数据类型；如不需要返回结果，那么应声明返回类型为 void。
- 方法名。
- 参数列表：包含在方法名后的一对括号中，表示方法的参数集合，对每个参数都应指定参数类型和参数名，参数之间用逗号分隔；没有任何参数则为空括号。
- 执行体：包含在一对大括号中的方法执行代码。

例如，下面的代码定义了一个名为 Max 的方法，其参数列表包含两个 int 类型的参数 x 和 y，方法的返回类型也为 int，而方法执行体中只有一行代码：

```
public int Max(int x, int y)
{
    return x >= y ? x : y;
}
```

方法是实现程序模块化的重要手段，它将对不同变量的操作统一抽象为对参数的操作，从而实现对方法代码的"一次定义、多次使用"。例如下面的代码中多次使用到了类似的条件表达语句：

```
int x = -10, y = -5;
Console.WriteLine("x 和 y 中较大的数为:{0}", x >= y ? x : y);
int a = x + y, b = x - y, c = y - x;
Console.WriteLine("a、b、c 中较大的数为:{0}", c>=(a>=b ? a:b) ? c : (a>=b ? a:b));
```

而使用上面定义的 Max 方法，就可以对代码进行有效的简化：
```
int x = -10, y = -5;
Console.WriteLine("x 和 y 中较大的数为:{0}", Max(x,y));
int a = x + y, b = x - y, c = y - x;
Console.WriteLine("a、b、c 中较大的数为:{0}", Max(c, Max(a,b)));
```

4.2.1 方法的返回值

如果方法的返回类型不为 void，那么方法的执行代码中必须通过 return 语句来返回一个值，而且 return 关键字后面的表达式类型必须与声明的返回类型一致（相同或是能进行隐式转换）。例如对于上面的 Max 方法，表达式"x >= y ? x : y"的类型就为 int；而对于前一小节中 Circle 类的 GetArea 方法，表达式"PI * r * r"的类型就为 double。

当方法的返回类型为 void 时，其执行代码要么没有 return 语句，要么只有不跟任何表达式的 return 语句。例如可以为 Circle 类增加下面的方法，用于实现圆的缩放：
```
public void Zoom(double rate)
{
    r *= rate;
}
```
进一步，如果要避免缩放比例为 0 或负数的情况，那么可对 Zoom 方法作如下修改：
```
public void Zoom(double rate)
{
    if (rate <= 0) return;
    r *= rate;
}
```
方法代码的执行过程中，只要遇到 return 语句就会返回，剩余的代码将被跳过。在编程过程中，要注意无条件的 return 语句之后的代码都是不可达的。

4.2.2 参数类型

1. 值传递和引用传递

方法声明的参数列表中的参数叫做形式参数（以下简称形参），而实际调用时传递给方法的参数叫做实际参数（以下简称实参），二者的名称不要求相同。例如对于 Circle 类的 Zoom 方法，其参数 rate 就是形参；下面的代码创建了一个 Circle 对象并调用了其 Zoom 方法，传递给该方法的参数 x 就是实参：
```
Circle c1 = new Circle();
double x = 0.5;
c1.Zoom(x);
```
调用方法时，程序会首先将实参的值传递给对应的形参，而后执行方法体中的代码。对于引用类型的参数，实参和形参会指向同一个对象；而对于值类型的参数，实参的值将被复制一份给形参，方法代码中对形参值的修改并不会影响到实参。例如下面的代码试图交换 x 和 y 的值，但实际上交换的只是形参的值，而不真正改变实参的值：
```
public void Swap(int x, int y)
{
    int z = x;
    x = y;
    y = z;
}
```
为了解决这一问题，C#中提供了方法参数的引用传递方式，此时形参是实参的"引用"，二者指向同一个变量；如果方法代码中修改了形参的值，实参的值也会发生相应的变化。引用传递的方法参数（以下简称引用型参数）通过 ref 关键字加以修饰，例如把 Swap 方法声明中的参数改为引用型，这样方法就能真正起到交换参数值的效果：
```
public void Swap(ref int x, ref int y)
```
而在方法调用时，引用型参数对应的实参之前也需要使用 ref 关键字。下面的程序对值传递和引用传递进行了比较：

```
//程序 P4_3
using System;
namespace P4_3
{
    class Program
    {
        static void Main()
        {
            int a = 10, b = 20;
            ValueSwap(a, b);
            Console.WriteLine("值传递: a = {0}, b = {1}", a, b);
            ReferenceSwap(ref a, ref b);
            Console.WriteLine("引用传递: a = {0}, b = {1}", a, b);
        }

        Public static void ValueSwap(int x, int y)
        {
            int z = x;
            x = y;
            y = z;
        }

        Public static void ReferenceSwap(ref int x, ref int y)
        {
            int z = x;
            x = y;
            y = z;
        }
    }
}
```

该程序的输出显示了值传递和引用传递的不同效果：

```
值传递: a = 10, b = 20
引用传递: a = 20, b = 10
```

2. 输出型参数

程序计算中有时会需要一个方法返回多个结果值，这时可通过 return 语句返回其中一个结果，并通过输出型参数返回其他结果。输出型参数使用 out 关键字加以修饰。例如，下面的方法将返回 x 元在 n 年后的本息合计金额，而输出型参数 interest 将返回最后一年的年息：

```
Public decimal Gain(decimal x, int n, out decimal interest)
{
    int i = 0;
    do {
        interest = x * 0.08M;
        x += interest;
        i++;
    }
    while (i < n);
    return x;
}
```

输出型参数也采用引用传递方式，但其形参要求在方法返回之前必须被赋值（以满足"输出"

的要求），而其所对应的实参在使用时可以不进行初始化，例如，下面代码中的变量 y 就被用作 Gain 方法的输出型实参：

```
decimal x = 1000;
decimal y;
decimal z = Gain(x, 10, out y);  //通过输出型参数对变量y进行初始化
Console.WriteLine("本息合计:{0}，最后一年利息:{1}", z, y);
```

但如果在调用 Gain 方法之前插入语句"Console.WriteLine(y)"就是错误的，因为此时变量 y 并没有被赋值。

3. 数组型参数

方法的参数类型也可以是数组，例如下面的方法可用于计算整数数组所有元素的和：

```
public int Sum(int[] array)   //使用普通数组作参数
{
    int s = 0;
    for (int i = 0; i < array.Length; i++)
        s += array[i];
    return s;
}
```

调用该方法的示例代码如下：

```
int[] a = { 1, 3, 5, 7, 9 };
int x = Sum(a);
```

而如果在参数前面加上关键字 params，该参数就成为了数组型参数：

```
public int Sum(params int[] array)   //使用数组型参数
{
    int s = 0;
    for (int i = 0; i < array.Length; i++)
        s += array[i];
    return s;
}
```

传递给数组型参数的实参既可以是一个数组，又可以是任意多个数组元素类型的变量。例如，下面调用 Sum 方法的代码都是合法的：

```
int[] a = { 1, 3, 5, 7, 9 };
int x = Sum(a);
int y = Sum(1, 3, 5, 7, 9);
int z = Sum(10, 20, 50, 100);
```

C#中对数组型参数的使用有着严格的规定：

- 方法中只允许定义一个数组型参数，而且该参数必须位于参数列表中的最后。
- 数组型参数所定义的数组必须是一维数组。
- 数组型参数不能同时作为引用型参数或输出型参数。

C#应用程序的主方法 Main 也可以使用数组型参数，但它额外规定数组元素的类型必须是字符串 string，此时的数组型参数表示程序所能接受的参数集合。看下面的例子：

```
//程序 P4_4
using System;
namespace P4_4
{
    class Program
    {
        static void Main(params string[] args)
        {
            if (args.Length == 0){
                Console.WriteLine("未指定输入参数");
                return;
```

```
        }
        double s = double.Parse(args[1]);
        if (args[0] == "SUM"){
            for (int i = 2; i < args.Length; i++)
                s += double.Parse(args[i]);
        }
        else if (args[0] == "PROD"){
            for (int i = 2; i < args.Length; i++)
                s *= double.Parse(args[i]);
        }
        Console.WriteLine(s);
    }
}
```

程序 P4_4 要求第一个参数必须为 "SUM" 或 "PROD"，对前者将计算剩余各参数之和，对后者则计算剩余各参数之积，其输出如图 4-1 所示。

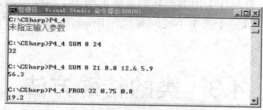

图 4-1　带参数的应用程序输出示例

4.2.3　方法的重载

有时会希望使用同一个方法名来表示多个操作，比如交换两个整数、两个实数、两个字符串的方法都可以叫做 Swap，而不是将其分别命名为 SwapInt、SwapDouble、SwapString 等。C#允许一个类中包含多个同名的方法，但要求它们的参数数量或类型不完全相同，这叫做方法的重载。例如下面的 Program 类中就定义了 3 个重载的 Swap 方法：

```
class Program
{
    Public static void Swap(ref int x, ref int y)
    {
        int z = x;
        x = y;
        y = z;
    }

    public static void Swap(ref double x, ref double y)
    {
        double z = x;
        x = y;
        y = z;
    }

    public static void Swap(ref string x, ref string y)
    {
        string z = x;
        x = y;
        y = z;
    }
}
```

在使用过程中，程序会根据传递给方法的实参类型来确定具体调用哪一个方法：

```
double x = 12.5, y = 20;
Program.Swap(ref x, ref y);   //调用第 2 种重载形式
string s1 = "Apple", s2 = "Orange";
Program.Swap(ref s1, ref s2); //调用第 3 种重载形式
```

不过，如果两个方法名称和参数列表都相同，而仅仅是返回类型不同，那么 C#认为它们不能构成重载，例如一个类中不能同时包含下面两个方法：

```
public int Add(int y)
{
    return 100 + y;
}

public double Add(int y)//错误：仅返回类型与前一方法不同，不能重载
{
    return 1.5* y;
}
```

此外，由于输出型参数只是在引用型参数的基础上增加了特殊限制，因此只有 ref 和 out 型参数的区别也不能构成重载，如一个类中不能同时包含 Swap(ref int x, ref int y)和 Swap(ref int x, out int y)两个方法。

4.3 类的特殊方法

4.3.1 构造函数和析构函数

对象都有自己的生命周期，对象的创建和销毁是其中两个重要的时间点，构造函数（Constructor）和析构函数（Destructor）就表示这两个时间点上的操作：构造函数用于创建对象，而析构函数会在对象销毁时自动执行。

1．构造函数

结构和类都可以有构造函数，函数名称与类型名称相同。和一般方法不同的是，构造函数不声明返回类型，其中的代码主要用于完成对象的初始化工作。例如下面 Student 类的构造函数将 grade 字段值设为 1：

```
Public class Student
{
  public string name;
  public int  grade;
  public Student() // 构造函数
  {
     grade = 1;
  }
}
```

如果希望通过不同的数据来创建不同的对象，那么可以使用带参数的构造函数，例如：

```
public class Student
{
  public string name;
  public int  grade;
  public Student(string s) // 带参构造函数
  {
     name = s;
  }
}
```

这样就可以指定学生姓名来创建不同的 Student 对象了：
```
Student s1 = new Student("王小红");
Student s2 = new Student("周军");
```
构造函数可以有多种重载形式，以便使用不同的方式来创建对象。但如果一个类只定义了带参数的构造函数，创建对象时就必须指定相应的参数。例如下面的 Student 类定义了两个重载的带参构造函数，那么创建 Student 对象时要么指定学生姓名，要么同时指定学生的姓名和年龄，而不允许使用 "new Student()" 这样的表达式来创建对象。
```
public class Student
{
   public string name;
   public int age;
   public byte grade;

   public Student(string n)   // 带参构造函数
   {
      name = n;
   }

   public Student(string n, int a)   // 带参构造函数
   {
      name = n;
      age = a;
   }
}
```

2. 对象初始化表达式

C# 从 3.0 开始提供了对象初始化表达式，即只要类具有无参构造函数，那么可以在创建对象的同时为其公有字段赋值。例如下面的类定义会包含一个默认的无参构造函数：
```
Public class Student
{
   public string name;
   public int age;
   public byte grade;
}
```
下面的代码是先创建对象，再为字段赋值：
```
Student s1 = new Student();
s1.name = "王小红";
s1.grade = 2;
```
而对象初始化表达式允许将上述语句合并在一起：
```
Student s1 = new Student() { name = "王小红", grade = 2 };
```
表达式大括号中的代码将在调用构造函数后被执行。例如在 Student 的无参构造函数中设置 grade 字段值为 1，使用上面这行代码创建的 Student 对象的 grade 字段值仍将是 2。

根据访问限制，声明为 protected 或 private 的构造函数不能用于对象创建。因此当一个类只定义了私有的构造函数，那么无法通过 new 关键字来创建其对象。此外，如果一个类没有定义任何构造函数，那么 C#编译器会为其自动生成一个默认的无参构造函数。

对于结构类型，C#编译器总是会为其创建一个公有的默认无参构造函数，而开发人员只能显式地为结构类型定义带参数的构造函数。

3. 静态构造函数

和其他成员一样，实例构造函数属于对象所有，而使用 static 修饰符的静态构造函数为类的

所有对象共享，并且只在首次使用该类时被调用。C#规定静态构造函数没有任何参数，且不声明访问限制（默认为 public）。

下面的 Student 类中定义了两个静态字段 objects 和 classes，前者在实例构造函数中加 1，后者则在静态构造函数中加 1：

```csharp
//程序 P4_5
using System;
namespace P4_5
{
    class Program
    {
        static void Main()
        {
            Student s1 = new Student("王小红");
            Student s2 = new Student("周军");
            s1 = new Student("Jerry");
        }
    }
    public class Student
    {
        public static int objects = 0, classes = 0;
        public string name;

        public Student(string n)//实例构造函数
        {
            name = n;
            Console.WriteLine("对象计数: {0}", ++objects);
        }
        static Student()//静态构造函数
        {
            Console.WriteLine("类计数: {0}", ++classes);
        }
    }
}
```

从该程序的输出中可以看到，class_count 的值只会增加一次，而 object_count 的值在每次创建对象都会增加：

```
类计数: 1
对象计数: 1
对象计数: 2
对象计数: 3
```

4. 析构函数

和构造函数不同，一个类中只能有一个析构函数，其名称是在类名前加上符号"~"。析构函数不能有参数和返回类型、不能是静态的，也不能有访问限制修饰符。例如 Student 类的析构函数可定义为：

```csharp
~Student()   // 析构函数
{
    Console.WriteLine("学生对象{0}销毁", name);
}
```

析构函数中的代码主要用于执行释放资源（如删除临时文件、断开与数据库的连接）等任务。在 C#程序中，用户不能显式地调用析构函数；公共语言运行时（CLR）负责管理内存中的所有对象，并使用垃圾收集器 GC 自动销毁无用的对象。通过静态类 GC 的 Collect 方法可强制程序进行垃圾回收，从而间接调用相关对象的析构函数：

```
GC.Collect();
```

4.3.2 属性

为了实现对数据的良好封装，C#为类提供了属性访问函数（简称属性，Property），主要用于控制对字段的访问。属性包括 get 访问函数和 set 访问函数，分别用于值的读取和设置。例如在 Student 类中可使用 Name 属性来封装对私有字段 name 的访问：

```
private string name;
public string Name  //属性
{
    get { return name; }
    set { name = value; }
}
```

作为类的特殊函数成员，get 和 set 访问函数需要包含在属性声明的内部，而函数声明只需要写出 get 和 set 关键字即可。其中 get 访问函数没有参数，默认返回类型就是属性的类型，表示属性的返回值；set 访问函数的默认返回类型为 void，且隐含了一个与属性类型相同的参数 value，表示要传递给属性的值。这样就可以通过属性来访问隐藏的字段，例如：

```
Student s1 = new Student("王小红");
Console.WriteLine(s1.Name);  //调用 get 访问函数访问 name 字段
Console.WriteLine("请输入新姓名:");
s1.Name = Console.ReadLine();  //调用 set 访问函数修改 name 字段
```

属性也可以只包含一个访问函数。如果只有 get 访问函数，那么属性的值不能被修改；如果只有 set 访问函数，则表明属性的值只能写不能读。例如希望 Student 对象在创建之后就不允许修改其姓名，那么 Name 属性的定义可改写为：

```
public string Name  //只读属性
{
    get { return name; }
}
```

使用只读属性时要注意：如果属性所封装的字段本身也是一个复合类型，那么并不能限制对其成员的修改。看下面的代码示例：

```
public class Address
{
    public string Province;
    public string City;
    public string Detail;
}

public class Student
{
    private string name;
    public string Name
    {
        set { name = value; }
    }

    private Address address;
    public Address Address
```

```
        get { return address; }
    }
    public Student(string n)
    {
        name = n;
        address = new Address();
    }
}
```

Student 类的私有字段 address 的类型为 Address 类,尽管该字段通过属性进行了只读封装,通过下面代码还是可以修改 address 对象自身的字段成员:

```
Student s1 = new Student("王小红");
s1.Address.Province = "江西";
s1.Address.City = "庐山";
```

不过,像下面这样直接设置 address 对象的代码仍是不合法的:

```
s1.Address = new Address();
s1.Address = null;
```

C# 从 3.0 开始还提供了"自动属性"的编程方式,它允许只写出属性及其访问函数的名称,编译器就会自动为其生成所要封装的字段以及访问函数的执行代码。下面定义的 Student 类中就使用了两个这样的自动属性:

```
class Student
{
    public string Name { get; set; }
    public int Age { get; set; }
}
```

其效果和下面这种传统定义方式是一样:

```
class Student
{
    Private string name;
    public string Name
    {
        get { return name; }
        set { name = value; }
    }

    private int age;
    public int Age
    {
        get { return age; }
        set { age = value; }
    }
}
```

自动属性的使用也存在一定的限制:首先它要求属性必须同时包含 get 和 set 访问函数;其次是自动生成的字段是完全隐藏的,即使在当前类的方法代码中也不能直接访问,而仍需通过属性进行读写。

在对象初始化表达式中,对可读写的公有属性也可以进行直接赋值,例如:

```
Student s1 = new Student() { Name = "王小红", Age = 20 };
```

属性的典型用法是一个公有属性封装一个私有或保护字段,但这并不是强制要求。属性在本质上是函数,在其代码中可以进行各种控制和计算。例如在学生类中有一个表示出生年份的私有字段 birthYear,那么表示年龄的属性 Age 的返回值就是当前年份(假定为 2012 年)减去出生年份:

```
private int birthYear;
public int Age
{
    get { return 2012 - birthYear; }
}
```

4.3.3 索引函数

程序设计过程中常常会遇到这样一些类,它们的主要作用是维护一组数据的集合。例如"同学录"类用于维护一组同学的信息;使用属性来访问数据集合的话,该类可定义为如下形式:

```
public class ClassmateList
{
    private string title;
    private Student[] classmates;

    public Student[] Classmates
    {
        get { return classmates; }
        set { classmates = value; }
    }

    public ClassmateList(string t)
    {
        title = t;
        classmates = new Student[100];
    }
}
```

通过对象的属性来访问集合元素的示例代码如下:
```
ClassmateList clist = new ClassmateList("我的大学同学录");
clist.Classmates[0] = new Student("王小红");
clist.Classmates[1] = new Student("周军");
clist.Classmates[2] = new Student("方小白");
for (int i = 0; i < 3; i++)
    Console.WriteLine(clist.Classmates[i].Name);
```

索引函数(Indexer)可看作是属性的扩展,它能够以数组的方式来访问对象内部的多项数据。下面是使用了索引函数改写后的 ClassmateList 类定义:
```
publicclass ClassmateList
{
    private string title;
    private Student[] classmates;

    public Student this[int index]    //索引函数
    {
        get { return classmates[index]; }
        set { classmates[index] = value; }
    }

    public ClassmateList(string t)
    {
        title = t;
        classmates = new Student[100];
    }
}
```

接下来就可以以"对象+下标"的方式来访问 ClassmateList 对象中的 classmates 数组元素了:
```
ClassmateList clist = new ClassmateList("我的大学同学录");
clist[0] = new Student("王小红");
```

```
clist[1] = new Student("周军");
clist[2] = new Student("方小白");
for (int i = 0; i < 3; i++)
    Console.WriteLine(clist[i].Name);
```

从中可以看出，索引函数和属性一样包含 get 和 set 访问函数，它使用 this 关键字加中括号[]进行定义，其中的参数表示所要访问的集合元素的索引下标。使用时只要在对象名后的中括号里指定下标就可以访问目标元素。很显然，一个类中只能定义一个索引函数。

索引函数的参数类型必须为整数或字符串；此外它所维护的集合对象不一定要是数组，也可以是多个离散的字段。例如在下面的 Teacher 类包含了 homePhone、officePhone、mobilePhone 这 3 个私有字段，并通过索引函数来控制它们的读写：

```
public class Teacher
{
    private string name;
    private string homePhone = "未知";
    private string officePhone = "未知";
    private string mobilePhone = "未知";

    public string this[string index]
    {
        get{
            if (index == "家庭电话")
                return homePhone;
            else if (index == "办公电话")
                return officePhone;
            else
                return mobilePhone;
        }
        set{
            if (index == "家庭电话")
                homePhone = value;
            else if (index == "办公电话")
                officePhone = value;
            else if(index == "手机")
                mobilePhone = value;
        }
    }

    public Teacher(string s)
    {
        name = s;
    }
}
```

这样的索引函数定义稍显复杂，但它能够在使用时大大提高代码的可读性，例如：

```
Teacher t1 = new Teacher("张大强");
t1["手机"] = "13011110234";
t1["办公电话"] = "88664321";
Console.WriteLine(t1["家庭电话"]);
```

4.3.4 操作符重载

考虑下面定义的复数类 ComplexNumber，其中定义了一个静态方法 Add，用于实现两个复数的相加操作：

```
Public class ComplexNumber
{
    public double a = 0, b = 0;
    public ComplexNumber(double x1, double x2)
    {
        a = x1;
        b = x2;
    }
    public static ComplexNumber Add(ComplexNumber c1, ComplexNumber c2)
    {
        return new ComplexNumber(c1.a + c2.a, c1.b + c2.b);
    }
}
```
通过该方法进行复数相加运算的示例代码如下：
```
ComplexNumber c1 = new ComplexNumber(-1.5, 3);
ComplexNumber c2 = new ComplexNumber(1, -1);
ComplexNumber c3 = ComplexNumber.Add(c1, c2);
```
不过上面这种表达方式显得很不自然，人们更习惯使用"c3 = c1 + c2"这样的语句。在本书 3.2 节中介绍了 C#语言中的各种预定义操作符，而操作符重载机制能够使这些操作符作用于用户自定义的类型。对于 ComplexNumber 类，只需将其中的方法名"Add"改为"operator +"，就能实现对"+"操作符的重载：
```
public static ComplexNumber operator +(ComplexNumber c1, ComplexNumber c2)
{
    return new ComplexNumber(c1.a + c2.a, c1.b + c2.b);
}
```
接下来就可以直接使用加号进行复数相加运算了：
```
ComplexNumber c3 = c1 + c2;
```
作为一种特殊的函数成员，被重载的操作符必须被声明为公有的和静态的，并通过在关键字 operator 后跟操作符来声明。此外，重载的一元操作符要求有一个参数，且参数类型应与当前类型一致（相同或是能进行隐式转换）；重载的二元操作符要求有两个参数，且至少有一个参数类型与当前类型一致。C#中的允许被重载的操作符有：

- 一元操作符：++, -, !, ~, (*T*), true, false。
- 二元操作符：+, -, *, /, %, &, |, ^, <<>>, ==, !=, ><=, <=。

其中(*T*)表示类型转换操作，它使得用户自定义的类型能够自动转换到其他类型，重载时还需要通过 explicit 或 implicit 关键字来说明要进行的是显式类型还是隐式类型转换。此外 C#中的布尔常量 true 和 false 也可以作为一元操作符来重载，它要求返回类型为 bool，这样就可以将对象本身作为一个布尔表达式来使用。

下面的程序 P4_6 定义了一个 Prime 类，它不仅重载了加法和减法操作符，而且还重载了类型转换操作符 uint（用于将 Prime 对象显式转换为一个整数），以及操作符 true 和 false（用于判断一个 Prime 对象是否为一个合法的素数）。借助这些操作符重载，程序的主方法直接输出了 50～100 之间的所有素数：

```
//程序 P4_6
using System;
namespace P4_6
{
    class Program
    {
        static void Main()
        {
```

```csharp
            for (uint i = 50; i <= 100; i++){
                Prime p1 = new Prime(i);
                if (p1)
                    Console.WriteLine((uint)i);
            }
        }
    }
    public class Prime
    {
        public uint x;
        public Prime(uint x1)
        {
            x = x1;
        }
        public static uint operator +(Prime p1, Prime p2)
        {
            return p1.x + p2.x;
        }
        public static int operator -(Prime p1, Prime p2)
        {
            return (int)(p1.x - p2.x);
        }
        public static explicit operator uint(Prime p)
        {
            return p.x;
        }
        public static bool operator true(Prime p)
        {
           for (uint i = 2; i <= p.x / 2; i++)
             if (p.x % i == 0)
                return false;
           return true;
        }
        public static bool operator false(Prime p)
        {
           for (uint i = 2; i <= p.x / 2; i++)
             if (p.x % i == 0)
                 return true;
           return false;
        }
    }
```

对于所有的复合赋值操作符，只要其左部的二元操作符已被重载，那么整个赋值操作符也会被自动重载。例如在ComplexNumber中重载了加法操作符后，对两个复数c1和c2就可以使用"c1 += c2"这样的表达式，它相当于"c1 = c1 + c2"。此外考虑到操作的对称性，一元操作符true和false，二元操作符==和!=、>和<，以及>=和<=都要求被成对重载。

在C#语言中，大多数操作符重载的返回类型都没有限制，例如两个复数相乘的结果可定义为一个复数，也可以定义为一个实数。但对于比较操作符而言，建议它们的返回类型都应当为 bool，这不仅符合人们的思维习惯（例如表达式"x>y"返回一个实数会让人很难理解），也有利于保持程序语义的一致性。

4.4　this 对象引用

类的代码中常常会需要访问到当前对象。例如对于一个银行账户类 Account，其 Remit 方法用于向其他账户汇款。显然一个账户不能给自己汇款，那么方法代码就要先进行判断：

```
public bool Remit(Account a1, decimal money)
{
    if (a1 == this){
        Console.WriteLine("不能向本账户汇款");
        return false;
    }
    else…//执行汇款操作
}
```

其中，this 关键字就表示对当前对象的引用，其类型就是当前类型。因此，在 this 后跟圆点连接符就可以访问当前对象的所有实例成员，例如：

```
public class ComplexNumber
{
    private double a = 0;
    public double A
    {
        get { return a; }
        set {
            if (this.a != value)
                this.a = value;
        }
    }

    private double b = 0;
    public double B
    {
        get { return b; }
        set {
            if (this.b != value)
                this.b = value;
        }
    }

    public void Write()
    {
        Console.WriteLine("{0}+{1}i", this.a, this.b);
    }
}
```

上面代码中的 "this." 其实都可以省略。不过 C#允许方法的参数名与类型的字段名相同，那么在方法的代码中就应当使用 this 关键字来区分的字段和形参变量，下面的构造函数定义就是这样一个例子：

```
public ComplexNumber(double a, double b)
{
    this.a = a;
    this.b = b;
}
```

4.5　常用类型

4.5.1　Object 类

Object 类（可简写为 object）是.NET 类库中最顶层的基类，它提供了以下 4 个公有成员方法。

- string ToString()：获得对象的字符串表示。
- Type GetType()：获得对象的数据类型。
- bool Equals(Object obj)：判断当前对象与对象 obj 是否相等。
- int GetHashCode()：获得对象的哈希函数值，适用于基于哈希表的数据结构类型。

这些方法都自动被所有其他类型所继承，因此对任何对象都可以调用。前面已经介绍过，预定义值类型的 ToString 方法将返回数值的字符串表示，这是因为它们对该方法进行了重载，否则该方法将直接返回对象类型的字符串表示。事实上，Console 类的 Write 和 WriteLine 方法输出的都是参数对象的字符串表示。例如对于一个不为空的 Student 对象 s1，下面两行代码的效果实际上是相同的：

```
Console.WriteLine(s1);
Console.WriteLine(s1.ToString());
```

GetType 方法将返回对象的类型，这里的 Type 类也是 .NET 类库中的一个类。类型表示的字符串格式为"命名空间" + "." + "类型名"，例如 object 类型的字符串表示就是"System.Object"。要注意 GetType 方法总是返回对象的实际类型，而不是声明类型。例如：

```
object o1 = new object();
Console.WriteLine(o1.GetType()); //输出"System.Object"
o1 = new Student();
Console.WriteLine(o1.GetType()); //输出"Student"
```

object 还提供了两个静态方法 Equals 和 ReferenceEquals，它们都接受两个 object 类型的参数。对于对象 o1 和 o2，调用实例方法 o1.Equals(o2)和静态方法 object.Equals(o1,o2)的结果是相同的。ReferenceEquals 则用于判断两个变量是否指向同一个内存对象，即判断二者的引用是否相同。

4.5.2 String 类

1. 字符串与字符数组

String 类（可简写为 string）表示字符串类型，它将一组字符视为一个整体进行处理，使用起来很像是一个字符数组，例如可以通过 Length 属性判字符串的长度，还可以通过索引下标来访问指定字符；不过 String 类的索引函数是只读的（只提供了 get 访问函数），因此不能像数组那样直接修改字符串中的某个字符，例如：

```
string s1 = "Visual C#";
char c1 = s1[2]; //c1='s'
int i = s1.Length; //i=9
s1[0] = 'W'; //错误:不能直接修改字符
```

空字符串和空值 null 是不同的，前者是一个不包含任何字符的 string 对象，而后者尚未分配存储空间。例如前者的 Length 属性值为 0，而访问后者的 Length 属性时会引发程序异常。

除了直接赋值外，也可以通过 String 类的构造函数来创建字符串对象，其中常用的重载形式有以下几种。

- String(char[] v)：将字符数组 v 转换为一个字符串。
- String(char c, int n)：将字符 c 重复 n 次来得到一个字符串。
- String(char[] v, int s, int n)：从字符数组 v 的第 s 个字符开始，取 n 个字符构成一个字符串。

下面示例了这些构造函数的用法：

```
char[] v = new char[] { 's', 'o', 'f', 't', 'w', 'a', 'r', 'e'};
string s1 = new string(v); //s1 = "software"
string s2 = new string('M', 3); //s2 = "MMM"
```

```
string s3 = new string(v, 0, 4); //s3 = "soft"
```
和构造函数相反，String 类的 ToCharArray 方法用于将字符串转换为字符数组，例如：
```
string s1 = "北京 2008";
char[] v1 = s1.ToCharArray();
Console.WriteLine(new string(v1)); //输出"北京 2008"
char[] v2 = s1.ToCharArray(3, 4); //从 s1 的第 3 个字符开始，取 4 个字符
s2 = new string(v2); //s2="2008"
```
String 类对相等和不等操作符进行了重载：如果两个字符串长度相同，且各个对应位置上的字符也都相同，那么认为这两个字符串是相等的。但它并没有重载">"、">="等比较操作符，而是提供了一个替代方法 CompareTo。当字符串 s1 通过该方法与字符串 s2 进行比较时，返回 0 表示二者相等，返回正数表示 s1 大于 s2，否则表示 s1 小于 s2。CompareTo 的具体比较规则为：

（1）如果 s1 和 s2 都为空字符串，那么返回整数 0。
（2）如果 s1 为空字符串而 s2 非空，那么返回整数-1；反之则返回整数 1。
（3）比较 s1 的第一个字符 a1 和 s2 的第一个字符 a2，如相等则继续比较 s1 和 s2 的剩余子串，否则返回 a1 和 a2 的比较结果（即比较 a1 和 a2 字符编码值）。
（4）如果 s1 和 s2 的所有字符都相等，那么返回整数 0。

数字字符的编码小于英文字母，小写字母的编码大于大写字母，汉字字符的编码则大于所有拉丁字符；对于两个汉字字符，系统默认的是比较它们的拼音排序。但不同国家或地区的文本排序规则可能有所不同，这在开发国际化应用软件时要加以注意。

下面给出了一些使用 CompareTo 方法进行字符串比较的示例代码：
```
string s1 = "北京 Olympics";
int i1 = s1.CompareTo(""); //i1 = 1
int i2 = s1.CompareTo("北京西客站"); //i2 = -1
```
此外 String 类还重载了加法操作符，用于实现字符串的连接，例如：
```
string s1 = "Beijing";
string s2 = s1 + " Rail" + "way"; //s2 = "Beijing Railway"
s2+= " Station"; //s2 = "Beijing Railway Station"
```

2．字符操作

和字符数组相比，String 类还通过 IndexOf、LastIndexOf、IndexOfAny 和 LastIndexOfAny 等方法提供了强大的字符查找功能，其中 IndexOf 方法有以下几种常用的重载形式。

- int IndexOf(char c)：查找字符 c 在字符串中首次出现的位置。
- int IndexOf(char c, int s)：从字符串的第 s 个字符开始，查找字符 c 首次出现的位置。
- int IndexOf(char c, int s, int n)：从字符串的第 s 个字符开始，在 n 个字符范围内查找字符 c 首次出现的位置。

方法的返回值为指定字符在字符串中的索引位置，如果未找到则返回-1；LastIndexOf 方法与之类似，只不过它是从后向前查找。看下面的代码示例：
```
string s1 = "apple";
int i = s1.IndexOf('p'); //i = 1
i = s1.IndexOf('l'); //i = 3
i = s1.IndexOf('l', 0, 3); //i = -1
i = s1.LastIndexOf('p'); //i = 2
```
和 IndexOf 方法相比，IndexOfAny 方法的第一个参数是一个字符数组，方法的返回值则为数组中任意一个字符在字符串中首次出现的位置；对应的，LastIndexOfAny 方法返回字符数组中任意一个字符在字符串中最后出现的位置。看下面的代码示例：
```
char[] v = { 'e', 'o', 't' };
```

```
string s1 = "Object-Oriented";
int i = s1.IndexOfAny(v); //i = 3
i = s1.LastIndexOfAny(v); //i = 13
i = s1.IndexOfAny(v, 0, 3); //i = -1
```

String 类中其他一些常用的字符操作方法（及其重载形式）如下。

- string ToLower()：将字符串中的所有大写字母转换为小写字母。
- string ToUpper()：将字符串中的所有小写字母转换为大写字母。
- string Replace(char c1, char c2)：将字符串中的所有 c1 字符都替换为 c2。
- string Trim()：删除字符串两端的所有空格。
- string Trim(char[] v)：将字符数组 v 中包含的所有字符从字符串两端删除。
- string PadLeft(int n)：向字符串左端填充 n 个空格。
- string PadLeft(int n, char c)：向字符串左端填充 n 个字符 c。
- string PadRight(int n)：向字符串右端填充 n 个空格。
- string PadRight(int n, char c)：向字符串右端填充 n 个字符 c。

但要注意这些方法都不修改当前字符串对象的内容，而是将修改后的内容作为一个新字符串对象返回，例如：

```
string s1 = "tea table~~~";
s1.Trim('~');
Console.WriteLine(s1);  //并未修改 s1 本身，仍然输出"tea table…"
string s2 = s1.Trim('~');
Console.WriteLine(s2); //输出"tea table"
```

3. 子串操作

字符串中任意一段连续的字符称为该字符串的子串。如果将字符视为长度为 1 的字符串，那么子串操作也是对字符操作的扩展。String 类的 IndexOf 和 LastIndexOf 方法不仅可查找指定字符，还可查找指定子串，其返回的结果是子串的首字符位置。例如：

```
string s1 = "第29届奥运会中国奥运代表团";
int i = s1.IndexOf("奥运"); //i=4
i = s1.IndexOf("奥运", 6); //i=9
i = s1.LastIndexOf("奥运"); //i=9
```

如果只需要知道字符串之间的包含关系，那么可使用 Contains 方法来判断字符串中是否包含指定子串，使用 StartsWith 方法和 EndsWith 方法来判断字符串是否以指定的子串开始和结束，例如：

```
string s1 = "中国工商银行北京海淀支行";
bool b = s1.Contains("银行"); //b = true
b = s1.StartsWith("中国"); //b = true
b = s1.EndsWith("银行"); //b = false
```

SubString 方法也是 String 类的一个常用方法，它有以下两种重载形式。

- string Substring(int s)：获得字符串从第 s 个字符开始直至结束的子串。
- string Substring(int s, int n)：获得字符串从第 s 个字符开始的连续 n 个字符的子串。

String 类还提供了向字符串中插入子串的 Insert 方法，以及从字符串中删除子串的 Remove 方法。

- string Insert(int s, string v)：在字符串的第 s 个字符前插入子串 v。
- string Remove(int s)：删除字符串第 s 个字符之后的子串。
- string Remove(int s, int n)：从字符串从第 s 个字符开始删除 n 个字符。

看下面的代码示例：

```
string s1 = "中国工商银行海淀支行";
```

```
string s2 = s1.Substring(2);  //s2 = "工商银行海淀支行"
string s3 = s2.Insert(4, "北京");  //s3 = "工商银行北京海淀支行"
string s4 = s4.Remove(4, 4);  //s4 = "工商银行海淀支行"
```

此外，String 类的 Split 方法用于将字符串分隔为一组子串，分隔的标记可存放在一个字符数组中。这一功能在文本处理中非常有用，例如：

```
string s1 = "83,45,19,100";
char[] sep1 = { ',' };  //分隔标记
//从字符串 s1 得到一组整数
string[] ss1 = s1.Split(sep1);
int[] x = new int[ss1.Length];
for (int i = 0; i < ss1.Length; i++)
    x[i] = int.Parse(ss1[i]);
```

最后，String 类的 Replace 方法也提供了用于子串替换的重载形式。Windows 应用程序 P4_7 利用了该方法来进行拼写检查，它将一组成语存放在一个字符串数组 rightWords 中，再将这些成语常见的错误拼写放在另一个数组 errorWords 中，那么只要将文本中包含的所有错误形式替换为正确形式，就可以实现成语纠错的功能（textBox1.Text 表示窗体文本框 textBox1 中的文本内容程序输出如图 4-2 所示）：

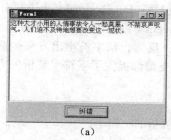
(a)
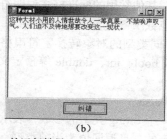
(b)

图 4-2 程序 P4_7 的运行结果

```
//程序 P4_7(Form1.cs)
using System;
using System.Windows.Forms;
namespace P4_7
{
    public partial class Form1 : Form
    {
        protected static string[] errors = new string[] {
            "哀声叹气", "重山峻岭", "大才小用", "甘败下风","留芳百世" ,"美仑美奂",
            "迫不急待", "人情事故", "食不裹腹", "谈笑风声", "一愁莫展", "再接再励"
        };

        protected static string[] rightWords = new string[] {
            "唉声叹气", "崇山峻岭", "大材小用", "甘拜下风", "流芳百世" ,"美轮美奂",
            "迫不及待", "人情世故", "食不果腹", "谈笑风生", "一筹莫展", "再接再厉"
        };

        private void Form1_Load(object sender, EventArgs e)
        {
            textBox1.Text = "这种大才小用的人情事故令人一愁莫展，不禁哀声叹气。人们迫不急待地想要改变这一现状。";
        }

        private void button1_Click(object sender, EventArgs e)
```

```
            {
                for (int i = 0; i < errorWords.Length; i++)
                    textBox1.Text = textBox1.Text.Replace(errors[i], rights[i]);
            }
        }
    }
```

4. 格式化和解析

String 类提供了一个非常实用的静态方法 Format，它能够对字符串进行参数格式化，这和 Console 类的静态方法 WriteLine 极为相似，即在方法的第一个字符串参数中包含形如{0}、{1}的指代标记，而使用随后的参数值来取代这些标记。例如：

```
string s1 = string.Format("{0}年级{1}班", 1, 3); //s1 = "1年级3班"
int x = 8;
s1 = string.Format("{0}年{1}月{1}日", 2008, x); //s1 = "2008年8月8日"
int y = 15;
s1 = string.Format("{0}*{1} = {2}", x, y, x * y); //s1 = "8*15=120"
```

Format 方法还可以针对数值类型使用特殊的输出格式，例如：

```
string s1 = String.Format("{0:C3}元", 9999); //s1 = "￥9,999.000元"
```

其中{0:C3}表示将后面的第 0 个参数按货币格式 C（Currency）输出，小数后保留 3 位（不写则默认为 2 位）。其他常用的标准格式说明符还包括：D（Decimal，十进制）、E（Exponential，指数）、Float（浮点数）、N（Number，数字）、Percent（百分比）和 X（Hex，十六进制）等。

格式化能将不同类型的对象转换字符串格式；反之，从字符串出发来构造其对象的过程叫做字符串解析。像 bool、int、double 等预定义值类型都提供了字符串解析的方法，它们有下列共同点：

- 方法名为 Parse。
- 方法只有一个 String 类型的参数，表示要解析的字符串。
- 方法是静态的，且返回值为当前类型，表示解析的结果。

看下面的代码示例：

```
bool b = bool.Parse("False");
int i = int.Parse("9999");
double d = double.Parse("4.14");
```

在使用上述类型的 Parse 方法时，如果字符串不符合格式要求就会发生异常。例如不能从字符串"Hello"中解析出一个整数值。为了提高程序的可靠性，一些类型还提供了另一种"尝试"类型解析的方法，其共同点为：

- 方法名为 TryParse。
- 方法有两个参数，一个为 String 类型，表示要解析的字符串；另一个输出参数为当前类型，表示解析的结果。
- 方法是静态的，且返回值为布尔类型，表示解析是否成功。

TryParse 的效率会低于 Parse 方法，但它在格式错误时不会引发程序异常，而是返回 false 值，而输出参数的值会变为类型的默认值（整数和实数均为 0）。例如：

```
decimal d; //可以先不赋值
decimal.TryParse("9,999", out d); //d = 9999
decimal.TryParse("0X100", out d); //d = 0
```

程序 P4_8 为类 ComplexNumber 定义了字符串解析和格式化的方法，这使得复数对象和其字符串表示"(a+bi)"之间能够相互转换。

```
//程序 P4_8
using System;
namespace P4_8
```

```csharp
{
    class Program
    {
        static void Main()
        {
            ComplexNumber c1 = ComplexNumber.Parse("(50+100i)");
            ComplexNumber c2;
            ComplexNumber.TryParse("(100,300i)", out c2);
            Console.WriteLine("{0} + {1} = {2}", c1, c2, c1 + c2);
            ComplexNumber.TryParse("(100+300i)", out c2);
            Console.WriteLine("{0} + {1} = {2}", c1, c2, c1 + c2);
        }
    }

    public class ComplexNumber
    {
        public double A = 0, B = 0;

        public ComplexNumber(double a, double b)
        {
            A = a;
            B = b;
        }

        public static ComplexNumber operator +(ComplexNumber c1, ComplexNumber c2)
        {
            return new ComplexNumber(c1.A + c2.A, c1.B + c2.B);
        }

        public static ComplexNumber operator -(ComplexNumber c1, ComplexNumber c2)
        {
            return new ComplexNumber(c1.A - c2.A, c1.B - c2.B);
        }

        public static ComplexNumber Parse(string s)
        {
            if (s == null)
                throw new ArgumentNullException();
            s = s.ToUpper();
            int pos1 = s.IndexOf('+');
            int pos2 = s.IndexOf('I');
            if (pos1 == -1 || pos2 == -1)
                throw new FormatException("输入的字符串格式不正确");
            double a = double.Parse(s.Substring(1, pos1 - 1));
            double b = double.Parse(s.Substring(pos1 + 1, pos2 - pos1 - 1));
            return new ComplexNumber(a, b);
        }

        public static bool TryParse(string s, out ComplexNumber c)
        {
            c = new ComplexNumber(0, 0);
            if (s == null)
                return false;

            s = s.ToUpper();
            int pos1 = s.IndexOf('+');
            int pos2 = s.IndexOf('I');
            if (pos1 == -1 || pos2 == -1)
                return false;
            string s1 = s.Substring(1, pos1 - 1);
            string s2 = s.Substring(pos1 + 1, pos2 - pos1 - 1);
```

```
            if (pos1 == -1'|| pos2 == -1  || !double.TryParse(s1, out c.A)
|| !double.TryParse(s2, out c.B))
                return false;
            else
                return true;
        }
        public override string ToString()
        {
            return String.Format("({0}+{1}i)", this.A, this.B);
        }
    }
}
```

程序 P4_8 的输出结果如下:

```
(50+100i) + (0+0i) = (50+100i)
(50+100i) + (100+300i) = (150+400i)
```

4.5.3 StringBuilder 类

前面介绍过，String 类型修改字符串的方法实际上都是返回一个新的 String 对象，原字符串仍然保留。那么当字符串较长或是操作频繁时就会消耗大量的资源。例如下面这段简单的代码在循环过程中将产生一共 200 个字符串对象:

```
string s = "";
for (int i = 1; i <= 100; i++)
{
    s += i.ToString();
    s += ",";
}
```

为了改善字符串的性能，.NET 类库的 System.Text 命名空间下专门定义了一个 StringBuilder 类，用于对字符串进行动态管理，而不是每次都生成新的字符串。StringBuilder 对象可以使用无参构造函数来创建，也可以基于一个现有字符串来创建，例如:

```
StringBuilder sb1 = new StringBuilder();    //包含空字符串的 StringBuilder 对象
StringBuilder sb2 = new StringBuilder("abc"); //字符串内容为"abc"
```

StringBuilder 与 String 类的用法有很多类似之处，如通过 Length 属性获取字符串长度，通过索引函数访问字符，但 StringBuilder 的索引函数是可读写的。StringBuilder 也提供了 Insert、Remove、Replace 这些操作字符串的方法；尽管它们的返回类型为 StringBuilder，但方法并没有创建新的对象，返回的仍是当前对象。此外，StringBuilder 的 ToString 方法直接返回所包含的字符串内容。

为了方便频繁的字符连接操作，StringBuilder 提供了 Append、AppendLine 和 AppendFormat 这三个方法。其中 Append 方法用于将一个新串加到字符串的尾端，其参数可以是字符串，也可以是其他基本值类型；AppendLine 方法会在追加新串后再增加一个换行符，而 AppendFormat 方法还能在追加新串的同时进行参数格式化。看下面的代码示例:

```
StringBuilder sb1 = new StringBuilder();
sb1.Append("公元");
sb1.AppendFormat("{0}年{1}月{1}日", 2008, 8);
Console.WriteLine(sb1.AppendLine());  //输出"2008年8月8日"并换行
```

4.5.4 Math 类

.NET 类库中的 Math 类提供了对初等数学计算的基本支持。它是一个静态类，其两个公有静态字段 E 和 PI 分别表示自然对数 e 和圆周率 π，例如：
```
Console.WriteLine("请输入圆的半径:");
double r = double.Parse(Console.ReadLine());
Console.WriteLine("周长:{0},面积:{1}", 2 * Math.PI * r, Math.PI * r * r);
```
Math 的一组静态方法则实现了指数、对数、三角等函数的计算功能，其中常用的有：
- Abs（求绝对值），Ceiling（求大于或等于指定数值的最小整数），Floor（求小于或等于指定数值的最大整数），Round（对数值进行四舍五入）。
- Exp（求 e 的指数幂），Pow（指数函数），Log（对数函数），Log10（求以 10 为底的对数），Sqrt（求平方根）。
- Sin（正弦函数），Cos（余弦函数），Tan（正切函数）。
- Sinh（双曲正弦函数），Cosh（双曲余弦函数），Tanh（双曲正切函数）。
- Asin（反正弦函数），Acos（反余弦函数），Atan（反正切函数）。

注意三角函数方法的参数和返回值都以弧度为单位，因此在使用角度时要注意单位换算（弧度值 = 角度值×π/180）。下面给出了一些函数计算的示例代码：
```
double a = 3, b = 4, c = 5;
double s = (a + b + c) / 2;
double A = Math.Sqrt(s * (s - a) * (s - b) * (s - c));
double x = Math.Asin(2 * A / b / c) * 180 / Math.PI;
double y = Math.Asin(2 * A / a / c) * 180 / Math.PI;
double z = Math.Asin(2 * A / a / b) * 180 / Math.PI;
```

4.5.5 DateTime 结构

程序中常常会使用到大量的日期和时间信息，.NET 类库中的结构类型 DateTime 对此进行了封装。该结构定义了多个重载的构造函数，其中最常用的是使用 3 个整数参数来指定时间的年月日，或是使用 6 个整数参数来指定时间的年月日和时分秒，例如：
```
DateTime dt1 = new DateTime(2008, 8, 8); //2008年8月8日
DateTime dt2 = new DateTime(2009, 1, 1, 12, 0, 0); //2009年1月1日12时
```
对于一个 DateTime 对象而言，其各个时间部分可通过 Year（年）、Month（月）、Day（日）、Hour（小时）、Minute（分）、Second（秒）、Millisecond（毫秒）这些实例属性获得，通过属性 DayOfWeek 和 DayOfYear 还能知道当天是一周和一年中的第几天。例如：
```
DateTime dt1 = new DateTime(2008, 12, 31);
Console.WriteLine(dt1.Year - 2000); //输出 8
Console.WriteLine(dt1.Day); //输出 31
Console.WriteLine(dt1.DayOfWeek); //输出 Wednesday
Console.WriteLine(dt1.DayOfYear); //输出 366
Console.WriteLine(dt1.Hour); //输出 0
```
此外，通过 DateTime 的静态属性 Now 可获得当前的系统时间，通过其静态属性 Today 则可获得当前时间的日期部分。不过这些实例属性和静态属性都是只读的，不能通过它们直接设置时间值。要修改时间，就要用到 DateTime 的一组形如 "AddXXXs" 的方法，比如 AddYears 用于修改年份，AddHours 用于修改小时等；这些方法并不改变当前对象，而是返回一个新的 DateTime 对象(这一点和 String 类型较为类似)；传递给这些方法的参数值为正时会使时间增加相应的部分，值为负时会使时间减少相应的部分。例如：
```
DateTime dt1 = new DateTime(2008, 12, 31);
DateTime dt2 = dt1.AddDays(1);
Console.WriteLine(dt2.AddMonths(1)); //输出"2009-2-1 00:00:00"
Console.WriteLine(dt2.AddSeconds(-1)); //输出"2008-12-31 23:59:59"
```

```
DateTime dt3 = dt2.AddHours(2450);
Console.WriteLine(dt2);  //输出"2009-4-11 02:00:00"
```
和基础值类型一样，DateTime 也提供了字符串解析的 Parse 和 TryPrase 方法，以及转换到字符串的 ToString 方法，此外它还提供了 ToLongDateString、ToShortDateString、ToLongTimeString、ToShortTimeString 这 4 个转换到特定格式字符串的方法。下面的代码说明了这些方法的不同转换效果：

```
DateTime dt1 = new DateTime(2008, 8, 8, 20, 0, 0);
Console.WriteLine(dt1);  //输出"2008-8-8 20:00:00"
Console.WriteLine(dt1.ToLongDateString());  //输出"2008年8月8日"
Console.WriteLine(dt1.ToShortDateString());  //输出"2008-8-8"
Console.WriteLine(dt1.ToLongTimeString());  //输出"20:00:00"
Console.WriteLine(dt1.ToShortTimeString());  //输出"20:00"
DateTime dt2 = DateTime.Parse("2008-12-31");
Console.WriteLine(dt2.ToLongDateString());  //输出"2008年12月31日"
```

DateTime 结构也重载了各种比较运算符，并提供了专用于时间比较的 CompareTo 方法。如果时间 dt1 在时间 dt2 之前，那么认为 dt1 小于 dt2。

4.6 案例研究——旅行社业务类的实现

4.6.1 省份、城市和景点类

这里对旅行社管理系统中的主要业务类进行细化和实现。首先考虑基础的 Province、City、Scene 等类型。除了名称之外，系统还希望提供每个省旅游信息的简要介绍；而除了直辖市之外，每个省都包含一组城市。那么 Province 类的定义如下：

```
public class Province
{
    private string _name;
    public string Name  //名称
    {
        get { return _name; }
    }

    public string Introduction {get; set;}  //简介
    public bool Municipal {get; set; }  //是否为直辖市
    public City[] Cities {get; set; }  //城市集合

    public City this[string name]  //通过名称索引城市
    {
        get {
            for (int i = 0; i <Cities.Length; i++)
                if (Cities[i].Name == name)
                    return Cities[i];
            return null;
        }
    }

    public Province(string name)  //构造函数
    {
        _name = name;
```

```csharp
    }
    public override string ToString()
    {
        return _name;
    }
}
```

其中，_name 字段值需要在构造函数中设定，而对应的 Name 属性为只读的；Introduction、Municipal 和 Cities 这些可读写的属性都采用自动属性的方式来定义；其中 Cities 属性可直接访问所有城市，而索引函数则能够根据名称来查询指定城市。类似的，City 包含一组景点集合；为了查询方便，该类还存储了所属省份的信息：

```csharp
public class City
{
    private string _name;
    public string Name //名称
    {
        get { return _name; }
    }

    public string Introduction {get; set; } //简介
    public Province Province {get; set; } //所属省份
    public Scene[] Scenes {get; set; } //城市景点集合

    public override string ToString()
    {
        return _name;
    }
}
```

Scene 类不仅定义了景点的名称和简介，还存储了景点类型、星级，以及各种票价和折扣等更丰富的信息：

```csharp
public class Scene
{
    private string _name;
    public string Name //名称
    {
        get { return _name; }
    }

    public string Introduction {get; set; } //简介
    public SceneType Type {get; set; } //景点类型
    public byte Star {get; set; }//景点星级
    public City City {get; set; }//所属城市

    public decimal Price {get; set; } //标准票价
    public decimal OffSeasonPrice {get; set; } //淡季票价
    public decimal ChlDiscount {get; set; } //儿童票折扣
    public decimal OldDiscount {get; set; } //老年票折扣
    public decimal StuDiscount {get; set; } //学生票折扣

    public override string ToString()
    {
        return City.Name + _name;
    }
}
```

4.6.2 旅游线路和方案类

接下来考虑 Line、Package、Tour 这些关键的业务类。为了方便管理，系统为这些类的每个对象都赋予了一个编号 Id，并要求在这些类的构造函数中赋值。Line 类还通过如下的字段和属性来维护旅游线路的相关信息：

```csharp
public class Line// 旅游线路类
{
    private int _id;
    public int Id //线路编号
    {
        get { return _id; }
    }

    private string _name;
    public string Name //线路名称
    {
        get { return _name; }
    }

    private Area _area;
    public Area Area //所属地区
    {
        get { return _area; }
    }

    public short Days {get; set; } //旅行天数
    public short Nights {get; set; } //旅行夜数
    public Scene[] Scenes {get; set; } //线路景点集合

    public Line(int id, string name, Area area, short days, short nights)
    {
        _id = id;
        _name = name;
        _area = area;
        Days= days;
        Nights= nights;
    }
}
```

系统一个常见的功能是查询旅游线路，这可以通过 Line 类的静态方法来实现。下面的 GetAll 方法用于返回系统中所有 Line 对象，而 Get 方法则返回指定编号的 Line 对象（实际应用中的数据可来源于文件或数据库）：

```csharp
public class Line// 旅游线路类续
{
    public static Line[] GetAll() //查询所有线路
    {
        Line l1 = new Line("10001", "长城十三陵一日游", Area.华北, 1, 0);
        Line l2 = new Line("10002", "云冈石窟恒山二日游", Area.华北, 2, 1);
        Line l3 = new Line("10003", "黄山九华山双卧五日游", Area.华东, 5, 5);
        return new Line[] {l1, l2, l3};
    }

    public static Line Get(int id) //查询指定线路
    {
```

```
            foreach (Line l in GetLines())
                if (l._id == id)
                    return l;
            return null;
        }
    }
```
下面的定义给出了组团方案所需的基本信息，注意 Package 类并没有 Name 属性，而其 ToString 方法对方案的字符串描述格式是"线路名(组团级别)"（如"长城十三陵一日游(经济)"），因为旅行社不会在同一线路上设置两个相同级别的组团方案：
```
public class Package// 组团方案类
{
    private int _id;
    public int Id  //组团方案编号
    {
        get { return _id; }
    }

    private Line _line;
    public Line Line  //旅游线路
    {
        get { return _line; }
    }

    private PackageType _type;
    public PackageType Type  //组团类型
    {
        get { return _type; }
    }

    private PackageGrade _grade;
    public PackageGrade Grade  //组团等级
    {
        get { return _grade; }
    }

    public string Introduction {get; set; }//旅行过程简介

    public short Number {get; set; }//总人数限制
    public short ChlNumber {get; set; }//儿童人数限制
    public decimal Price {get; set; } //成人价格
    public decimal ChlPrice {get; set; } //儿童价格

    public HotelGrade HotelGrade {get; set; } //酒店等级
    public DinnerGrade DinnerGrade {get; set; } //餐饮等级

    public Agent Agent {get; set; } //负责业务员

    public Package(int id, Line line, PackageType type, PackageGrade grade)
    {
        _id = id;
        _line = line;
        _type = type;
        _grade = grade;
    }

    public override string ToString()
```

```
        return string.Format("{0}({1})", _line.Name, _grade);
    }
}
```
和 Line 类似，Package 也可定义查询组团方案的方法 Get 和 GetAll，具体代码此处略。

4.6.3 旅行团和游客类

Tour 类的基本定义代码如下，其构造函数在创建旅行团时设置的初始状态为"报名"，而其字符串描述格式是"组团方案名-旅行团编号"：

```
public class Tour// 旅行团类
{
    private int _id;
    public int Id //旅行团编号
    {
        get { return _id; }
    }

    private Package _package;
    public Package Package //组团方案
    {
        get { return _package; }
    }

    public DateTime RegDue {get; set; } //报名截止时间
    public DateTime StartTime {get; set; } //开始时间
    public DateTime EndTime {get; set; } //结束时间
    public string StartAddress {get; set; } //出发地点
    public Guide Guide {get; set; } //随团导游
    public TourState State {get; set; } //旅行团状态
    public Customer[] Customers {get; set; } //游客集合

    public Tour(int id, Package package, DateTime regDue, DateTime startTime)
    {
        _id = id;
        _package = package;
        RegDue = regDue;
        StartTime = startTime;
        State = TourState.报名;
    }

    public bool IsFull()
    {
        return _customers.Count >= _package.Number;
    }
    public override string ToString()
    {
        return string.Format("{0}-{1}", _package, _id);
    }
}
```

Tour 类使用了一个 Customer[]来存储游客集合。下面给出了游客类的基本定义，其中 E-mail 属性使用的不是 string 类型，而是.NET 类库提供的 MailAddress 类型（在 System.Net.Mail 命名空间中定义），它能够自动分析电子邮件地址的有效性：

```csharp
public class Customer  // 游客类
{
    protected int _id;
    public int Id //编号
    {
        get { return _id; }
    }
    private string _name;
    public string Name //姓名
    {
        get { return _name; }
    }

    private CustomerType _type = CustomerType.Common;
    public CustomerType Type //游客类型
    {
        get { return _type; }
    }

    public bool Gender{get; set; } //性别
    public DateTime Birthday{get; set; } //生日
    public IDCard IdCard{get; set; } //身份证件

    public string Address{get; set; } //通信地址
    public string Phone{get; set; } //电话
    public MailAddress Email{get; set; } //电子邮件

    public Customer(int id, string name, bool gender, CustomerType type)
    {
        _id = id;
        _name = name;
        Gender = gender;
        _type = type;
    }
}
```

游客类还需要实现两个重要方法：报名参团的 Enroll 方法和取消报名的 Cancel 方法。为了避免 Customer 对象去过多地操作 Tour 对象，这里引进一个 Enrollment 类来描述游客报名的信息，其定义代码如下：

```csharp
public class Enrollment// 报名类
{
    private Tour _tour;
    public Tour Tour //旅行团
    {
        get { return _tour; }
    }

    private DateTime _regTime;
    public DateTime RegTime //报名时间
    {
        get { return _regTime; }
    }

    private Customer _applier;
    public Customer Applier //报名者
    {
```

```csharp
        get { return _applier; }
    }

    public Customer[] Customers{get; set; }   //报名游客集合
    public int Priority{get; set; }   //优先级
    public EnrollmentState State{get; set; }   //报名状态

    // 单个游客报名
    internal Enrollment(Tour tour, Customer applier)
    {
        _tour = tour;
        _applier = applier;
        Customers = new Customer[] { applier };
        _time = DateTime.Now;
        State = EnrollmentState.待确认;
    }

    // 一组游客报名
    internal Enrollment(Tour tour, Customer applier, Customer[] customers)
    {
        _tour = tour;
        _regTime= DateTime.Now;
        _applier = applier;
        Customers = customers;
        State = EnrollmentState.待确认;
    }
}
```

其中，表示报名状态的枚举类型定义为：

```csharp
public enum EnrollmentState{待确认, 接受, 拒绝}
```

这样在 Tour 类中就还应增加一个 Enrollment[] 数组类型的属性，以描述旅行团的所有报名信息：

```csharp
public Enrollment[] Enrollments{get; set; }   //报名集合
```

接下来就可以为 Customer 类实现 Enroll 和 Cancel 方法，它们作用的结果分别是向 Tour 对象的 Enrollments 属性中加入和删除一个 Enrollment 对象（Enrollment 对象只能由报名者本人取消）：

```csharp
public Enrollment Enroll(Tour t, Customer[] customers)   //报名参加旅行团
{
    Enrollment enr = new Enrollment(t, this, customers);
    if (t.Enrollments == null)
        t.Enrollments = new Enrollment[] { enr };
    else{
        Enrollment[] enrollments = new Enrollment[t.Enrollments.Length + 1];
        for (int i = 0; i < t.Enrollments.Length; i++)
            enrollments[i] = t.Enrollments[i];
        enrollments[t.Enrollments.Length] = enr;
        t.Enrollments = enrollments;
    }
    return enr;
}

public void Cancel(Enrollment enr)   //取消报名
{
    Tour t = enr.Tour;
    if (t.Enrollments.Length == 1 && t.Enrollments[0] == enr&& t.Enrollments[i].Applier == this)
        t.Enrollments = null;
    else
```

```
            for (int i = 0; i < t.Enrollments.Length; i++)
                if (t.Enrollments[i]==enr&& t.Enrollments[i].Applier==this) {
                    Enrollment[]  enrollments=new Enrollment[t.Enrollments.Length-1];
                    for (int j = 0; j < i; j++)
                        enrollments[j] = t.Enrollments[j];
                    for (int j = i; j < t.Enrollments.Length - 1; j++)
                        enrollments[j] = t.Enrollments[j + 1];
                    t.Enrollments = enrollments;
                }
    }
```

4.7 小　　结

　　类和结构中可以定义用于描述状态的字段成员，以及用于行为实现的方法成员，还可以在其中定义嵌套的类或结构。这些成员的访问权限可以通过访问限制修饰符进行控制，其中只有 public 成员才允许外部对象访问，private 成员只能由对象本身进行访问，而 protected 成员允许对象本身及其派生对象访问。

　　方法是最基本的函数成员，是对象消息通信的基本手段，也是实现程序模块化的基础之一。方法声明中的参数叫做形参，而实际调用时传递给方法的参数叫做实参。构造函数、析构函数、属性、索引函数和操作符都可以视为特殊的方法成员。对象的生命周期从构造函数开始，到析构函数结束。利用属性和索引函数提供的访问方法，可以隐藏数据处理的细节，更好地实现对象的封装性。操作符重载则能使 C# 预定义的操作符直接作用于各种自定义类型。

4.8 习　　题

1. 比较静态字段、常量字段、只读字段和只读属性的相同点和不同点。
2. 程序 P4_5 中，Student 的静态字段 objects 将记录被创建的学生对象的总数。试修改该程序，使该字段记录的是当前内存中学生对象的总数。
3. 使用可读写的属性来封装一个私有字段，与直接将该字段定义为公有字段，二者之间有何区别？
4. 定义完整的 ComplexNumber 类，在其中重载加减乘除等基本算术操作符，相等和不等操作符（比较复数的实部和虚部是否均相等），以及大小比较操作符（比较复数模的大小）。
5. 在类的静态方法代码中能否使用 this 对象引用，为什么？
6. 在从字符串中解析值类型时，哪些情况下应使用 String 类的 Parse 方法，哪些情况下则适合使用 TryParse 方法？
7. 使用 StringBuilder 类改写程序 P4_7。
8. 知道一个人的生日，如何计算出他（她）的年龄？写出相应的代码，要求严格按照周岁计算（例如某人的生日为 2000 年 10 月 1 日，那么在 2010 年 9 月 30 日计算出的年龄应为 9 岁，而在 2010 年 10 月 1 日计算出的年龄应为 10 岁）。
9. 为旅行社管理系统中的 Province 和 City 类添加静态的 GetAll 方法，通过它创建一些常见的省份和城市对象。再编写控制台应用程序，测试已定义的旅行社业务类。

第 5 章 继承和多态

类有自己的成员字段和方法，其他类通过访问这些成员来享受其提供的服务，这就形成了类和类之间的一般关联关系。本章将深入讨论面向对象中另一种重要关系——继承，并介绍如何通过多态性来处理基类和派生类的对象行为。

5.1 继 承

5.1.1 基类和派生类

继承是面向对象的关键要素之一。在程序设计过程中，类的层次结构有两种基本的构造方式。

- 自顶向下：从基础类型开始向下分解，不断得到新的派生类型。例如从"图形"中细化出"椭圆"和"多边形"等具体图形，"多边形"又可进一步细分为"三角形"、"四边形"等更为具体的类型。
- 自底向上：对现有的一组具体类型进行抽象，得到新的基础类型。例如对"飞机"、"轮船"、"汽车"等概念进行归类，得到更为抽象的"交通工具"。

在复杂软件设计中，上述两种方式往往要综合运用。.NET 类库就是一个典型的面向对象的实现：Object 类是其他所有类的基类，而所有的 C#数据类型都是从类中衍生而来。图 5-1 给出了.NET 类库中的继承结构示意图。

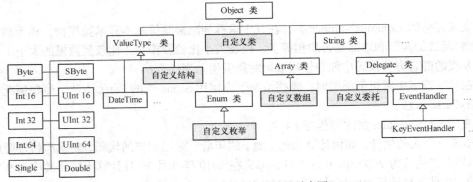

图 5-1 .NET 类库中的继承示意图

派生类能够继承基类的所有成员，并增加自己的成员来进行功能扩展。程序 P5_1 中定义了一个汽车类 Automobile，它提供了 Speed 和 Weight 两个属性以及 Run 方法；其派生类 Bus 和 Truck 则分别增加了 Passengers 和 Load 属性：

```csharp
//程序 P5_1
using System;
namespace P5_1
{
    class Program
    {
        public static void Main()
        {
            Bus b1 = new Bus();
            Console.WriteLine("客车行驶1000公里需{0}小时", b1.Run(1000));
            Truck t1 = new Truck();
            Console.WriteLine("卡车行驶1000公里需{0}小时", t1.Run(1000));
        }
    }

    public class Automobile
    {
        protected float speed;
        public float Speed
        {
            get { return speed; }
        }

        private float weight;
        public float Weight
        {
            get { return weight; }
            set { weight = value; }
        }

        public float Run(float distance)
        {
            return distance / speed;
        }
    }

    public class Bus : Automobile
    {
        private int passengers;
        public int Passengers
        {
            get { return passengers; }
            set { passengers = value; }
        }

        public Bus()
        {
            passengers = 20;
            speed = 60;
            Weight = 10;
        }
    }

    public class Truck : Automobile
    {
        private float load;
        public float Load
        {
            get { return load; }
```

```
            set { load = value; }
        }

        public Truck()
        {
            load = 30;
            speed = 50;
            Weight = 15;
        }
    }
}
```

上面两个派生类都定义了自己的构造函数,并在其中访问了基类的保护字段 speed,但基类的私有字段 weight 需要通过其公有属性 Weight 间接访问。程序 P5_1 的输出如下:

> 客车行驶 1000 公里需 16.66667 小时
> 卡车行驶 1000 公里需 20 小时

还有一点要注意的是:基类的保护成员是指允许派生类的方法代码访问,而不是指通过派生类的对象访问。例如在下面 Main 方法中的第 3 行代码是错误的,因为其所属的 Program 类无权访问 Automobile 类的保护字段 speed:

```
class Program
{
  static void Main()
  {
     Bus b1 = new Bus();
     b1.speed = 80; //错误:不能通过对象访问基类的保护成员
     Console.WriteLine("客车行驶 1000 公里需{0}小时", b1.Run(1000));
   }
}
```

而将上述代码放在 Automobile 派生类的某个方法中就是正确的:

```
public class Bus : Automobile
{
    public static void CreateInstance()
    {
       Bus b1 = new Bus();
       b1.speed = 80;
       Console.WriteLine("客车行驶 1000 公里需{0}小时", b1.Run(1000));
    }
}
```

提示

如果基类中的字段通过公有且可读写的属性进行了封装,那么建议将字段定义为私有的,这样包括其派生类在内的所有其他类型都必须通过属性进行访问。

在 4.1.2 小节中介绍了访问限制修饰符,其中 public 和 internal 修饰符可用于控制外部程序集对当前程序集中类型的访问。由于派生类和基类是"is a"的关系,因此基类的可访问性不应低于派生类。换句话说,如果外部程序集能够访问某个类,那么同时就能访问其基类。例如上面的代码中定义 Bus 和 Truck 为公有的,那么就不能将 Automobile 定义为内部的。

5.1.2 隐藏基类成员

有时候派生类会使用和基类中相同的成员,但希望这些成员提供与基类不同的服务。例如可为程序 P5_1 中的 Truck 类也定义一个 Run 方法,但它计算的行驶时间不仅与距离和速度相关,还受卡车载货量的影响：
```
public class Truck
{
    //…
    public float Run(float distance)
    {
        return (1 + load / Weight / 2) * distance /speed;
    }
}
```

此时，程序仍能通过编译，但编译器会警告"Truck.Run 隐藏了继承的成员 Automobile.Run"。也就是说，如果派生类中定义了与基类相同的成员（字段名称和类型相同，或是方法名称、参数列表和返回类型相同），默认情况下基类的成员在派生类中会被隐藏，这也称之为派生类成员覆盖了基类成员。为了提高代码的可读性，C#建议使用 new 关键字来明确修饰派生类中的成员。那么 Truck 类的 Run 方法应定义为：

```
public new float Run(float distance)
{
    return (1 + load / Weight / 2) * distance /speed;
}
```

　　隐藏基类成员所用的 new 关键字是一种修饰符，它和创建对象所用的 new 操作符是不同的。此外，如果派生类中的成员与基类中的私有成员同名，那么无须使用 new 修饰符。

属性和索引函数这样的特殊方法也可以被覆盖。下面的程序中，Truck 类就同时覆盖了基类的 Speed 属性和 Run 方法：

```
//程序 P5_2
using System;
namespace P5_2
{
    class Program
    {
        static void Main()
        {
            Truck t1 = new Truck();
            Console.WriteLine("卡车速度{0}公里/小时", t1.Speed);
            Console.WriteLine("卡车行驶 1000 公里需{0}小时", t1.Run(1000));
            Automobile a1 = t1;
            Console.WriteLine("汽车速度{0}公里/小时", a1.Speed);
            Console.WriteLine("汽车行驶 1000 公里需{0}小时", a1.Run(1000));
        }
    }

    public class Automobile
    {
        protected float speed;
        public float Speed
        {
            get { return speed; }
        }

        private float weight;
        public float Weight
        {
            get { return weight; }
            set { weight = value; }
```

```csharp
        }
        public float Run(float distance)
        {
            return distance / speed;
        }
    }

    public class Truck : Automobile
    {
        private float load;
        public float Load
        {
            get { return load; }
            set { load = value; }
        }
        public new float Speed
        {
            get { return speed / (1 + load / Weight / 2); }
        }
        public Truck()
        {
            load = 30;
            speed = 50;
            Weight = 15;
        }

        public new float Run(float distance)
        {
            return (1 + load / Weight / 2) * base.Run(distance);
        }
    }
}
```

在出现成员隐藏的情况下，程序中究竟调用哪一个成员取决于对象的声明类型。所以在程序 P5_2 的主方法中，尽管 t1 和 a1 实际上是同一个对象，但对前者调用的是 Truck 类中定义的方法和属性，对后者调用的则是 Automobile 的方法和属性，那么程序的输出为：

```
卡车速度 25 公里/小时
卡车行驶 1000 公里需 40 小时
汽车速度 50 公里/小时
汽车行驶 1000 公里需 20 小时
```

5.1.3　base 关键字

在 4.4 节中，介绍了 this 关键字表示当前对象引用。例如对于上面的 Automobile 类，其方法代码中出现"speed"和"this.speed"是等价的。C#中还提供了一个 base 关键字，通过它可以访问基类的成员。例如在 Bus 和 Truck 类的方法代码中，"Weight"和"base.Weight"是等价的，它们都表示基类 Automobile 的 Weight 属性。

 和 this 不同，base 关键字不能作为单个对象变量使用，例如使用 Console.WriteLine (base)这样的代码是错误的。

不过，当派生类成员隐藏了基类成员时，base 关键字就能发挥其特有的作用：直接写出的成

员名表示派生类的成员,而增加了 base 引用的成员表示被隐藏的基类成员。看下面的代码示例:
```
public class Truck : Automobile
{
    //...
    public void ShowSpeed()
    {
        Truck t1 = new Truck();
        //访问基类的属性和方法
        Console.WriteLine("空载速度:{0}", base.Speed);
        Console.WriteLine("行驶1000公里需{0}小时", base.Run(1000));
        //访问自身的属性和方法
        Console.WriteLine("满载速度:{0}", Speed);  //自身属性
        Console.WriteLine("行驶1000公里需{0}小时", Run(1000));  //自身方法
    }
}
```
如果派生类的方法是在基类方法的基础上增加功能,那么通过 base 关键字能够有效减少派生类中的代码量。例如 Truck 类的 Run 方法代码还可写成如下形式:
```
public new float Run(float distance)
{
    return (1 + load / Weight / 2) * base.Run(distance);
}
```

5.1.4 对象的生命周期

在 4.3.1 小节中介绍过对象在创建时调用构造函数,在销毁时调用析构函数。对于派生类对象而言,它在创建时将自顶向下地调用各级基类的构造函数,最后调用自身的构造函数;销毁时首先调用自身的析构函数,而后自底向上地调用各级基类的析构函数。下面的程序说明了这一点:

```
//程序 P5_3
using System;
namespace P5_3
{
    class Program
    {
        public static void Main()
        {
            Son s1 = new Son();
            System.GC.Collect();
        }
    }

    public class Grandsire
    {
        public Grandsire()
        {
            Console.WriteLine("调用Grandsire的构造函数");
        }

        ~Grandsire()
        {
            Console.WriteLine("调用Grandsire的析构函数");
        }
    }

    public class Father : Grandsire
    {
```

```csharp
    public Father()
    {
        Console.WriteLine("调用 Father 的构造函数");
    }

    ~Father()
    {
        Console.WriteLine("调用 Father 的析构函数");
    }
}
public class Son : Father
{
    public Son()
    {
        Console.WriteLine("调用 Son 的构造函数");
    }

    ~Son()
    {
        Console.WriteLine("调用 Son 的析构函数");
    }
}
```

程序在创建 Son 对象时会依次调用 Grandsire、Father 及 Son 的构造函数；而通过垃圾收集器 GC 调用析构函数时则是相反的顺序。程序 P5_3 的输出如下：

```
调用 Grandsire 的构造函数
调用 Father 的构造函数
调用 Son 的构造函数
调用 Son 的析构函数
调用 Father 的析构函数
调用 Grandsire 的析构函数
```

如果基类中定义了多个重载的构造函数，那么派生类也会重载这些构造函数。对于派生类的构造函数而言，在其定义中可以通过 base 关键字来指定创建对象时要调用的基类构造函数。例如假设 Automobile 类定义了下面 3 个构造函数：

```csharp
public Automobile()
{ }

public Automobile(float speed)
{
    this.speed = speed;
}

public Automobile(float speed, float weight)
{
    this.speed = speed;
    this.weight = weight;
}
```

那么对于如下的派生类 Truck 的构造函数，使用其创建对象时将调用基类的第 2 个构造函数：

```csharp
public Truck(float speed) : base(speed)
{
    load = 30;
```

```
        Weight = 15;
}
```
而如果改为下面这种写法，调用的就是基类的第 3 个构造函数：
```
public Truck(float speed) : base(speed, 15)
{
    load = 30;
}
```
此时 base 的作用是指代基类的构造函数，这种方式只能出现在派生类的构造函数定义中，而不允许出现在方法代码中。如果派生类的构造函数定义中没有出现 base 关键字，那么默认调用的是基类的无参构造函数。而如果基类中只定义了带参数的构造函数，那么派生类中的构造函数就必须通过 base 关键字来调用基类的带参构造函数。

5.2 多 态 性

5.2.1 虚拟方法和重载方法

根据上一节的介绍，如果派生类的成员方法隐藏了基类的成员方法，那么程序会根据对象的声明类型而非实际类型来决定调用哪一个方法。例如 Bus 和 Truck 都隐藏了 Automobile 的 Run 方法，为了准确计算某汽车对象 a1 的行驶时间，程序中就必须明确判断其实际类型：
```
if (a1 is Bus)
    Console.WriteLine("客车行驶 1000 公里需{0}小时", ((Bus)a1).Run(1000));
else if (a1 is Truck)
    Console.WriteLine("卡车行驶 1000 公里需{0}小时", ((Truck)a1).Run(1000));
else
    Console.WriteLine("汽车行驶 1000 公里需{0}小时", a1.Run(1000));
```
而按照多态性的思想，客户并不一定要去了解服务器对象的详细信息，而是由这些对象根据自己的方式来提供服务。在 C#程序中，只要将基类的方法定义为虚拟方法（使用关键字 virtual 修饰），将派生类中的对应方法定义为重载方法（使用关键字 override 修饰），那么程序就能够根据对象的实际类型来决定调用哪一个方法。看下面的程序示例：
```
//程序 P5_4
using System;
namespace P5_4
{
    class Program
    {
        static void Main()
        {
            foreach (Automobile a in GetAutos()){
                a.Speak();
                Console.WriteLine("{0}行驶 1000 公里需{1}小时", a.Name, a.Run(1000));
            }
        }

        static Automobile[] GetAutos()
        {
            Automobile[] autos = new Automobile[4];
            autos[0] = new Bus("客车", 20);
            autos[1] = new Truck("东风卡车", 30);
            autos[2] = new Truck("黄河卡车", 45);
            autos[3] = new Automobile("汽车", 80, 3);
```

```csharp
        return autos;
    }
}

public class Automobile
{
    private string name;
    public string Name
    {
        get { return name; }
    }

    private float speed;
    public float Speed
    {
        get { return speed; }
    }

    private float weight;
    public float Weight
    {
        get { return weight; }
        set { weight = value; }
    }

    public Automobile(string name, float speed, float weight)
    {
        this.name = name;
        this.speed = speed;
        this.weight = weight;
    }

    public virtual float Run(float distance)   //虚拟方法
    {
        return distance / speed;
    }

    public virtual void Speak()   //虚拟方法
    {
        Console.WriteLine("汽车鸣笛……");
    }
}

public class Bus : Automobile
{
    private int passengers;
    public int Passengers
    {
        get { return passengers; }
        set { passengers = value; }
    }

    public Bus(string name, int passengers) : base(name, 60, 10)
    {
        this.passengers = passengers;
    }

    public override void Speak()   //重载方法
    {
        Console.WriteLine("嘀……嘀……");
    }
```

}
```csharp
public class Truck : Automobile
{
    private float load;
    public float Load
    {
        get { return load; }
        set { load = value; }
    }

    public Truck(string name, int load) : base(name, 50, 15)
    {
        this.load = load;
    }

    public override float Run(float distance)  //重载方法
    {
        return (1 + load / Weight / 2) * base.Run(distance);
    }

    public override void Speak()  //重载方法
    {
        Console.WriteLine("叭……叭……");
    }
}
```
}

在实际程序设计中，GetAutos 方法可能来自另一个类乃至另一个程序集，其调用者难以确定每个汽车对象的实际类型。但从程序的输出中可以看到，尽管声明类型都是 Automobile，真正的客车对象仍调用 Bus 类中的定义的方法，而真正的卡车对象仍调用 Truck 类中的方法：

```
嘀……嘀……
客车行驶 1000 公里需 16.66667 小时
叭……叭……
东风卡车行驶 1000 公里需 40 小时
叭……叭……
黄河卡车行驶 1000 公里需 50 小时
汽车鸣笛……
汽车行驶 1000 公里需 12.5 小时
```

在多层继承结构中，重载方法可以继续被派生类重载，也就是说重载方法在本质上也是一种虚拟方法。例如 Truck 的派生类 Light-Truck 和 Heavy-Truck 也可以使用 overirde 修饰符来重载 Run 和 Speak 方法；而如果再定义 Automobile 的基类 Vehicle，其中使用了 virtual 修饰符来定义 Run 方法，那么 Automobile 中的 Run 方法的修饰符就应改为 overirde。

提示

在 4.2.3 小节中介绍的方法重载是指方法的名称相同而标识不同，此处的重载方法则是指对基类中标识相同的虚拟方法的重载；但二者有着类似的含义，即"同样的操作在不同的条件下可以产生不同的结果"。在面向对象的语义中，重载方法特指使用 override 关键字定义的方法。一个类中如果定义了重载方法，那么其基类中就必须有相同标识的虚拟方法，且二者的访问权限应当相同。此外，new 关键字可用于隐藏基类的字段和方法，而 virtual 和 override 关键字仅限于修饰方法。

5.2.2 抽象类和抽象方法

现实生活中有很多抽象概念，它们本身不与具体的对象相联系，但可以为其派生类提供一个公共的界面。这类概念在 C#中可定义为抽象类，它通过 abstract 关键字来进行修饰。例如"图形"就可作为一个抽象类，因为每一个图形对象实际上都是其派生类的实例，如"椭圆"、"三角形"、"四边形"等；但"三角形"不应是一个抽象类，因为其对象既可以是派生的"等腰三角形"、"直角三角形"等特殊三角形，也可以是一般的三角形。再如"交通工具"也可定义为一个抽象类，并通过字段和方法来描述其派生类共有的特性：

```
public abstract class Vehicle
{
    protected float speed;
    public float Speed
    {
        get { return speed; }
    }
    public float Run(float distance)
    {
        return distance / speed;
    }
}
```

抽象类不能使用 new 关键字来直接创建对象，但可以将其与派生类的实例相关联，例如：

```
Vehicle v = new Vehicle();    //错误：不允许创建抽象类的实例！
Vehicle v1 = new Bus();       //正确
```

抽象类中常常会出现这种情况：其所有派生类都应提供某个方法，但无法为这些方法定义一个统一的实现形式。例如所有的图形对象都应有计算面积的 GetArea 方法，但在抽象类"图形"中无法给出该方法的实现代码。一种解决办法是在抽象类中将该方法定义为虚拟的，方法的代码简单地返回 0 值，而由各个派生类的重载方法提供具体的计算代码。不过这种方式显得很不自然，而且如果开发人员在某个派生类中忘了重载该方法，那么对此类图形计算面积的结果就总是 0。

C#提供的解决方式是：同样使用 abstract 关键字将此类方法定义为抽象方法。抽象方法没有实现代码，但该抽象类的派生类都必须重载该方法：

- 如果抽象类的派生类是非抽象的，那么它必须重载基类中的所有抽象方法，并为这些方法提供具体实现。
- 如果抽象类的派生类也是抽象的，那么它必须重载基类中的所有抽象方法，且重载方法要么提供具体实现，要么也是抽象的（同时使用 override 和 abstract 修饰符）。

下面的程序 P5_5 就定义了抽象的 Vehicle 类、它的非抽象派生类 Train、抽象派生类 Automobile，以及 Automobile 的非抽象派生类 Truck，请注意其中抽象方法、虚拟方法和重载方法的不同定义方式：

```
//程序 P5_5
using System;
namespace P5_5
{
    class Program
    {
        static void Main()
        {
            Vehicle v1 = new Train();
            v1.Speak();
            Console.WriteLine("行驶 1000 公里需{0}小时", v1.Run(1000));
```

```csharp
        v1 = new Truck(16, 24);
        v1.Speak();
        Console.WriteLine("行驶1000公里需{0}小时", v1.Run(1000));
    }
}

public abstract class Vehicle
{
    private float speed;
    public float Speed
    {
        get { return speed; }
    }

    public virtual float Run(float distance)   //虚拟方法
    {
        return distance / speed;
    }

    public abstract void Speak();   //抽象方法:无执行代码

    public Vehicle(float speed)
    {
        this.speed = speed;
    }
}

public class Train : Vehicle
{
    public Train() : base(160)
    { }

    public override void Speak()   //重载
    {
        Console.WriteLine("呜……");
    }
}

public abstract class Automobile : Vehicle
{
    public Automobile(float speed) : base(speed)
    { }

    public overrideabstract void Speak();   //重载+抽象
}

public class Truck : Automobile
{
    private float weight;
    public float Weight
    {
        get { return weight; }
    }

    private float load;
    public float Load
    {
        get { return load; }
```

```
        }
        public Truck(int weight, int load) : base(50)
        {
            this.weight = weight;
            this.load = load;
        }

        public override float Run(float distance)  //重载
        {
            return (1 + load / Weight / 2) * base.Run(distance);
        }

        public override void Speak()  //重载
        {
            Console.WriteLine("叭……叭……");
        }
    }
}
```

在上面的程序中,如果非抽象的 Train 和 Truck 类不包含 Speak 方法,那么就不能通过编译。从该程序中还可以看到:尽管抽象类不能创建实例,但它仍可以定义构造函数(但不能使用 abstract 修饰符),其作用是供派生类的构造函数重载之用。程序 P5_5 的输出结果如下:

> 呜……
> 行驶 1000 公里需 6.25 小时
> 叭……叭……
> 行驶 1000 公里需 35 小时

属性和索引函数也可以是抽象的,此时它们也不提供实现代码,但要声明访问器的类型,例如:
```
public abstract int X
{
    get;
}

public abstract int this[int index]
{
    get; set;
}
```
而重载的索引函数和属性必须提供一致的访问器类型,例如重载上面的 X 属性时只能提供 get 访问器,重载上面的索引函数则必须同时提供 get 和 set 访问器(这和隐藏基类的成员是不一样的)。

抽象类和静态类都不能创建实例,但一个类不能既是抽象的又是静态的。同样,抽象方法、虚拟方法和重载方法也不能是静态的。

5.2.3 密封类和密封方法

还有一些类型不允许或是不需要再有派生类型,那么在 C#中可使用 sealed 修饰符将其定义为密封类。例如下面的代码就定义了一个密封类 Circle,程序如果再试图定义 Circle 的派生类就会发生错误:
```
public abstract class Shape
{
    public abstract double GetArea();
```

```
    public sealed class Circle : Shape    //密封类
    {
        private double r;
        public Circle(double r)
        {
            this.r = r;
        }
        public override double GetArea()
        {
            return Math.PI * r * r;
        }
    }

    public class Ellipse : Circle    //错误：不允许继承密封类！
    { }
```

sealed 修饰符还可用于类的成员方法，其含义是该方法在派生类中不能被重载。也就是说，密封类是对类继承的"截止"，而密封方法是对类继承中方法重载的"截止"。注意密封方法必然也是重载方法，即 sealed 和 override 修饰符应同时出现。例如上面 Circle 类的 GetArea 方法也可定义为密封的：

```
    public sealed override double GetArea()
    {
        return Math.PI * r * r;
    }
```

实际上，上面代码中的 sealed 可以省略，因为 Circle 本身是密封类，其成员方法不可能再被重载。只有对于非密封类，sealed 修饰符才能真正对方法起到"密封"的效果：对声明为该类型的任何对象调用该方法，都能确保当前方法中的代码被执行（但这并不能阻止其派生类使用 new 关键字来隐藏该方法）。

在下面的程序 P5_6 中，Student 类就重载了 Object 类的 ToString 和 Equals 方法，并对后者进行了密封。Equals 方法是通过比较学号 id 来判断两个学生对象是否相等，这种判断方式也适用于 Student 的所有派生类。Graduate 类可以重载 Student 的 ToString 方法，但不能再重载其 Equals 方法。

```
//程序 P5_6
using System;
namespace P5_6
{
    class Program
    {
        static void Main()
        {
            Student s1 = new Student(101, "王小红");
            Console.WriteLine(s1);
            Student s2 = new Graduate(101, "王晓红", "张大伟");
            Console.WriteLine(s2);
            Console.WriteLine(s1.Equals(s2));
        }
    }

    public class Student
    {
        private string name;
        public string Name
```

```csharp
            {
                get { return name; }
            }
            private int id;
            public int ID
            {
                get { return id; }
            }
            public Student(int id, string name)
            {
                this.id = id;
                this.name = name;
            }
            public override string ToString()
            {
                return string.Format("学号{0},姓名{1}", id, name);
            }
            public sealed override bool Equals(object obj)
            {
                if (obj is Student && ((Student)obj).id == this.id)
                    return true;
                else
                    return false;
            }
        }
        public class Graduate : Student
        {
            private string supervisor;
            public string Supervisor
            {
                get { return supervisor; }
                set { supervisor = value; }
            }
            public Graduate(int id, string name) : base(id, name)
            { }
            public Graduate(int id, string name, string supervisor)
                : base(id, name)
            {
                this.supervisor = supervisor;
            }
            public override string ToString()
            {
                return base.ToString() + ",导师:" + supervisor;
            }
        }
    }
```

程序 P5_6 的输出结果如下：

```
学号101，姓名王小红
学号101，姓名王晓红，导师：张大伟
True
```

5.3 案例研究——旅行社业务类的实现和精化

5.3.1 会员类

本节将继续实现旅行社管理系统中的业务类。游客和会员都属于旅行社的客户，但游客主要提供参加旅行团所必需的信息，而会员属于旅行社长期发展的客户。下面定义了从 Customer 类派生的 Member 类，其中维护了会员的用户名、密码和积分：

```
public class Member : Customer //会员类
{
    private int _id;
    public int Id //会员号
    {
        get { return _id; }
    }

    private string _username;
    public string Username //用户名
    {
        get { return _username; }
    }

    private string _password;
    public string Password //密码
    {
        get { return _password; }
    }

    public int Credit { get; set; } //积分

    public string Interests { get; set; } //兴趣爱好

    public Member(int id, string name, bool gender, CustomerType type, string username, string password)
        : base(id, name, gender, type)
    {
        _username = username;
        _password = password;
    }
}
```

和一般游客相比，会员报名参加旅行团具有更高的优先级，这实现起来也很简单。首先修改 Customer 类的 Enroll 方法，在声明中加入一个 virtual 修饰符使其成为一个虚拟方法，而后在 Member 类中增加下面的重载方法：

```
public override Enrollment Enroll(Tour t)
{
    Enrollment enr = base.Enroll(t);
    enr.Priority = 1 + Credit / 1000;
    return enr;
}
```

也就是说，会员的报名优先级至少是 1，且每多 1000 个积分优先级就会增加 1。

5.3.2 职员类

职员类和游客类存在许多重复的信息，如姓名、年龄、性别等。但它们是系统中不同的研究对象，因此不再定义它们的共同基类。下面给出了职员类 Staff 的基本定义代码，其中只定义了保护型的构造函数，那么就不能为其直接创建对象，而只能创建其派生类对象：

```csharp
public class Staff  //职员类
{
    private int _id;
    public int Id  //职员编号
    {
        get { return _id; }
    }

    private string _name;
    public string Name  //姓名
    {
        get { return _name; }
    }

    private bool _gender;
    public bool Gender  //性别
    {
        get { return _gender; }
    }

    private DateTime _birthday;
    public DateTime Birthday  //生日
    {
        get { return _birthday; }
    }

    protected DateTime _joinday;
    public DateTimeJoinday  //聘用时间
    {
        get { return _joinday; }
    }

    public string Birthplace { get; set; }  //籍贯
    public string IdCard { get; set; }  //身份证号码
    public Degree Degree { get; set; }  //学历
    public StaffState State { get; set; }  //状态
    public string Phone { get; set; }  //联系电话
    public MailAddress Email { get; set; }  //电子邮件

    protected Staff(int id, string name, bool gender, DateTime birthday, DateTimejoinday)
    {
        _id = id;
```

```csharp
            _name = name;
            _gender = gender;
            _birthday = birthday;
            _joinday = joinday;
            State = StaffState.在岗;
        }

        public override string ToString()
        {
            return string.Format(_id.ToString() + _name);
        }

        public static Staff Get(int id)    //根据编号查询指定职员
        {
            foreach (Staff s in Staff.GetAll())
                if (s._id == id)
                    return s;
            return null;
        }

        public static Staff[] GetAll()    //获取所有职员（模拟）
        {
            Manager s1 = new Manager(105001, "程学兵", true, new DateTime(1972,2,20), new DateTime(2008, 7, 9));
            Director s2 = new Director(105003, "张文强", true, new DateTime(1980,9,7), new DateTime(2009, 11, 15));
            Agent s3= new Agent(105007, "马秋萍", false, new DateTime(1979,1,30), new DateTime(2010, 3, 23));
            Guide s4= new Guide(105009, "何艳", false, new DateTime(1985,10,9), new DateTime(2011, 1, 18));
            return new Staff[] { s1, s2, s3, s4 };
        }
}
```

注意 Staff 的 IdCard 属性是 string 类型而非 IDCard 结构类型，因为职员只使用身份证号码，而不像游客那样可以有多种证件。

再看 Staff 的派生类。导游需要持证才能上岗，因此 Guide 类在 Staff 的基础上增加了导游证的相关信息：

```csharp
public class Guide : Staff    //导游类
{
    public string GuideCard { get; set; }    //导游证号

    public DateTime GuideDue { get; set; }    //导游证有效期

    public Guide(int id, string name, bool gender, DateTime birthday, DateTime joinday)
        : base(id, name, gender, birthday, joinday)
    { }
}
```

Agent 类则着重提供处理游客报名的操作，具体实现代码如下：

```csharp
public class Agent : Staff    //业务员类
{
```

```csharp
public Agent(int id, string name, bool gender, DateTime birthday, DateTime joinday)
    : base(id, name, gender, birthday, joinday)
{ }

public void Accept(Enrollment enr)  //接受报名
{
    Tour t = enr.Tour;
    if (enr.State != EnrollmentState.待确认 || t.State != TourState.报名)
        return;
    Customer[] customers = new Customer[t.Customers.Length + enr.Customers.Length];
    for (inti = 0; i<t.Customers.Length; i++)
        customers[i] = t.Customers[i];
    for (inti = 0; i<enr.Customers.Length; i++)
        customers[t.Customers.Length + i] = enr.Customers[i];
    t.Customers = customers;
    enr.State = EnrollmentState.接受;
    if (t.IsFull())
        t.State = TourState.满员;
    if (enr.Applier.Email != null)//发送通知邮件
    {
        string subject = string.Format("欢迎参加'{0}'旅行", t);
        StringBuilder sb1 = new StringBuilder("尊敬的客户:\n");
        sb1.Append("欢迎您参加我社组织的");
        sb1.Append(t.Package.Line.Name);
        sb1.Append("\n 本次旅行的发团时间为: ");
        sb1.Append(t.StartTime);
        sb1.Append("\n 更详细的相关信息可登录我社网站查询: ");
        sb1.Append("\n 希望您能够充分享受这次美好的旅行生活! ");
        sb1.Append("\n 如您有任何意见或建议，请及时反馈给我们。");
        sb1.Append("\n\n\n 碧水丹山旅行社");
        sb1.Append(DateTime.Now);
        TravelMail.SendMail(enr.Applier.Email, subject, sb1.ToString());
    }
}

public void Reject(Enrollment enr, string reason)  //拒绝报名
{
    enr.State = EnrollmentState.拒绝;
    if (enr.Applier.Email != null)//发送通知邮件
    {
        string subject = enr.Tour.Package.Line.Name + "报名结果";
        StringBuilder sb1 = new StringBuilder("尊敬的客户:\n");
        sb1.Append("感谢您报名参加我社组织的");
        sb1.Append(enr.Tour.Package.Line.Name);
        sb1.Append("\n 很遗憾，由于 ");
        sb1.Append(reason);
        sb1.Append("\n, 您的报名没有成功。");
        sb1.Append("\n 在此我们深表歉意，希望下次能为您提供更优质的服务。");
```

```
            sb1.Append("\n 如您有任何意见或建议，请及时反馈给我们。");
            sb1.Append("\n\n\n 碧水丹山旅行社");
            sb1.Append(DateTime.Now);
            TravelMail.SendMail(enr.Applier.Email, subject, sb1.ToString());
        }
    }
}
```

TravelMail 通过 System.Net.Mail 命名空间中提供的 MailAddress、MailMessage 和 SmtpClient 等类型来进行邮件收发，其功能实现方式可参考 MSDN，具体定义代码则可参考教案源程序中的 TravelMail.cs 文件。

接下来考虑职员管理中的请假功能，"休假"这个概念也可以抽象为一个类，其属性包括请假原因、起止时间、请假人和批准人，以及请假结果，具体定义代码如下：

```
public class Vocation// 休假类
{
    private string _reason;
    public string Reason //请假原因
    {
        get { return _reason; }
    }

    public DateTime From { get; set; } //请假开始时间
    public DateTime To { get; set; } //请假截止时间

    private Staff _applicant;
    public Staff Applicant //请假人
    {
        get { return _applicant; }
    }

    public Staff Approver { get; set; } //批准人
    public bool Approved { get; set; } //是否已被批准

    internal Vocation(Staff applicant, string reason, DateTime from, DateTime to, Director approver)
    {
        _applicant = applicant;
        _reason = reason;
        From = from;
        To = to;
        Approver = approver;
        Approved = false;
    }
}
```

Vocation 的构造函数使用了 internal 修饰符，那么外部程序集就不能直接创建该对象。从业务角度分析，请假必须由某个职员提出，那么可以为 Staff 类增加如下的成员方法，该方法生成请假条邮件并发送给某个主管，然后返回一个 Vocation 对象：

```csharp
        public virtual Vocation Apply(DateTime from, DateTime to, string reason, Director approver)   //请假
        {
            StringBuilder sb1 = new StringBuilder("请假时间: ");
            sb1.Append(from);
            sb1.Append(" 至 ");
            sb1.Append(to);
            sb1.Append("\n事由: ");
            sb1.Append(reason);
            TravelMail.SendMail(approver.Email, _name + "请假条", sb1.ToString());
            return new Vocation(this, reason, from, to, approver);
        }
```

主管只能向经理请假，那么可在 Director 类中重载 Staff 的 Apply 方法，并在 Director 和 Manager 类中实现批准请假的 Approve 方法；此外 Director 还要实现安排业务员和导游的方法。这两个类的基本定义代码如下：

```csharp
    public class Director : Agent// 主管类
    {
      public Director(int id, string name, bool gender, DateTime birthday, DateTime joinday)
            : base(id, name, gender, birthday, joinday)
      { }

      public override Vocation Apply(DateTime from, DateTime to, string reason, Director approver)   //请假
      {
          if (!(approver is Manager))
              return null;
          return base.Apply(from, to, reason, approver);
      }

      public virtual void Approve(Vocation v)  //批假
      {
          if (v.Applicant is Director)
              return;
          v.Approver = this;
          v.Approved = true;
          v.Applicant.State = StaffState.休假;
      }

      public virtual void Approve(Vocation v, DateTime from, DateTime to)   //有条件批假
      {
          if (v.Applicant is Director)
              return;
          v.Approver = this;
          v.From = from;
          v.To = to;
          v.Approved = true;
          v.Applicant.State = StaffState.休假;
      }

      public void Assign(Package p, Agent a)   //安排业务员
      {
          if (a.State == StaffState.试用 || a.State == StaffState.在岗)
```

```
            p.Agent = a;
    }

    public void Assign(Tour t, Guide g)   //安排导游
    {
        if (g.State == StaffState.试用 || g.State == StaffState.在岗)
            t.Guide = g;
    }
}

public class Manager : Director// 经理类
{
    public Manager(int id, string name, bool gender, DateTime birthday, DateTime joinday)
        : base(id, name, gender, birthday, joinday)
    { }

    public override void Approve(Vocation v)   //批假
    {
        v.Approver = this;
        v.Approved = true;
        v.Applicant.State = StaffState.休假;
    }
    public override void Approve(Vocation v, DateTime from, DateTime to)//有条件批假
    {
        v.Approver = this;
        v.From = from;
        v.To = to;
        v.Approved = true;
        v.Applicant.State = StaffState.休假;
    }

    public void Hire(Staff s, bool bTry)   //雇佣员工
    {
        if (bTry)
            s.State = StaffState.试用;
        else
            s.State = StaffState.在岗;
    }

    public void Fire(Staff s)   //解雇员工
    {
        s.State = StaffState.离职;
    }
}
```

5.4 小　　结

　　封装、继承和多态性是面向对象程序设计的 3 个基本要素。通过继承，派生类能够自动获得基类中已定义的功能特性，而对基类代码的修改也会自动反映到派生类中，从而实现良好的可重用性和可维护性。

　　派生类中还可以定义与基类中相同的成员，并对基类成员进行覆盖和重载，而对基类虚拟成

员的重载则是实现多态性的一个关键技术。这样客户就可以以统一的方式来处理基类对象和派生类对象，而不同的对象会自行选择执行不同的操作，从而产生不同的结果。

5.5 习　　题

1. 简述公有、保护和私有成员的继承规则。
2. 简要叙述一下各种特殊方法成员在继承过程中要注意的事项。
3. 举例说明密封类和密封方法的作用。
4. 定义磁盘类 Disk 及其派生类硬盘 HardDisk、闪盘 Flash 和光盘 CDROM，在其中定义记录磁盘容量的字段，并通过虚拟方法和重载方法来模拟对磁盘内容的写入和删除。
5. 设计基本几何图形的继承层次结构，并编程实现其中的主要类型，要求通过抽象方法、虚拟方法和重载方法来计算各种图形的面积和周长。
6. 结构类型不支持继承，那么有哪些办法可以重用一个现有结构的功能呢？
7. 在旅行社管理系统中，如果游客报名已经成功，那么要在其取消报名时发生邮件通知负责的业务员，请在 Customer 类的 Cancel 方法中加入此功能。
8. 编写控制台应用程序，测试旅行社管理系统中的旅行团报名和职员请假功能。

第6章
委托和事件

面向对象的核心思想之一就是将数据和对数据的操作封装为一个整体。前面我们已经掌握了将数据对象作为方法参数进行传递，而使用委托（Delegate）则能够将方法本身作为参数进行传递。本章将介绍委托的实现机制，并介绍其在匿名方法和事件中的应用。

6.1 委托和方法

6.1.1 通过委托来封装方法

委托是一种特殊的引用类型，它将方法也作为特殊的对象封装起来，从而将方法作为变量或参数进行传递。看下面的程序示例：

```
//程序 P6_1
using System;
namespace P6_1
{
    delegate void DualFunction(double x, double y);
    class Program
    {
        static void Main()
        {
            DualFunction fun1;
            double a = 2.5, b = 2;
            Console.Write("请选择函数(加 0 减 1 乘 2 除 3)：");
            int i = int.Parse(Console.ReadLine());
            if (i == 1)
                fun1 = new DualFunction(Sub);
            else if (i == 2)
                fun1 = new DualFunction(Mul);
            else if (i == 3)
                fun1 = new DualFunction(Div);
            else
                fun1 = new DualFunction(Add);
            fun1(a, b);
        }

        static void Add(double x, double y)
        {
```

```
            Console.WriteLine("{0} + {1} = {2}", x, y, x + y);
        }
        static void Sub(double x, double y)
        {
            Console.WriteLine("{0} - {1} = {2}", x, y, x - y);
        }
        static void Mul(double x, double y)
        {
            Console.WriteLine("{0} * {1} = {2}", x, y, x * y);
        }
        static void Div(double x, double y)
        {
            Console.WriteLine("{0} / {1} = {2}", x, y, x / y);
        }
    }
}
```

该程序首先定义了一个委托类型 DualFunction，注意 delegate 关键字之后要说明委托的返回类型（没有返回类型则为 void）和参数类型：

```
delegate void DualFunction(double x, double y);
```

程序主方法的第一行代码则创建了一个该类型的变量 fun1：

```
DualFunction fun1;
```

之后就可以将指定的方法封装到该委托对象中，不过方法的参数列表和返回类型都必须和委托的定义保持一致，例如：

```
fun1 = new DualFunction(Sub);
```

注意在 new 表达式的括号中是要封装的方法名，而不包括方法的参数和返回值。和其他类型的对象一样，委托对象的声明和初始化也可以合并在一行语句中，例如：

```
DualFunction fun1 = new DualFunction(Sub);
```

程序 P6_1 是通过委托调用当前类的静态方法，因此在创建表达式中只需写出方法名；如果是调用外部类型的静态方法，那么应写出方法所属的类型，如"DualFunction fun1 = new DualFunction(Program.Sub)"；如果调用的是非静态方法，那么还需指出方法所属的对象名。

C#还允许将方法名直接写在委托赋值表达式的等号右边，而不必写出完整的委托，例如：

```
DualFunction fun1 = Sub;
```

最后，在程序中可以通过委托对象来调用指定的方法，其效果和直接调用原始方法的效果是一样的，例如：

```
fun1(a, b);
```

从中可以看到，委托的使用过程一般可分为 3 步：类型定义、对象创建和方法绑定，以及方法调用。除了最后的方法调用外，委托变量的使用和一般对象变量没有本质的区别。比如可以将一组委托对象放在一个数组中，那么程序 P6_1 的主方法可以改写为如下内容：

```
static void Main()
{
    DualFunction[] funs = new DualFunction[] {Add, Sub, Mul, Div};
    double a = 2.5, b = 2;
    Console.Write("请选择函数(加 0 减 1 乘 2 除 3)：");
    int i = int.Parse(Console.ReadLine());
    funs[i](a, b);
}
```

6.1.2 委托的加减运算

一个委托对象还可以封装多个方法,这是通过委托对象的相加(也叫合并)来实现的。例如下面的委托对象 fun3 就能够连续输出两个数相乘和相除的结果:

```
DualFunction fun1 = new DualFunction(Mul);
DualFunction fun2 = new DualFunction(Div);
DualFunction fun3 = fun1 + fun2;
```

委托的合并还可以使用复合赋值操作符,例如:

```
DualFunction fun1 = Mul;
fun1 += Div;
```

对应的,减法操作符能够将方法从已合并的委托对象中删除,例如:

```
fun3 = fun3 -fun2;
```

当然,参与加减运算的委托对象必须属于同一个委托类型,也就是说参与合并或删除的方法的参数和返回类型必须完全一致。

不过,如果要减去的委托未包含在作为被减数的委托对象中,减法运算不会产生任何效果。而如果将所包含的委托对象全部删除之后,最后得到的是一个空的委托对象 null,通过其进行方法调用会引发程序异常。很显然,委托对象减去自身后就等于 null,而加减 null 值对委托对象不会产生任何效果,例如:

```
DualFunction fun1 = null;
DualFunction fun2 = Mul + fun1;
fun2(2.5, 2);
fun2 -= fun2;
Console.WriteLine(fun1 == fun2);  //输出 True
fun2(2.5, 2);  //错误:不能调用空委托!
```

因此,在通过某个委托对象进行方法调用时,如果不能确定其他对象是否对委托进行了删除操作,那么安全的做法是先判断该委托对象是否为空,例如:

```
if (fun2 != null)
    fun2(2.5, 2);
```

6.1.3 传递委托对象

C#的委托机制将方法视为特殊的对象,并允许将其作为其他方法的参数或返回值进行传递。以程序 P6_1 中定义的 DualFunction 为例,我们可以为 Program 类添加如下的静态方法,它根据整数 i 来返回不同的 DualFunction 委托对象:

```
static DualFunction GetDualFunction(int i)
{
    if (i == 1)
        return new DualFunction(Sub);
    else if (i == 2)
        return new DualFunction(Mul);
    else if (i == 3)
        return new DualFunction(Div);
    else
        return new DualFunction(Add);
}
```

那么 P6_1 主方法中的代码就可以改写为如下内容:

```
static void Main()
{
    double a = 2.5, b = 2;
    Console.Write("请选择函数(加 0 减 1 乘 2 除 3): ");
    int i = int.Parse(Console.ReadLine());
```

```
        DualFunction fun1 = GetDualFunction(i);
        fun1(a, b);
```

上述最后两行代码还可以合并为一句：

```
GetDualFunction(i)(a, b);
```

程序 P6_2 则给出了一个以委托对象作为方法参数的示例：Student 类的静态方法 SortAndPrint 用于对一组学生进行排序而后输出，而排序的依据由 CompareFunction 委托对象（其类型被定义为 Student 的嵌套类型）决定。主程序类 Program 的 3 个静态方法 CompareName、CompareAge 和 CompareGrade 分别按照姓名、年龄和年级来比较两个学生对象，那么程序就可以不同的排序方式来输出学生信息。

```
//程序 P6_2
using System;
namespace P6_2
{
    class Program
    {
        static void Main()
        {
            Student[] students = new Student[5];
            students[0] = new Student("王小红", 20, 1);
            students[1] = new Student("周军", 23, 2);
            students[2] = new Student("方小白", 21, 2);
            students[3] = new Student("高强", 25, 3);
            students[4] = new Student("王浩", 22, 3);
            Student.CompareFunction compare;
            Console.WriteLine("请选择排序方式：A 姓名 B 年龄 C 年级");
            char ch = Console.ReadKey().KeyChar;
            if (ch == 'B' || ch == 'b')
                compare = CompareAge;
            else if (ch == 'C' || ch == 'c')
                compare = CompareGrade;
            else
                compare = CompareName;
            Student.SortAndPrint(students, compare);
        }

        static int CompareName(Student s1, Student s2)
        {
            return 0 - s1.Name.CompareTo(s2.Name);
        }

        static int CompareAge(Student s1, Student s2)
        {
            return s1.Age - s2.Age;
        }

        static int CompareGrade(Student s1, Student s2)
        {
            return s1.Grade - s2.Grade;
        }
    }

    public class Student
    {
        private string name;
```

```csharp
    public string Name
    {
        get { return name; }
    }

    private int age;
    public int Age
    {
        get { return age; }
    }

    public int Grade { get; set; }
    public Student(string name, int age, int grade)
    {
        this.name = name;
        this.age = age;
        this.Grade = grade;
    }

    public static void SortAndPrint(Student[] students, CompareFunction compare)
    {
        for (int i = students.Length - 1; i > 0; i--)
            for (int j = 0; j < i; j++)
                if (compare(students[j], students[j + 1]) > 0){
                    Student s = students[j];
                    students[j] = students[j + 1];
                    students[j + 1] = s;
                }
        foreach (Student s in students)
            Console.WriteLine(s);
    }

    public override string ToString()
    {
        return string.Format("{0} {1}岁 {2}年级", name, age, Grade);
    }

    public delegate int CompareFunction(Student s1, Student s2);
}
```

SortAndPrint 方法中使用的排序算法叫做冒泡排序，即先通过相邻元素的两两比较将数组中最大的元素交换到数组的最后，再将次大的元素交换到数组倒数第二的位置……最后就得到了有序的数组。

6.1.4 Delegate 类型成员

C#中定义的所有委托类型实际上都是.NET 类库中 Delegate 类的派生类。不过，Delegate 是一个抽象类，不能创建实例；而且 C#编译器不允许显式定义 Delegate 的派生类，而是必须通过 delegate 关键字来创建委托类型。

和一般对象相比，委托对象的特殊性在于它能够封装和调用方法，具有方法的过程性；当然它也同样具有变量的说明性，可以访问 Delegate 类中定义的方法成员。如 Delegate 类重载了相等和不等操作符，通过它们可以判断两个委托对象所包含的内容是否相等。而其成员方法 DynamicInvoke 也可调用委托对象所封装的方法，例如下面两行代码的运行结果是一样的：

```
fun1(2.5, 2);
fun1.DynamicInvoke(2.5, 2);
```

不过 DynamicInvoke 方法的使用的是一个 params object[]类型的数组型参数，因此无论将什么参数传递给该方法都能通过编译，但运行时可能发生错误。例如下面第一行代码将不能通过编译，而第二行代码直到程序运行时才会引发异常，因此在通常情况下应采用前一种方式来进行方法调用：

```
fun1(2.5, 2, 1.5);                  //编译错误
fun1.DynamicInvoke(2.5, 2, 1.5);    //程序异常
```

Delegate 类的 GetInvokationList 方法返回一个 Delegate[]数组，通过它可以得到所包含的委托对象列表。此外，Delegate 的公有属性 Method 表示所封装的方法，其类型为 System.Reflection 命名空间下的 MethodInfo 类，通过其可进一步获取方法的详细信息。

Delegate 类还定义了两个静态方法 Combine 和 Remove，它们实际上等效于委托的加减运算，只不过方法的参数和返回类型都是基类 Delegate，使用时通常需要进行类型转换。例如下面两行代码的作用是一样的：

```
DualFunction fun3 = fun1 + fun2;
DualFunction fun3 = (DualFunction)Delegate.Combine(fun1, fun2);
```

6.2 匿名方法和 Lambda 表达式

6.2.1 匿名方法

如果一个方法只是通过委托进行调用，那么在 C#中允许不写出该方法的定义，而是将方法的执行代码直接写在委托对象的创建表达式中，此时被封装的方法叫做匿名方法（对应的，使用常规方式定义的方法成员叫做命名方法）。例如对于程序 P6_1，将 Add 方法封装到 DualFunction 对象 fun1 中的代码可以换为：

```
DualFunction fun1 = delegate (double x, double y)
{
    Console.WriteLine("{0} + {1} = {2}", x, y, x + y);
};
```

该语句的等号右边是一个匿名方法表达式，它不使用 new 关键字来创建委托对象，而是在 delegate 关键字之后直接写出方法的参数列表和执行代码。其中参数列表应和委托类型的定义保持一致；如果执行代码中不出现参数变量，那么参数列表还可以省略。如果委托有返回类型，那么在执行代码中也应通过 return 语句来返回该类型的值。

 匿名方法并不是没有名称，而是指开发人员无须在源代码中为其命名。C#在编译程序时会自动生成一个方法定义，该方法实际上是当前类型的一个私有静态方法。

使用匿名方法来改写程序 P6_2 的话，可以不再定义 CompareName、CompareAge 和 CompareGrade 方法，而是将程序主方法中的排序代码改写为如下内容：

```
static void Main()
{
    Student[] students = new Student[5];
    students[0] = new Student("王小红", 20, 1);
    students[1] = new Student("周军", 23, 2);
    students[2] = new Student("方小白", 21, 2);
    students[3] = new Student("高强", 25, 3);
```

```csharp
    students[4] = new Student("王浩", 22, 3);
    Student.CompareFunction compare;
    Console.WriteLine("请选择排序方式：A姓名 B年龄 C年级");
    char ch = Console.ReadKey().KeyChar;
    if (ch == 'B' || ch == 'b')
        compare = delegate(Student s1, Student s2){
            return s1.Age - s2.Age;
        };
    else if (ch == 'C' || ch == 'c')
        compare = delegate(Student s1, Student s2){
            return s1.Grade - s2.Grade;
        };
    else
        compare = delegate(Student s1, Student s2){
            return 0 - s1.Name.CompareTo(s2.Name);
        };
    Student.SortAndPrint(students, compare);
}
```

6.2.2 Lambda 表达式

从 C# 3.0 中开始可以使用 Lambda 表达式来替换匿名方法表达式。例如：

```
Student.CompareFunctioncompare = (s1, s2) => s1.Age - s2.Age;
```

该语句的等号右边就是一个 Lambda 表达式，它表示"给定两个对象 s1 和 s2，返回 s1.Age - s2.Age"，其中参数 s1 和 s2 的类型可从委托类型的定义中推断出来。Lambda 表达式有多种书写形式，上面采用的是最简洁的一种。实际上，下面这些 Lambda 表达式都是等价的：

```
(Student s1, Student s2) =>{ return s1.Age - s2.Age; }
(Student s1, Student s2) => s1.Age - s2.Age;
(s1, s2) =>{ returns1.Age - s2.Age; }
(s1, s2) => s1.Age - s2.Age;
```

在 Lambda 表达式中，符号"=>"的左侧为输入参数，右侧为计算体。如果只有一个参数，那么左侧的括号还可以省略，例如：

```
x =>x++;
```

如果方法执行代码中用不到某个输入参数，那么可在 Lambda 表达式的参数列表中使用下划线来表示该参数，例如：

```
_ =>DateTime.Now.Year;
```

6.2.3 外部变量

由于匿名方法和 Lambda 表达式总是定义在另一个方法的执行代码中，那么匿名方法的参数名和变量名就可能和外部代码发生冲突。为此 C#规定：匿名方法的参数名不能和已有的外部变量名相同；如果匿名方法执行体中的局部变量名和外部变量名相同，那么它们代表同一个变量，此时称外部变量被匿名方法所"捕获"。例如下面的代码是错误的，因为匿名方法的参数名和外部变量名相冲突：

```csharp
double x = 2.5, y = 2;
DualFunction fun1 = delegate(double x, double y) //错误：不能和外部变量名相同
{
    Console.WriteLine("{0} + {1} = {2}", x, y, x + y);
};
fun1(x, y);
```

下面的代码则将两次输出"2.5 + 2 = 4.5",不过第一次输出属于参数传递,第二次属于捕获外部变量:

```
double a = 2.5, b = 2;
DualFunction fun1 = delegate(double x, double y)
{
    Console.WriteLine("{0} + {1} = {2}", x, y, x + y);
    Console.WriteLine("{0} + {1} = {2}", a, b, a + b); //捕获外部变量
};
fun1(a, b);
```

通过捕获外部变量,匿名方法就能够实现与外部程序代码的状态共享。下面的程序演示了命名方法和匿名方法对外部变量的不同处理方式:

```
//程序 P6_3
using System;
namespace P6_3
{
    public delegate int OneDelegate(int x);

    class Program
    {
        static void Main()
        {
            int x = 0;
            OneDelegate dg1 = new OneDelegate(Increment); //封装命名方法
            Console.WriteLine(dg1(x));
            Console.WriteLine(x);
            Console.WriteLine(dg1(x));
            Console.WriteLine(x);
            dg1 = delegate { return ++x; }; //封装匿名方法
            Console.WriteLine(dg1(x));
            Console.WriteLine(x);
            Console.WriteLine(dg1(x));
            Console.WriteLine(x);
        }

        public static int Increment(int x)
        {
            return ++x;
        }
    }
}
```

当委托对象 dg1 封装的是命名方法 Increment 时,变量 x 只是作为形参传递,调用之后并不修改 x 的值,因此程序前四次输出分别是 1、0、1、0;而当 dg1 封装的是匿名方法时,x 被其捕获,方法代码能够修改 x 的值,因此程序后四次输出的分别是 1、1、2、2。

6.3 事件处理

6.3.1 委托发布和订阅

由于委托能够封装方法,而且能够合并或删除其他委托对象,那么就能够通过委托来实现"发布者/订阅者(Publisher/Subscriber)"的设计模式。具体实现步骤为:

(1)定义委托类型,并在发布者类中定义一个该类型的公有成员。

（2）在订阅者类中定义委托处理方法。
（3）订阅者对象将其事件处理方法合并到发布者对象的委托成员上。
（4）发布者对象在特定的情况下激发委托操作，从而自动调用订阅者对象的委托处理方法。

以交通红绿灯为例，如果希望车辆对交通灯的颜色变化作出响应（红灯停，绿灯行），那么可以定义一个委托类型 LightEvent，其参数 color 表示交通灯的颜色（布尔类型，false 为绿灯，true 为红灯）：

```
public delegate void LightEvent(bool color);
```

而后在交通灯类 TrafficLight 中定义一个 LightEvent 类型的公有成员 OnColorChange，并在其成员方法 ChangeColor 中激发委托操作：

```
public class TrafficLight
{
    private bool color = false;
    public bool Color
    {
        get { return color; }
    }

    public LightEvent OnColorChange;  //委托发布

    public void ChangeColor()
    {
        color = !color;
        Console.WriteLine(color ? "红灯亮" : "绿灯亮");
        if(OnColorChange != null)
            OnColorChange(color);
    }
}
```

接下来在车辆类 Car 中定义事件处理方法 LightColorChange，并在需要时将其合并到某个 TrafficLight 对象的 OnColorChange 委托上，例如：

```
public class Car
{
    private bool bRun = true;

    public void Enter(TrafficLight light)
    {
        light.OnColorChange += LightColorChange;  //通过委托合并进行订阅
    }

    public virtual void LightColorChange(bool color)  //委托处理方法
    {
        if (bRun && color){
            bRun = false;
            Console.WriteLine("{0}停车", this);
        }
        else if (!bRun && !color){
            bRun = true;
            Console.WriteLine("{0}启动", this);
        }
    }
}
```

这样当 TrafficLight 对象使用其 ChangeColor 方法改变交通灯颜色时，就会自动调用相关 Car 对象的 LightColorChange 方法，例如：

```
TrafficLight light = new TrafficLight();
```

```
Car car1 = new Car();
car1.Enter(light);             //委托合并
light.ChangeColor();           //绿灯变红灯,输出"Car 停车"
light.ChangeColor();           //红灯变绿灯,输出"Car 启动"
```

图 6-1 给出了这种基于委托的发布和订阅模式的示意图。

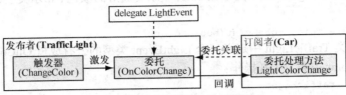

图 6-1　基于委托的发布和订阅模式

这种模式下,一个发布者可以对应多个订阅者对象,这些对象还可以属于不同的类型,并采用不同的委托处理方法。比如对于救护车类 Ambulance,它在一般情况下采用与基类 Car 相同的方法来处理交通灯的颜色变化,但在紧急情况下(emergent 字段值为 true)允许闯红灯:

```
public class Ambulance : Car
{
    private bool emergent = false;
    public bool Emergent
    {
        get { return emergent; }
        set { emergent = value; }
    }
    public override void LightColorChange(bool color)
    {
        if (emergent)
            Console.WriteLine("{0}紧急行驶", this);
        else
            base.LightColorChange(color);
    }
}
```

在不需要的情况下,订阅者对象可以通过委托删除来取消订阅。例如 Car 可以通过如下方法来取消对交通灯的响应:

```
public void Leave(TrafficLight light)
{
    light.OnColorChange -= LightColorChange;  //取消订阅
}
```

6.3.2　事件发布和订阅

通过委托发布和订阅,可使不同的对象对特定的情况作出反应。但这种机制存在一个问题,即外部对象可以任意修改已发布的委托,这就会影响到其他对象对委托的订阅。比如一个新类型 Truck 也可以订阅 TrafficLight 的 OnColorChange 委托,但它没有使用复合赋值操作符"+=",而是直接使用了赋值操作符"=":

```
public class Truck
{
    public void Enter(TrafficLight light)
    {
        light.OnColorChange= LightColorChange;
    }
```

```csharp
        public virtual void LightColorChange(bool color)
        {
            Console.WriteLine(color ? "停车" : "启动");
        }
```

那么当某个 Truck 对象完成订阅后，Car 等其他对象的订阅就会被删除，TrafficLight 的 OnColorChange 委托将只调用 Truck 对象的 LightColorChange 方法。那么当交通灯变化时，其他对象不作出响应。将"+="写成"="很可能是程序员的一个疏忽，但也有可能是其他程序的故意破坏，更极端的情况是订阅者可能直接取消整个发布：

```csharp
light.OnColorChange= null;
```

为了解决这一问题，C#提供了专门的事件以实现可靠的订阅发布，其做法是在发布的委托定义中加上 event 关键字：

```csharp
public event LightEvent OnColorChange; //事件发布
```

经过这个简单的修改之后，其他类型再使用 OnColorChange 委托时，就只能将其放在在复合赋值操作符"+="或"-="的左侧，而类似于下面这样的代码都不能通过编译：

```csharp
light.OnColorChange = LightColorChange; //错误
light.OnColorChange = null; //错误
```

也就是说，事件是一种特殊的委托类型，发布者在发布一个事件之后，订阅者针对它只能进行自身的订阅或取消，而不能干涉其它订阅者。图 6-2 描述了基于事件的发布和订阅模式，程序 P6_4 则给出了交通灯事件处理的完整代码。

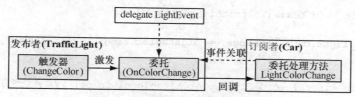

图 6-2　基于事件的发布和订阅模式

```csharp
//程序 P6_4
using System;
namespace P6_4
{
    class Program
    {
        static void Main()
        {
            TrafficLight light = new TrafficLight();
            Car car1 = new Car();
            car1.Enter(light);
            Ambulance amb1 = new Ambulance();
            amb1.Enter(light);
            light.ChangeColor();
            light.ChangeColor();
            amb1.Emergent = true;
            light.ChangeColor();
        }
    }

    public delegate void LightEvent(bool color);

    public class TrafficLight
    {
```

```csharp
        private bool color = false;
        public bool Color
        {
            get { return color; }
        }
        public event LightEvent OnColorChange;  //事件发布
        public void ChangeColor()
        {
            color = !color;
            Console.WriteLine(color ? "红灯亮" : "绿灯亮");
            if(OnColorChange != null)
                OnColorChange(color);
        }
    }
    public class Car
    {
        private bool bRun = true;

        public void Enter(TrafficLight light)
        {
            light.OnColorChange += LightColorChange;  //事件订阅
        }

        public void Leave(TrafficLight light)
        {
            light.OnColorChange -= LightColorChange;  //取消事件订阅
        }

        public virtual void LightColorChange(bool color)  //事件处理方法
        {
          if (bRun && color){
             bRun = false;
             Console.WriteLine("{0}停车", this);
          }
          else if (!bRun && !color){
             bRun = true;
             Console.WriteLine("{0}启动", this);
          }
       }
    }

    public class Ambulance : Car
    {
       private bool emergent = false;
       public bool Emergent
       {
          get { return emergent; }
          set { emergent = value; }
       }

       public override void LightColorChange(bool color)  //事件处理方法
       {
          if (emergent)
             Console.WriteLine("{0}紧急行驶", this);
          else
```

```
        base.LightColorChange(color);
    }
}
```

程序 P6_4 的输出结果如下：

红灯亮
P6_5.Car 停车, 60 秒后启动
绿灯亮
P6_5.Car 启动, 30 秒内通过

事件也是类的一种特殊方法成员；即使是公有事件，除了其所属类型之外，其他类型只能对其进行订阅或取消，其他任何操作都是不允许的，因此事件具有特殊的封装性。和一般委托成员不同，某个类型的事件只能由自身触发。例如在 Car 的成员方法中使用如下代码来直接调用 light 对象的 OnColorChange 事件是不允许的：

```
light.OnColorChange(true);   //错误
```

6.3.3 使用 EventHandler 类

在事件发布和订阅的过程中，定义事件的原型委托类型常常是一件重复性的工作。为此，.NET 类库中定义了一个 EventHandler 委托类型，并建议尽量使用该类型作为事件的原型。该委托的定义为：

```
public delegate void EventHandler(object sender, EventArgs e);
```

其中 object 类型的参数 sender 表示引发事件的对象。由于事件成员只能由类型本身触发，因此在触发时传递给该参数的值通常应为 this。例如可将 TrafficLight 类的 OnColorChange 事件定义为 EventHandler 类型，那么触发事件的代码就是"OnColorChange(this, null);"：

```
public event EventHandler OnColorChange;

public void ChangeColor()
{
    color = !color;
    Console.WriteLine(color ? "红灯亮" : "绿灯亮");
    if (OnColorChange != null)
        OnColorChange(this, null);
}
```

事件的订阅者可以通过 sender 参数来了解是哪个对象触发的事件，不过在访问对象时通常要进行拆箱转换。例如 Car 类对 TrafficLight.OnColorChange 事件的处理方法可修改为：

```
public virtual void LightColorChange(bool color)
{
    TrafficLight light = (TrafficLight)sender;
    Console.WriteLine(light.color ? "停车" : "启动");
}
```

如果事件的发布者是静态类，那么触发事件时传递给 EventHandler 委托的 sender 参数值可以为空（null），而订阅者总是可以通过静态类本身来访问发布者的信息。

EventHandler 委托的第二个参数 e 表示事件中包含的数据。如果发布者还要向订阅者传递额外的事件数据，那么就需要定义 EventArgs 类型的派生类。例如要描述交通灯变化后持续的时间，那么可以定义如下的 LightEventArgs 类：

```csharp
public class LightEventArgs : EventArgs
{
    private int seconds;
    public int Seconds
    {
        get { return seconds; }
    }
    public LightEventArgs(int seconds)
    {
        this.seconds = seconds;
    }
}
```

而 TrafficLight 在触发 OnColorChange 事件时,就可以将该数据作为参数传递给 EventHandler 委托:

```csharp
public event EventHandler OnColorChange;
public void ChangeColor(int seconds)
{
    color = !color;
    Console.WriteLine(color ? "红灯亮" : "绿灯亮");
    if (OnColorChange != null)
        OnColorChange(this, new LightEventArg(seconds));
}
```

由于 EventHandler 原始定义中的参数类型是 EventArgs,那么订阅者在读取参数内容时同样需要进行拆箱转换,例如:

```csharp
public virtual void LightColorChange(object sender, EventArgs e)
{
    if (((TrafficLight)sender).Color){
        bRun = false;
        Console.WriteLine("{0}停车, {1}秒后启动", this, ((LightEventArgs)e).Seconds);
    }
        else{
        bRun = true;
        Console.WriteLine("{0}启动, {1}秒内通过", this, ((LightEventArgs)e).Seconds);
    }
}
```

6.3.4 在事件中使用匿名方法

匿名方法和 Lambda 表达式同样可以作为事件的处理方法,例如程序 P6_4 中 Car 的事件订阅代码可以改写为如下内容:

```csharp
light.OnColorChange += delegate(bool color) //匿名方法
{
    if (bRun && color){
        bRun = false;
        Console.WriteLine("{0}停车", this);
    }
    else if (!bRun && !color){
        bRun = true;
        Console.WriteLine("{0}启动", this);
    }
};
```

程序 P6_5 中就使用了 EventHandler 类型和匿名方法来处理交通灯事件:

```csharp
//程序 P6_5
```

```csharp
using System;
namespace P6_5
{
    class Program
    {
        static void Main()
        {
            TrafficLight light = new TrafficLight();
            Car car1 = new Car();
            car1.Enter(light);
            light.ChangeColor(60);
            light.ChangeColor(30);
        }
    }

    public class LightEventArgs : EventArgs
    {
        private int seconds;
        public int Seconds
        {
            get { return seconds; }
        }

        public LightEventArgs(int seconds)
        {
            this.seconds = seconds;
        }
    }

    public class TrafficLight
    {
        private bool color = false;
        public bool Color
        {
            get { return color; }
        }

        public event EventHandler OnColorChange;  //事件发布

        public void ChangeColor(int seconds)
        {
            color = !color;
            Console.WriteLine(color ? "红灯亮" : "绿灯亮");
            if (OnColorChange != null)
                OnColorChange(this, new LightEventArgs(seconds));
        }
    }

    public class Car
    {
        private bool bRun = true;

        public void Enter(TrafficLight light)
        {
            light.OnColorChange += delegate(object sender, EventArgs e)
            {
                if (light.Color){
                    bRun = false;
                    Console.WriteLine("{0}停车, {1}秒后启动", this, ((LightEventArg)e).
```

```
Seconds);
                }
                else{
                    bRun = true;
                    Console.WriteLine("{0}启动，{1}秒内通过", this, ((LightEventArgs)e).Seconds);
                }
            };
        }
    }
```

从上面 Car 类的事件处理方法中可以看到，在使用基于 EventHandler 类型的事件时，匿名方法的另一个好处是可以直接访问外部变量（如 Enter 方法中的 light 对象），这就能有效地减少或避免拆箱转换。

 不过匿名方法不适合需要取消事件订阅的场合。因为 C#编译器会为每个匿名方法创建一个类型成员，那么向事件中合并和删除的就不是同一个成员，起不到取消订阅的效果。

6.4 Windows 控件事件概述

事件在 Windows 这样的图形界面程序中有着极其广泛的应用，事件响应是程序与用户交互的基础。用户的绝大多数操作，如移动鼠标、单击鼠标、在文本框中输入内容、选择菜单命令等，都可以触发相关的控件事件。以按钮控件 Button 为例，其成员 Click 就是一个 EventHandler 类型的事件：

```
public event EventHandler Click;
```

用户单击按钮时，Button 对象就会调用其保护成员方法 OnClick，并通过它来触发 Click 事件：

```
protected virtual void OnClick(EventArgs e)
{
    if(Click != null)
        Click(this, e);
}
```

此时如果在程序中定义了相应的事件处理方法，那么单击按钮就能够执行其中的代码。假设窗体 Form1 包含一个名为 button1 的按钮，那么可以在窗体构造函数中关联事件处理方法，并在方法代码中执行所需的功能：

```
public Form1()
{
    InitializeComponent();
    button1.Click += new EventHandler(button1_Click);  //关联事件处理方法
}

private void button1_Click(object sender, EventArgs e)  //事件处理方法
{
    this.Close();
}
```

在下面的示例程序 P6_6 中，Windows 窗体中包含了 3 个按钮控件和 1 个文本框控件，3 个按钮的 Click 事件共用一个事件处理方法 button_Click：

```
//程序 P6_6
using System;
```

```
using System.Windows.Forms;
namespace P6_6
{
    public partial class Form1 : Form
    {
        public Form1()
        {
            InitializeComponent();
            this.button1.Click += new EventHandler(button_Click);
            this.button2.Click += new EventHandler(button_Click);
            this.button3.Click += new EventHandler(button_Click);
        }
        void button_Click(object sender, EventArgs e)
        {
            textBox1.Text = "您按下了" + ((Button)sender).Text;
        }
    }
}
```

其中TextBox控件和Button控件的Text属性分别表示文本框中和按钮上显示的文字，那么当用户在窗体上按下某个按钮后，文本框中就会显示"您按下了buttonX"，如图6-3所示。

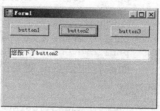

图6-3 程序P6_6的运行结果

在Visual Studio开发环境中，只要在Windows窗体的设计视图中选择某个控件，而后单击"属性"窗口中的 按钮，窗口中就会显示该控件的事件列表；双击某个事件项，Visual Studio就会自动创建并关联其事件处理方法，并切换到代码视图中，以便开发人员输入事件处理代码。例如在设计视图中选择整个窗体，在属性窗口中看到的Form类的事件列表如图6-4所示，其中有加载窗体时触发的Load事件（在构造窗体对象之后发生），在窗体上单击鼠标时触发的MouseClick事件，双击鼠标时触发的MouseDoubleClick事件等。而在设计视图中选择并双击某个控件，Visual Studio会自动创建并关联其最常用的事件处理方法，如按钮的Click事件，窗体的Load事件等。

 如果是通过Visual Studio的设计视图来添加控件的事件处理方法，那么将处理方法关联（合并）到事件的代码并不位于窗体的源程序文件（如Form1.cs）中，而是位于窗体的设计器源程序文件（如Form1.Designer.cs）中，这两个源文件共同组成了窗体类的完整定义。

图6-4 属性窗口中的控件事件列表

6.5 案例研究——旅行团基本事件处理

6.5.1 旅行团事件发布

下面分析旅行社业务的一些事件需求。每个旅行团都会有一组对象与其相关联,那么在旅行团信息发生变化时就需要通知这些对象。特别的,旅行团准备出发应提醒随团导游以及所有已确认报名的游客;旅行团成功返回后,系统应向游客发送致谢邮件。如果一个旅行团因故取消,那么所有已确认报名的游客都应得到通知。那么可以为 Tour 类增加如下的事件成员,分别表示旅行团信息变化、准备出发、出发、结束和取消时所触发的事件:

```csharp
public event EventHandler OnChange;
public event EventHandler OnPreSend;
public event EventHandler OnSend;
public event EventHandler OnComplete;
public event EventHandler OnCancel;
```

而旅行团信息改变要说明内容,取消也要说明原因,下面定义的事件参数类中就封装了相应的消息:

```csharp
public class TourEventArgs : EventArgs// 旅行团事件参数类
{
    private string _description;
    public string Description
    {
        get { return _description; }
    }
    public TourEventArgs(string description)
    {
        _description = description;
    }
}
```

Tour 类的下列方法分别用于旅行团的准备出发、出发、完成和取消,它们各自触发了相应的事件:

```csharp
public void PreSend() //准备发团
{
    if (this.OnPreSend != null)
        OnPreSend(this, null);
    StringBuilder sb1 = new StringBuilder(this.ToString());
    sb1.AppendLine("即将发团");
    sb1.Append("时间: ");
    sb1.AppendLine(this.StartTime.ToString());
    sb1.Append("地点: ");
    sb1.AppendLine(this.StartAddress);
    TravelMail.SendMail(this.Guide.Email, this.ToString() + "发团提醒", sb1.ToString());
}

public void Send() //发团
{
    State = TourState.发团;
    Guide.State = StaffState.随团;
    if (this.OnSend != null)
```

```
        OnSend(this, null);
        TravelMail.SendMail(this.Guide.Email, this.ToString() + "发团消息",
this.ToString() + "已发团");
    }

    public void Complete()   //完成旅行
    {
        State = TourState.完成;
        Guide.State = StaffState.在岗;
        if (this.OnComplete != null)
            OnComplete(this, null);
    }

    public void Cancel(string reason)   //取消组团
    {
        State = TourState.取消;
        if (this.OnCancel != null){
            TourEventArgs e = new TourEventArgs(reason);
            OnCancel(this, e);
        }
    }
```
而 Tour 类的 StartTime 和 StartAddress 的 set 属性访问器也应作如下修改：
```
privateDateTime _startTime;
public DateTime StartTime
{
    get { return _startTime; }
    set {
        _startTime = value;
      if (this.OnChange != null)
            OnChange(this, new TourEventArgs("出发时间改为" + value));
    }
}

private string _startAddress;
public string StartAddress
{
    get { return _startAddress; }
    set{
            _startAddress = value;
            if (this.OnChange != null)
                OnChange(this, new TourEventArgs("出发地点改为" + value));
    }
}
```

6.5.2 旅行团事件处理

在 Customer 类的报名方法 Enroll 中，可以订阅旅行团的一系列事件：
```
public Enrollment Enroll(Tour t)   //报名参加旅行团
{
    Enrollment enr = new Enrollment(t, this);
    if (t.Enrollments == null)
        t.Enrollments = new Enrollment[] { enr };
    else {
        Enrollment[] enrollments = new Enrollment[t.Enrollments.Length + 1];
        for (int i = 0; i < t.Enrollments.Length; i++)
            enrollments[i] = t.Enrollments[i];
```

```csharp
        enrollments[t.Enrollments.Length] = enr;
        t.Enrollments = enrollments;
    }
    t.OnChange +=new EventHandler(tour_OnChange);
    t.OnPreSend += new EventHandler(tour_OnPreSend);
    t.OnComplete += new EventHandler(tour_OnComplete);
    t.OnCancel += new EventHandler(tour_OnCancel);
    return enr;
}
```

下面给出了 Customer 类中相应的事件处理方法:

```csharp
protected void tour_OnChange(object sender, EventArgs e)
{
   if (Email != null){
      Tour t = (Tour)sender;
      string subject = t.Package.Line.Name + "重要通知";
      StringBuilder sb1 = new StringBuilder("尊敬的客户:\n");
      sb1.Append("您报名参加的");
      sb1.Append(t.Package.Line.Name);
      sb1.AppendLine("有如下信息变化:\n");
      sb1.Append(((TourEventArgs)e).Description);
      sb1.Append("\n 给您带来的不便请多谅解。");
      sb1.Append("\n 如您有任何意见或建议,请及时反馈给我们。");
      sb1.Append("\n\n\n 碧水丹山旅行社");
      sb1.Append(DateTime.Now);
      TravelMail.SendMail(Email, subject, sb1.ToString());
   }
}

protected void tour_OnPreSend(object sender, EventArgs e)
{
   if (Email != null){
      Tour t = (Tour)sender;
      string subject = t.Package.Line.Name + "出发提醒";
      StringBuilder sb1 = new StringBuilder("尊敬的客户:\n");
      sb1.Append("欢迎您参加我社组织的");
      sb1.Append(t.Package.Line.Name);
      sb1.Append("\n 本次旅行团将于");
      sb1.Append(t.StartTime);
      sb1.Append("从");
      sb1.Append(t.StartAddress);
      sb1.Append("准时出发,请您及时做好相关准备工作");
      sb1.Append("\n 如您有任何意见或建议,请及时反馈给我们。");
      sb1.Append("\n\n\n 碧水丹山旅行社");
      sb1.Append(DateTime.Now);
      TravelMail.SendMail(Email, subject, sb1.ToString());
   }
}

protected virtual void tour_OnComplete(object sender, EventArgs e)
{
   if (Email != null){
      Tour tour = (Tour)sender;
      string subject = "感谢您参与我们的旅行";
      StringBuilder sb1 = new StringBuilder("尊敬的客户:\n");
```

```
        sb1.Append("您参加了我社组织的");
        sb1.Append(tour.Package.Line.Name);
        sb1.Append("\n 希望这次旅行给您留下了美好的回忆。\n");
        sb1.Append("如您有任何意见或建议,请及时反馈给我们。");
        sb1.Append("\n\n\n 碧水丹山旅行社");
        sb1.Append(DateTime.Now);
        TravelMail.SendMail(Email, subject, sb1.ToString());
    }
}

protected void tour_OnCancel(object sender, EventArgs e)
{
    if (Email != null){
        Tour tour = (Tour)sender;
        string subject = tour.Package.Line.Name + "取消通知";
        StringBuilder sb1 = new StringBuilder("尊敬的客户:\n");
        sb1.Append("很抱歉,本次旅行由于");
        sb1.Append(((TourEventArgs)e).Description);
        sb1.Append("而取消。\n");
        sb1.Append("\n 在此我们深表歉意,希望下次能为您提供更优质的服务。");
        sb1.Append("如您有任何意见或建议,请及时反馈给我们。");
        sb1.Append("\n\n\n 碧水丹山旅行社");
        sb1.Append(DateTime.Now);
        TravelMail.SendMail(Email, subject, sb1.ToString());
    }
}
```

显然,在游客取消报名或是业务员拒绝游客报名后,应取消这些事件订阅。为此可以在 Customer 类中的专门定义如下的成员方法:

```
public void Unsubscribe(Tour t)
{
    t.OnChange -= new EventHandler(tour_OnChange);
    t.OnPreSend -= new EventHandler(tour_OnPreSend);
    t.OnComplete -= new EventHandler(tour_OnComplete);
    t.OnCancel -= new EventHandler(tour_OnCancel);
}
```

那么 Customer 的 Cancel 方法可作如下修改:

```
public void Cancel(Enrollment enr) //取消报名
{
    Tour t = enr.Tour;
    if (t.Enrollments.Length == 1 && t.Enrollments[0] == enr)
        t.Enrollments = null;
    int i;
    for (i = 0; i < t.Enrollments.Length; i++){
        if (t.Enrollments[i] == enr && t.Enrollments[i].Applier == this){
            Enrollment[] enrollments = new Enrollment[t.Enrollments.Length - 1];
            for (int j = 0; j < i; j++)
                enrollments[j] = t.Enrollments[j];
            for (int j = i; j < t.Enrollments.Length - 1; j++)
                enrollments[j] = t.Enrollments[j + 1];
            t.Enrollments = enrollments;
        }
    }
    if (i != t.Enrollments.Length)
        this.Unsubscribe(t);
}
```

而在 Agent 的 Reject 方法的第一行也应插入下面的语句：
enr.Applier.Unsubscribe(enr.Tour);

6.6 小　　结

委托是一种特殊的引用类型，它将方法作为特殊的对象进行封装、传递和调用。只通过委托进行调用的方法可以定义为匿名方法。事件是类的特殊成员，它利用委托机制来使对象对外界发生的情况作出自动响应，这在 Windows 窗体等图形界面中有着广泛的应用。下一章将对 Windows 基本控件的用法作更为详细的介绍。

6.7 习　　题

1. 简述通过委托来调用对象方法的基本过程。
2. 为程序 P6_2 添加多重排序的功能，比如先按年级排序，年级相同再按年龄排序，年龄仍相同再按姓名排序。
3. 定义一个原型为 void WriteMethodInfo(Delegate dg1) 的方法，它能够在控制台依次输出委托对象 dg1 所封装的各个方法的返回类型、参数数量和类型等信息（提示：使用 Delegate 类型的 GetInvokationList 方法和 Method 属性）。
4. 写出下面代码的运行结果：

```
public delegate int IntFunction(int x);
class Program
{
    static void Main()
    {
        int x = 1;
        IntFunction fun = delegate(int y){
            y += x++;
            return y;
        };
        for (int i = 0; i < 3; i++)
            Console.WriteLine(fun(1));
    }
}
```

5. 通过匿名方法来处理事件有什么优缺点？
6. 什么是事件的触发方法，什么又是事件的处理方法？二者之间有何联系？
7. 在 6.5 节中，旅行团状态改变时，为什么给导游和负责业务员发邮件是通过 Tour 的成员方法来实现，而给游客发邮件要通过事件来完成？
8. 编写控制台应用程序，测试旅行团的各项事件功能。

第 7 章
Windows Form 应用程序设计

Windows 窗体和 Web 浏览器是两种最为常见的图形用户界面。目前主流的程序设计语言都采用面向对象的方式来处理各种 GUI 元素，并在集成开发环境中支持以"所见即所得"的方式来设计程序界面。针对 Windows 窗体开发，.NET Framework 提供了标准 Windows Form 和 WPF 两种风格的界面编程方式。本章将介绍 Windows Form 应用程序设计的基本要素，WPF 应用程序开发将在第 13 章中介绍。

7.1 图形用户界面概述

现代应用程序越来越多地使用图形用户界面（GUI）与用户进行可视化的交互。良好的界面设计是优秀应用程序不可或缺的一个要素。即使软件具有高效的算法、良好的面向对象设计、很高的可靠性，但如果不能提供良好的用户体验，也很难被用户所认可。以下简要地总结了设计 GUI 界面的一些基本原则。

- 界面一致性：整个界面应给人一种协调一致的感觉，尽量使用标准的控件和信息表现方法，相似的功能应使用相似的界面和操作和完成。
- 布局合理化：合理安排界面上的各个元素，把重要的元素放在醒目的位置，控件按照其使用顺序进行放置（一般是从上到下和从左至右）。
- 操作简便性：减少用户的工作量，避免重复操作，为频繁执行的操作设计快捷键，使用 Tab 键帮助用户在控件之间快速导航，尽可能地提供默认的或候选的输入内容。
- 操作容错性：为用户的误操作提供恢复功能，而在执行不可恢复的或有其他危险性的操作之前要求用户确认。
- 响应时间：避免程序停止响应，在需要用户等待时显示沙漏状光标或进度条。
- 帮助和提示：提供有效的用户帮助，为固定处理流程提供向导界面，在必要时给出错误和警告信息，且提示信息要便于用户理解。

提高用户界面设计能力的最佳方法就是参考他人的成功范例，特别是借鉴优秀商业软件的创意和经验，并认真采纳最终用户的意见和建议，使界面真正满足用户需求。

早期的 Windows 程序需要直接调用 Windows 底层 API 来绘制图形用户界面，工作量十分繁重。后来，Microsoft 逐步为 Windows 开发了 User32、GDI（Graphical Device Interface）、GDI+、DirectX 等编程接口，它们对底层 API 进行了良好的封装，从而大大简化了图形界面设计的工作。在.NET Framework 提供的两种 Windows 界面风格中，标准 Windows Form 使用的是 GDI/GDI+编程接口，WPF 使用的则是 DirectX 编程接口，如图 7-1 所示。

.NET 类库以面向对象的方式封装了大量 GUI 元素。在绝大多数情况下，我们只需要在程序

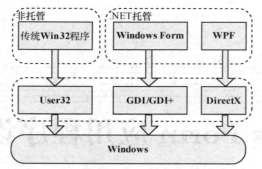

图 7-1　Windows 图形界面编程模型

中使用现有的类型，而不必去直接调用 GDI/GDI+ 或 DirectX 编程接口。在 Visual Studio 开发环境中，我们能够以"所见即所得"的方式来方便地创建窗体和拖放控件。这些都大大简化了界面设计的难度，从而使我们能够把更多的注意力放到程序业务逻辑上来。

7.2　位置、坐标、颜色和字体

在详细讲解 Windows Form 控件之前，首先介绍 Windows Form 中几个常用的基础类型，它们都在 System.Drawing 命名空间下定义。

7.2.1　Size 和 SizeF 结构

Size 结构表示二维平面上的矩形尺寸，它存储了一个有序的整数对，并通过属性 Width 和 Height 来分别表示矩形的水平宽度和垂直高度，可指定这两个值来构造 Size 对象。该结构还对加法、减法、相等和不等操作符进行了重载，其中加法和减法就是分别叠加两个 Size 对象的宽度和高度，相等和不等操作符则是同时比较两个 Size 对象的宽度和高度。例如：

```
Size s1 = new Size(10, 6);
Size s2 = new Size(5, 3);
Size s3 = s1 + s2; //s3.Width = 15, s3.Height = 9
Size s4 = s1 - s2; //s4.Width = 5, s4.Height = 3
bool b1 = (s2 == s4); //b1 = true
```

SizeF 结构和 Size 结构的用法基本相同，只不过它采用浮点数类型 float 来表示宽度和高度。Size 结构可以隐式转换为 SizeF 结构，而通过 SizeF 结构的 ToSize 方法可以得到取整后的 Size 结构，例如：

```
Size s1 = new Size(10, 8);
SizeF s2 = s1;
s2 += new SizeF(0.6F, 1.2F);
Size s3 = s2.ToSize(); //s3.Width = 10, s3.Height = 9
```

7.2.2　Point 和 PointF 结构

Point 结构也存储了一个有序的整数对，但它表示的是二维平面上的坐标点，它的两个属性 X 和 Y 分别表示点的横坐标和纵坐标。使用时要注意：Windows 屏幕坐标的 X 轴向右为正方向，而 Y 轴向下为正方向。

Point 结构也重载了加法、减法、相等和不等操作符，只不过其加法和减法的右操作数是 Size 类型，分别表示向右下和左上移动指定尺寸的水平和垂直分量。例如：

```
Point p1 = new Point(2, 3);
```

```
Point p2 = new Point(8, 12);
p1+= new Size(3, 4);//p1.X = 5, p1.Y = 7
p2-= new Size(3, 5);//p2.X = 5, p2.Y = 7
bool b1 = (p1 == p2); //b1 = true
```

对应的，PointF 结构使用浮点数来表示坐标值，从 Point 结构可以隐式转换到 PointF 结构。此外，在 Point 结构和 Size 结构、PointF 结构和 SizeF 结构之间还存在显式类型转换，例如：

```
Size s1 = new Size(5, 3);
Point p1 = (Point)s1;
p1 += new Size(3, 1);
Size s2 = (Size)p1;//s2.Width = 8, s2.Height = 4
```

7.2.3 Color 结构

Color 结构封装了窗体界面中的颜色信息，其基础表示是基于 ARGB 颜色结构的，即任何一种颜色都可看作是红（R）、绿（G）、蓝（B）这三种单色的组合。通过 Color 结构的静态方法 FromArgb 可指定 RGB 分量来创建颜色对象，通过 R、G、B 属性则可以访问这些分量的值，这些值都在 0~255 之间变化。例如：

```
Color c1 = Color.FromArgb(0, 0, 0); // 黑色
Color c2 = Color.FromArgb(255, 255, 255); // 白色
Color c3 = Color.FromArgb(255, 0, 0); // 红色
```

非专业用户可能不是很了解颜色的分量概念，这时可通过 Color 结构的一组静态属性来获取指定颜色，这些属性名都是具体颜色的英文单词名，例如：

```
Color c1 = Color.Black;
Color c2 = Color.White;
Color c3 = Color.Red;
```

7.2.4 Font 和 FontFamily 类

Font 类封装了窗体界面中的字体信息，每个 Font 对象又属于一个字体族 FontFamily。例如我们常说的"宋体"、"黑体"、"Times New Roman"都属于字体族，而在文档中使用的具体字体，如"五号宋体粗体"和"小五号宋体加下划线"，则属于不同的 Font 对象。

如果知道字体族的名称，那么就可以创建对应的 FontFamily 对象，例如：

```
FontFamily ff1 = new FontFamily("华文行楷");
```

FontFamily 的 Name 属性表示字体族的名称，而通过其静态属性 Families 可得到当前系统所有可用字体族的集合。下面的代码就试图得到一个"华文行楷"FontFamily 对象，如果该字体族不存在则使用默认的"宋体"：

```
FontFamily ff1 = null;
foreach (FontFamily ff in FontFamily.Families)
    if (ff.Name == "华文行楷")
        ff1 = ff;
if(ff1 == null)
    ff1 = new FontFamily("宋体");
```

在字体族的基础上可以方便地创建字体对象，而且在 Font 的构造函数中既可以使用 FontFamily 对象，也可以使用字体族的名称，例如下面创建的对象 f1 和 f2 都是 10 号宋体：

```
FontFamily ff1 = new FontFamily("宋体");
Font f1 = new Font(ff1, 10);
Font f2 = new Font("宋体", 10);
```

Font 类的构造函数有多种重载形式，上述代码创建的都是常规字体对象；如果再使用一个 FontStyle 枚举类型的参数，还可指定字体的样式。FontStyle 的枚举取值包括 Regular（常规）、

Bold（粗体）、Italic（斜体）、Strikeout（删除线）和 Underline（下划线）。例如下面的代码就分别创建了"10 号新罗马斜体"和"9 号宋体粗体加下划线"两个 Font 对象：

```
Font f1 = new Font("Times New Roman", 10, FontStyle.Italic);
Font f2 = new Font("宋体", 9, FontStyle.Bold | FontStyle.Underline);
```

创建了 Font 对象后，可从其 FontFamily 属性获得所属的字体族，从 Size 属性获得字体大小，以及从布尔类型的属性 Bold、Italic、Strikeout 和 Underline 来判断字体样式。不过上述属性都是只读的，也就是说不能在创建一个字体对象后再去改变其风格。

7.3 窗体、消息框和对话框

Windows Form 中的主要控件都在 System.Windows.Forms 命名空间下定义。窗体 Form 本身也是控件，但它又可以作为其他控件的容器。Windows 应用程序的开发通常都是先创建一个 Windows 主窗体、在其上加入各种控件、设置相关属性、再通过第 6 章中介绍的事件处理模型来响应用户操作，根据需要还可创建其他窗体和自定义控件，从而实现各种程序功能。

7.3.1 窗体

Form 类是对 Windows 窗体的抽象，在 Visual Studio 中创建的每一个窗体都是 Form 类的派生类。掌握了 Form 的基本用法，就能熟练地操控各种类型的 Windows 窗体。图 7-2 中显示了一个简单的窗体实例。

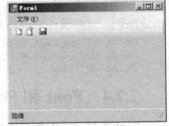

图 7-2 Windows 窗体实例

1. 窗体的常用属性

Form 类提供了一系列属性来设置窗体的可视化特征，如大小、位置、颜色和字体等。除了通过代码修改之外，常用属性也可以在 Visual Studio 的"属性"窗口中直接设置。

从图 7-2 中可以看到，窗体上方是一个标题栏，从左至右依次为窗体的图标、标题文本，以及最小化、最大化和关闭 3 个小按钮。通过 Form 类的 Text 属性可读取或设置窗体的标题文本，例如：

```
this.Text = "欢迎光临";  //this 为当前窗体对象
```

Form 类有 3 个与标题栏按钮相关的属性：MaximizeBox、MinimizeBox 和 ControlBox，它们均为布尔类型，且默认值都是 true。如果把 MaximizeBox 或 MinimizeBox 的值设为 false，可以取消窗体的最大化或最小化按钮；而如果把 ControlBox 的值设为 false，则 3 个按钮都被取消。

如果窗体的 ControlBox 属性值为 false，同时 Text 属性值又为空字符串，那么窗体将不显示标题栏。在 ControlBox 属性值为 false 时，要注意通过其他控件或快捷键来为用户提供关闭窗体的方式。

Form 还有 3 个与窗体尺寸相关的属性：Size、MaximumSize 和 MinimumSize，分别表示窗体正常显示、最大化和最小化时的尺寸，其类型均为 Size 结构。而窗体在屏幕上的位置则可以使用 Location 属性来设置，其类型为 Point 结构，表示窗体左上角在屏幕上的坐标（屏幕左上角为坐标原点）。下面的代码演示了如何设置当前窗体的大小和位置：

```
this.Size = new Size(320, 240);
this.Location = new Point(0, 50);
```

Form 类的 Font 属性表示窗体所使用的字体，ForeColor 和 BackColor 属性则分别表示窗体的前景色和背景色。例如下面的代码设置当前窗体字体为 12 号宋体，且使用蓝底黄字：

```
this.Font = new Font("宋体", 12);
this.ForeColor = Color.Yellow;
this.BackColor = Color.Blue;
```

Form 类还有如下一些常用的属性。

- ShowInTaskBar：bool 类型，表示是否在 Windows 任务栏中显示窗体说明。
- TopMost：bool 类型，表示窗体是否总位于最前方。
- Opacity：double 类型，表示窗体的透明度，取值在 0～1 之间；默认值 1 表示不透明，0 表示完全透明（窗体不可见）。
- AcceptButton：Button 类型，表示当窗体获得焦点时按下 Enter 键所触发的按钮。
- CancelButton：Button 类型，表示当窗体获得焦点时按下 Esc 键所触发的按钮。
- FormBorderStyle：FormBorderStyle 枚举类型，表示窗体的边框样式，枚举取值包括 None（无边框）、FixedSingle（固定单行边框）、Fixed3D（固定的三维边框）、FixedDialog（固定对话框式的粗边框）、Sizable（可调整大小的边框）、FixedToolWindow（不可调整大小的工具窗口式边框）、SizableToolWindow（可调整大小的工具窗口式边框）。
- WindowState：FormWindowState 枚举类型，表示窗体的显示状态，枚举取值包括 Normal（常规）、Minimized（最小化）、Maximized（最大化）。

2. 窗体的常用方法

一个程序中可能需要显示一个或多个窗体，显示方式可以分为模态和非模态两种。模态显示是指程序中只有当前窗体能与用户交互，在关闭该窗体之前无法切换到其他窗体；而非模态显示的窗体可以被切换到后台。Form 类的 ShowDialog 和 Show 方法分别用于模态和非模态显示窗体，例如：

```
Form frm1 = new Form();
frm1.ShowDialog(); //以模态方式显示窗体
(new Form()).Show(); //以非模态方式显示窗体
```

Show 方法没有返回类型，而 ShowDialog 方法的返回类型为 DialogResult 枚举，表示对话框的返回值，枚举取值包括 None、OK、Cancel、Abort、Retry、Ignore、Yes 和 No。此外，Form 类的 Close 方法用于关闭窗体，Activate 方法用于激活窗体（使窗体具有焦点），而 Update 方法则用于重新绘制窗体界面。

3. 窗体的常用事件

窗体在首次启动时将引发 Load 事件，在关闭过程中和关闭后将分别引发 FormClosing 和 FormClosed 事件。在使用过程中，窗体得到和失去焦点后将分别引发 Activated 和 Deactivate 事件。此外，而窗体的一些属性被修改时也会引发相应的事件，如标题文本被改变时引发 TextChanged 事件，窗体位置被改变时引发 LocationChanged 事件，窗体大小被改变时引发 SizeChanged 事件，字体被改变时引发 FontChanged 事件等。

下面的示例程序就为窗体 Form1 添加了 LocationChanged 事件和 SizeChanged 事件的处理方法，并在窗体的标题栏上分别显示窗体的当前位置和尺寸大小信息，其输出结果如图 7-3 所示。

```
//程序 P7_1(Form1.cs)
using System;
using System.Drawing;
using System.Windows.Forms;
namespace P7_1
{
    public partial class Form1 : Form
    {
```

```
public Form1()
{
    InitializeComponent();
    LocationChanged += new EventHandler(Form1_LocationChanged);
    SizeChanged += new EventHandler(Form1_SizeChanged);
}

void Form1_SizeChanged(object sender, EventArgs e)
{
    this.Text = "窗体尺寸: " + this.Size.ToString();
}

void Form1_LocationChanged(object sender, EventArgs e)
{
    this.Text = "窗体位置: " + this.Location.ToString();
}
```

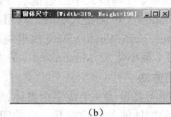

(a)　　　　　　　　　　　　　　(b)

图 7-3　程序 P7_1 的输出结果示例

7.3.2　消息框

命名空间 System.Windows.Forms 下有一个 MessageBox 类，它表示系统预定义的消息框。该类没有提供公有的构造函数，因此不能创建实例。事实上，除了从 Object 继承的成员外，MessageBox 的公有成员只有 Show 方法，用于向用户显示消息框。不过该方法有多种重载形式，从而能够向用户显示文本、标题、按钮、图标等不同的提示内容。

Show 方法最简单的用法只有一个参数，表示在消息框中显示的文本，例如：

`MessageBox.Show("欢迎光临！");`

此时显示的消息框如图 7-4（a）所示，可以看到其中包含消息文本和一个"确定"按钮，而标题栏上只有一个关闭按钮。如果向 Show 方法传递两个 string 类型的参数，那么第二个参数的字符串内容就会显示在消息框的标题栏上，例如下面的语句所显示的消息框将如图 7-4（b）所示：

`MessageBox.Show("欢迎光临！", "您好");`

消息框中还可以显示不同的按钮，这是通过向 Show 方法传递一个 MessageBoxButtons 枚举类型的参数来实现的，其枚举取值包括 OK（显示"确定"按钮）、OKCancel（"确定"和"取消"按钮）、YesNo（"是"和"否"按钮）、YesNoCancel（"是"、"否"和"取消"按钮）、RetryCancel（"重试"和"取消"按钮），以及 AbortRetryIgnore（"终止"、"重试"和"取消"按钮。例如下面的语句所显示的消息框如图 7-4（c）所示：

`MessageBox.Show("是否继续？", "警告", MessageBoxButtons.YesNoCancel);`

在 Show 方法中再加入一个 MessageBoxIcon 枚举类型的参数，就能够在消息框中显示具有不同含义的图标。例如下面的语句将显示将如图 7-4（d）所示的消息框：

`MessageBox.Show("发生错误！", "警告", MessageBoxButtons.RetryCancel, MessageBoxIcon.Error);`

MessageBoxIcon 有 9 个枚举取值，但实际上只能提供 4 种不同的图标，其中枚举值 None 表示不显示图标，其他枚举值和图标的对应关系如表 7-1 所示。

表 7-1　　　　　　　　　　　　　MessageBoxIcon 的枚举值

枚举取值	含义	图标	枚举取值	含义	图标
Asterisk	星号提示	ⓘ	Information	信息提示	ⓘ
Error	错误提示	⊗	Question	提问	❓
Exclamation	惊叹号	⚠	Stop	停止	⊗
Hand	禁止标志	⊗	Warning	警告	⚠

在显示多个按钮时，消息框总是以第一个按钮为默认按钮（类似于 Form 类的 AcceptButton）。如果要使其他按钮成为默认按钮，那么可以在 Show 方法最后加入一个 MessageBoxDefaultButton 枚举类型的参数，其枚举值 Button1、Button2 和 Button3 分别表示以第一、第二和第三个按钮作为默认按钮（消息框中最多一次显示 3 个按钮）。例如下面的语句将使消息框以第三个"取消"按钮作为默认按钮，如图 7-4（e）所示。

```
MessageBox.Show("是否继续? ", "警告", MessageBoxButtons.YesNoCancel, MessageBoxIcon.
Question, MessageBoxDefaultButton.Button3);
```

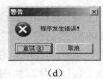
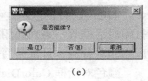

（a）　　　　（b）　　　　（c）　　　　　　　　（d）　　　　　　　　（e）

图 7-4　通过 MessageBox.Show 方法显示不同风格的消息框

和 Form 类的 ShowDialog 方法类似，MessageBox 的 Show 方法也返回一个 DialogResult 类型的值，各枚举取值分别表示用户单击了对应的按钮，例如 OK 表示单击"确定"按钮，Abort 表示单击"取消"按钮，等等。

在下面的程序 P7_2 中，窗体启动前将显示如图 7-5（a）所示的消息框，用户单击"确定"按钮才会进入主窗体，单击"取消"按钮将直接退出程序；而在关闭窗体前将显示如图 7-5（b）所示的消息框，用户单击"是"按钮才退出程序，否则仍留在主窗体中（对于 FormClosing 事件的参数 FormClosingEventArgs，设置其 Cancel 属性值为 true 表示放弃关闭窗体）。

```
//程序 P7_2(Form1.cs)
using System;
using System.Windows.Forms;
namespace P7_2
{
    public partial class Form1 : Form
    {
        public Form1()
        {
            InitializeComponent();
            this.Load += Form1_Load;
        }

        void Form1_Load(object sender, EventArgs e)
        {
            if (MessageBox.Show("要进入程序吗? ", "提示", MessageBoxButtons .OKCancel,
MessageBoxIcon.Question) == DialogResult.Cancel)
```

```
            this.Close();
        else
            this.FormClosing += Form1_FormClosing;
    }
    void Form1_FormClosing(object sender, FormClosingEventArgs e)
    {
        if (MessageBox.Show("确认要退出吗？", "提示", MessageBoxButtons .YesNo,
MessageBoxIcon.Question, MessageBoxDefaultButton.Button2) == DialogResult.No)
            e.Cancel = true;
    }
}
```

 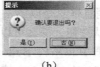

　　　　　　　　　　（a）　　　　　　　　　　（b）

图 7-5　程序 P7_2 启动和关闭时所显示的消息框

7.3.3　对话框

　　.NET 类库中定义了一个抽象类 CommonDialog，它是颜色对话框、字体对话框、文件对话框、打印对话框等一系列通用对话框的基类。之所以叫"通用对话框"，是因为它们在 Windows 中的应用非常普遍，且都是通过 CommonDialog 的 ShowDialog 方法来模态显示的。

　　颜色对话框 ColorDialog 的显示内容如图 7-6（a）所示，它最重要的一个成员就是 Color 属性，表示在对话框中选择的颜色。下面的代码就用于显示一个颜色对话框，用户在其中选择颜色并单击"确定"按钮后，当前窗体的前景色就被设为选定的颜色：

```
ColorDialog dlg1 = new ColorDialog();
if (dlg1.ShowDialog() == DialogResult.OK)
    this.ForeColor = dlg1.Color;
```

　　ColorDialog 的 Color 属性是可读写的。有时候希望对话框具有"记忆"功能，比如要使对话框一开始就显示当前窗体的前景色，那么可在上面第一行代码后插入下面这条语句：

```
dlg1.Color = this.ForeColor;
```

　　类似地，字体对话框 FontDialog 的显示内容如图 7-6（b）所示，它最重要的一个成员就是 Font 属性，表示在对话框中选择的字体。下面的代码用于显示一个字体对话框，用户选择字体并单击"确定"按钮后，当前窗体的字体就被设为选定的字体：

```
FontDialog dlg1 = new FontDialog();
dlg1.Font = this.Font;
if (dlg1.ShowDialog() == DialogResult.OK)
    this.Font = dlg1.Font;
```

　　有趣的是，字体对话框中不仅可以选择字体，还可以选择颜色。FontDialog 与颜色相关的两个属性是 Color 和 ShowColor，前者为 Color 结构类型，表示与对话框相关联的颜色；后者为 bool 类型，表示对话框是否显示颜色选项。下面的代码就能通过字体对话框来一次性设置当前窗体的字体和前景色：

```
FontDialog dlg1 = new FontDialog();
dlg1.ShowColor = true;
if (dlg1.ShowDialog() == DialogResult.OK){
    this.Font = dlg1.Font;
```

```
        this.ForeColor = dlg1.Color;
}
```

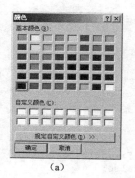
(a)

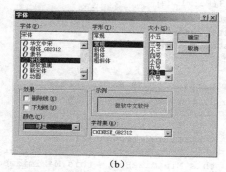
(b)

图 7-6　颜色对话框和字体对话框

7.4　常用 Windows 控件

7.4.1　Control 类

Control 类是包括 Form 类在内的所有其他 Windows 控件的基类。前面介绍的一些 Form 类的成员，如 Text、Location、Font、ForeColor 等属性，也是从 Control 类继承而来的。

Control 的许多属性在各种控件中都是通用的，其中 Name 属性表示控件的名称，Enable 属性表示控件是否可用，Visible 属性表示控件是否可见，AutoSize 属性表示控件是否随其内容而自动调整大小，Focused 属性则表示控件是否获得焦点。例如把一个按钮的 Visible 属性值设为 False，该按钮就不再显示；将其 Enable 属性值设为 False，该按钮就不能被按下，也就不能响应用户的鼠标和键盘事件；而如果将一个窗体的 Enable 属性值设为 False，窗体上的所有控件都会变得不可用。

此外，Control 的 Width 和 Height 属性分别表示控件的宽度和高度（等于其 Size 属性的 Width 和 Height 分量），Left 和 Top 属性分别表示控件左上角的横坐标和纵坐标（等于其 Location 属性的 X 和 Y 分量），Right 和 Bottom 属性分别表示控件左上角的横坐标和纵坐标（Right = Left + Width，Bottom = Top + Height），ClientSize 属性则表示控件工作区的尺寸（除去边框、滚动条等部分后的有效尺寸）。

Control 类还有一个 Controls 属性，表示控件中包含的其他控件的集合（类型为 ControlCollection），这通常只在 Form、Panel 等容器控件中使用。比如从工具箱中拖放一个 Label 控件到窗体上，那么在 Visual Studio 自动生成的设计源文件 Form1.Designer.cs 中可以找到如下代码，表示将标签控件 label1 加入到窗体的子控件集中：
```
            this.Controls.Add(this.label1);
```
鼠标移到某个控件上时，所显示的光标形状可由 Control 的 Cursor 属性来控制，其类型为 System.Drawing 命名空间下的 Cursor 类。不过大多数情况下开发人员不需要去创建一个 Cursor 对象，而是可以通过另一个 Cursors 类的一系列静态属性来得到所需的 Cursor 对象，比如 Cursors 的 Arrow 属性表示箭头状光标，Cross 属性表示十字状光标，WaitCursor 表示等待（沙漏状）光标等。

下面的程序就在窗体中加入了 9 个 Panel 控件，每一个控件都设置了不同的光标属性，并分别通过一个子标签控件来进行说明，其输出结果示例如图 7-7 所示（这种批量处理控件数组的方式在 Visual Studio 的设计视图中是难以做到的，只能通过代码动态实现）：

```
//程序 P7_3(Form1.cs)
using System;
using System.Drawing;
using System.Windows.Forms;
namespace P7_3
{
    public partial class Form1 : Form
    {
        public Form1()
        {
            InitializeComponent();
            this.FormBorderStyle = FormBorderStyle.FixedSingle;
            this.MinimizeBox = this.MaximizeBox = false;
            this.Size = new Size(278, 296);
            Panel[] panels = new Panel[9];
            Label[] labels = new Label[9];
            Cursor[] cursors = { Cursors.Arrow, Cursors.Cross, Cursors.Hand, Cursors.Help, Cursors.IBeam, Cursors.No, Cursors.PanEast, Cursors.SizeAll, Cursors.WaitCursor };
            for (int i = 0; i < 9; i++){
                panels[i] = new Panel();
                panels[i].Size = new Size(90, 90);
                panels[i].Location = new Point((i % 3) * 90, (i / 3) * 90);
                panels[i].Cursor = cursors[i];
                labels[i] = new Label();
                labels[i].Text = cursors[i].ToString();
                panels[i].Controls.Add(labels[i]);
                this.Controls.Add(panels[i]);
            }
        }
    }
}
```

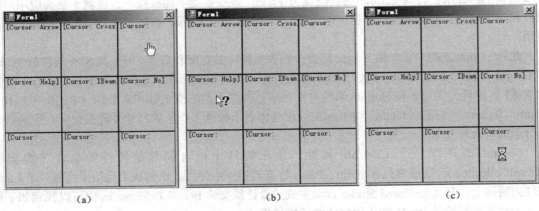

图 7-7　程序 P7_3 的示例输出结果

作为 Windows 控件的基类，Control 类的基本事件可以分为以下几类。

● **鼠标事件**：包括控件的鼠标单击事件 MouseClick、双击事件 MouseDoubleClick、按下鼠标按键事件 MouseDown、释放鼠标按键事件 MouseUp、移动鼠标事件 MouseMove、鼠标滚轮事件 MouseWheel 等。

● **键盘事件**：包括 KeyDown（按下某个键）、KeyUp（放开某个键）和 KeyPress（按某个键）；如果一次性按下而后放开某个键，那么将依次发生 KeyDown、KeyPress 和 KeyUp 事件。

- 焦点事件：包括得到焦点事件 GotFocus 和失去焦点事件 LostFocus。
- 属性改变事件：包括可见性改变事件 VisibleChanged、可用性改变事件 EnabledChanged 等。
- 其他控件行为事件：包括移动事件 Move、绘制事件 Paint、重绘事件 Invalidated 等。

Control 类同时提供了单击事件 Click 和鼠标单击事件 MouseClick。当使用鼠标单击控件时，将先后引发 Click 事件和 MouseClick 事件；而当控件获得焦点并按下 Enter 键时，将只引发 Click 事件。

7.4.2 标签、文本框和数值框

1. Label 控件

Label 控件表示文本标签，文本内容就是其 Text 属性。该控件主要用于显示静态文本，它不能获得焦点，实际应用中也很少进行事件处理。通过从 Control 继承的 Font、ForeColor 和 BackColor 等属性，能够为文本设置丰富的显示格式。例如设置其 BackColor 属性为 Color.Transparent 时，文本将透明显示。

Label 控件中还可以显示图像，图像的内容由 Image 属性指定（属性值为 null 时表示没有图像），文本和图像的位置则分别由 TextAlign 和 ImageAlign 属性指定，其类型为枚举 ContentAlignment，枚举取值包括 TopLeft（左上方）、TopCenter（正上方）、TopRight（右上方）、MiddleLeft（左方）、MiddleCenter（正中央）、MiddleRight（右方）、BottomLeft（左下方）、BottomCenter（正下方）、BottomRight（右下方）。例如下面的代码表示在标签 label1 的文本左方显示一个图像，图像内容来源于 C 盘根目录下 support.bmp 文件：

```
label1.Text = "文本和图像显示示例";
label1.TextAlign = ContentAlignment.MiddleRight;
label1.Image = Image.FromFile("C:\\support.bmp");
label1.ImageAlign = ContentAlignment.MiddleLeft;
```

2. TextBox 控件

Label 文本标签不可编辑，TextBox 控件则可供用户输入文本内容。其 MaxLength 属性表示允许输入的最大长度（默认值 2048），通过它可以限制用户输入的字符个数。如果要支持多行输入，可将其 MultiLine 属性值设置为 true，则最多可输入 64KB 的文本；此时其 Lines 属性将返回一个 string 数组，表示文本行的集合；通过其 WordWrap 属性还可设置文本是否自动换行。

在 Visual Studio 的窗体设计视图中，只有设置其 MultiLine 属性值为 true，才能使用鼠标纵向拖曳 TextBox 控件的大小。

将 TextBox 控件的 ReadOnly 属性值设为 true，能够禁止用户对文本框的输入，但通过代码来修改文本框中的文本仍是可行的。而如果把 TextBox 控件的 Enable 属性值设为 false，那么用户和程序代码都无法改变其中的文本内容。

很多时候还会用到 TextBox 类的 PasswordChar 属性（字符类型），其默认值为空，表示文本正常显示；如果设置该属性为某个字符，那么不管输入什么内容，控件上显示的都是该字符。当把 TextBox 控件作为密码框使用时，常常将该属性值设为星号"*"。

文本框经常需要处理键盘事件，其中 KeyDown 和 KeyUp 使用的事件参数为 KeyEventArgs 类型，而 KeyPress 使用的事件参数为 KeyPressEventArgs 类型，通过它们可以获取有关按键的信息。下面的示例程序就在窗体中放置了一个文本标签 label1 和一个文本框 textBox1；当用户在文本框中输入常规 ASCII 字符时，文本标签将逆序显示输入的内容（例如在文本框中输入"BJ2008"，

那么文本标签上将显示"8002JB")。

```csharp
//程序 P7_4(Form1.cs)
using System;
using System.Windows.Forms;
namespace P7_4
{
    public partial class Form1 : Form
    {
        public Form1()
        {
            InitializeComponent();
            this.textBox1.KeyPress += textBox1_KeyPress;
        }
        void textBox1_KeyPress(object sender, KeyPressEventArgs e)
        {
            if (e.KeyChar >= 32 && e.KeyChar < 127)
                label1.Text = e.KeyChar.ToString() + label1.Text;
            if (e.KeyChar == '\b' && label1.Text.Length > 0)
                label1.Text = label1.Text.Remove(0, 1);
        }
    }
}
```

3. RichTextBox 控件

如果希望文本框的格式更加丰富,可以使用功能更为强大的 RichTextBox 控件。RichTextBox 和 TextBox 都继承自抽象类 TextBoxBase,它们共有的主要特性有:

- 支持从键盘输入文本,设置 ReadOnly 属性值为 false 则可禁止文本编辑。
- 能够在文本框中选取指定的内容,这对应于 TextBoxBase 的 Select 方法,而其 SelectedText 属性表示所选取的文本内容。
- 支持文本内容的剪切、复制和粘贴(分别对应于 TextBoxBase 的 Cut、Copy 和 Paste 方法)。
- 能够使用快捷键 Ctrl+Z 撤销上一步操作,这对应于 TextBoxBase 的 Undo 方法。

而 RichTextBox 还有下面一些特有的功能。

- 默认支持多行显示,而且能够在输入超出控件尺寸时自动显示滚动条。
- 能够使用快捷键 Ctrl+Y 恢复已撤销的操作(对应 RichTextBox 的 Redo 方法)。
- 没有像 TextBox 那样的 64K 字符长度限制,因而能够支持大容量文本操作。
- 能够对不同的文本部分应用不同的字体和颜色格式,还能够显示超链接和图像。

下面的代码就依次选择了一段文本的不同部分,并通过 RichTextBox 的 SelectionFont 和 SelectionColor 属性来设置所选内容的字体和颜色,应用效果如图 7-8 所示:

```csharp
richTextBox1.Text = "北京2008年8月8日\nwww.beijing.com.cn";
richTextBox1.Select(0, 2);
richTextBox1.SelectionFont = new Font("宋体", 12, FontStyle.Bold);
richTextBox1.Select(2, 5);
richTextBox1.SelectionFont = new Font("黑体", 10, FontStyle.Italic);
richTextBox1.SelectionColor = Color.Red;
richTextBox1.Select(7, 4);
richTextBox1.SelectionFont = new Font("楷体_GB_2312", 10);
richTextBox1.SelectionColor = Color.Blue;
```

图 7-8 RichTextBox 的示例输出结果

4. NumericUpDown 控件

TextBox 中可以输入任何内容；如果要限制输入的内容只能是数字，那么可使用数值框 NumericUpDown。该控件可看作是一个文本框和一对上下箭头的组合，用户可在文本框部分直接输入数字（但不能输入其他字符），还可以单击向上或向下箭头来增大或减小数值。

NumericUpDown 的 Value 属性表示控件中的数值（默认值为 0），Minimum 和 Maximum 属性则分别表示允许输入的最小值和最大值（默认值分别为 0 和 100），这 3 个属性的类型都是 decimal。控件数值达到最小值或最大值后，单击向下或向上箭头不会再有效果。还有一个 Increment 属性也是 decimal 类型（默认值为 1），它表示一次单击箭头所增大或减小的数量。如果 NumericUpDown 的 Readonly 属性值为 true，那么只能通过箭头来调整数字，而不能在文本框部分进行输入。

NumericUpDown 控件中的数值显示格式还和下面 3 个属性有关。
- DecimalPlaces（整数类型）：允许输入的小数位数，为 0（默认值）时只能输入整数。
- ThousandsSeparator（布尔类型）：是否显示千位分隔符（默认值为 false）。
- Hexadecimal（布尔类型）：是否以十六进制显示数字（默认值为 false）。

以数值 1002.5 为例，图 7-9（a）中的 NumericUpDown 控件是以默认方式显示，图 7-9（b）中设置了 DecimalPlaces 属性值为 3，图 7-9（c）中设置了 DecimalPlaces 属性值为 2、ThousandsSeparator 属性值为 true，而图 7-9（d）中则设置了 Hexadecimal 属性值为 true。

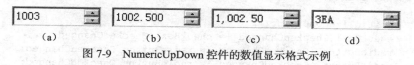

图 7-9 NumericUpDown 控件的数值显示格式示例

NumericUpDown 最常用的事件则是 ValueChanged，通过任何一种方式改变控件中的数值时都将引发该事件。NumericUpDown 控件还应当注意的是：用户的输入或调整的数值范围总是在 Minimum 和 Maximum 属性值之间；而如果通过程序代码设置 NumericUpDown 的 Value 值超出了指定范围，就会引发程序异常。

NumericUpDown 还有一个布尔类型的 InterceptArrowKeys 属性，其值为 true 时通过键盘的向上和向下方向键也能调整数值。NumericUpDown 还提供了 UpButton 和 DownButton 方法，它们分别模拟向上和向下调整一次数值。

7.4.3 按钮、复选框和单选框

作为最常用的 Windows 控件之一，Button 类一般只使用 Click 事件；在获得焦点时，可以使用鼠标、Enter 键或空格键单击按钮来引发事件。默认情况下，Button 控件不处理 DoubleClick 事件，而是将鼠标双击等同为两次单击。

CheckBox 控件又叫复选框，主要用于指示是否选中了某个指定条件。复选框控件包含一个小方框和一行文本说明，小方框中打勾时表示条件选中，否则表示未选中。通过单击复选框可以改变其选取状态。复选框可以单独使用，也可以成组使用。

RadioButton 控件又叫单选框，通常成组使用，供用户在一组互斥选项中选取其中一项。每个单选框控件包含一个小圆框和一行文本说明，小圆框中打点时表示选中，此时同组中的其他单选框将不被选中。如果将多个 RadioButton 控件放在同一个窗体上，它们会自动成为一组，那么一次最多只能选中其中的一个；如果要分为多组，那么可将其放在不同的 Panel 或 GroupBox 等容器控件中。

CheckBox 和 RadioButton 都提供了布尔类型的 Checked 属性，表示控件是否被选中；在改变控件的选取状态时将引发 CheckedChanged 事件。它们通常也不处理 DoubleClick 事件，而是由鼠标双击引发两次 CheckedChanged 事件。

Button、CheckBox 和 RadioButton 都继承自抽象类 ButtonBase，它们都可以通过 Image 属性中来显示图像，并由 TextAlign 和 ImageAlign 属性来指定文本和图像的位置，这一点和 Label 控件是类似的。ButtonBase 还提供了一个 FlatStyle 属性，用于设置按钮的外观，其类型就是 System.Windows.Form 命名空间下的 FlatStyle 枚举，枚举取值包括 Standard（标准三维）、Flat（平面）、Popup（鼠标移至控件时突出显示）和 System（由操作系统决定）。

程序 P7_5 的 Windows 窗体上放置了一个 Button 控件、一个 CheckBox 控件和 3 个 RadioButton 控件，下面的程序代码演示了通过 CheckBox 的选项来切换按钮的平面和三维外观，以及通过 RadioButton 的选项来切换按钮中的文本和图像位置：

```csharp
//程序 P7_5(Form1.cs)
using System;
using System.Drawing;
using System.Windows.Forms;
namespace P7_5
{
    public partial class Form1 : Form
    {
        public Form1()
        {
            InitializeComponent();
            checkBox1.CheckedChanged += checkBox1_CheckedChanged;
            radioButton1.CheckedChanged += radioButton_CheckedChanged;
            radioButton2.CheckedChanged += radioButton_CheckedChanged;
            radioButton3.CheckedChanged += radioButton_CheckedChanged;
        }
        void checkBox1_CheckedChanged(object sender, EventArgs e)
        {
            if (checkBox1.Checked) button1.FlatStyle = FlatStyle.Standard;
            else button1.FlatStyle = FlatStyle.Flat;
        }

        void radioButton_CheckedChanged(object sender, EventArgs e)
        {
            if (radioButton1.Checked){
                button1.TextAlign = ContentAlignment.MiddleCenter;
                button1.ImageAlign = ContentAlignment.MiddleCenter;
            }
            else if (radioButton2.Checked){
                button1.TextAlign = ContentAlignment.MiddleLeft;
                button1.ImageAlign = ContentAlignment.MiddleRight;
            }
            else if (radioButton3.Checked){
                button1.TextAlign = ContentAlignment.MiddleRight;
                button1.ImageAlign = ContentAlignment.MiddleLeft;
            }
        }
    }
}
```

这里 3 个 RadioButton 控件共用了一个 CheckedChanged 事件处理方法，这也是使用单选框事件的一个常用技术。程序 P7_5 的示例输出效果如图 7-10 所示。

(a)
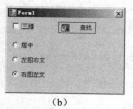
(b)

图 7-10　程序 P7_5 的示例输出结果

7.4.4　组合框和列表框

组合框 ComboBox 和列表框 ListBox 都可用于显示一组集合的元素，它们也有一个共同的基类——ListControl 类，其 Items 属性表示列表中的元素集合，元素类型为 Object，因此可以将任何对象加入到列表集合中。在 Visual Studio 的窗体设计视图中选择 ComboBox 或 ListBox 控件，而后在属性窗口中找到 Items 项，单击右侧的 ... 按钮即可打开一个"字符串集合编辑器"，在其中可输入集合内容，每一行代表一个列表项。不过，如果列表项是特殊类型的对象，或是集合元素具有明显的规律，那么通过代码来设置列表项会更为方便，例如下面的代码就将整数 0～99 加入到一个组合框 comboBox1 当中：

```
for (int i = 0; i < 100; i++)
    comboBox1.Items.Add(i);
```

ListControl 通过 SelectedIndex 或 SelectedItem 属性来读取或设置列表中的当前选项，只不过前者为 int 类型，表示选取项的索引（从 0 开始计数）；而后者为 object 类型，表示元素项本身。如果未选中任何项，那么这两个属性值分别为 -1 和 null。ComboBox 和 ListBox 最常用的事件是选项改变事件 SelectedIndexChanged，单击 ComboBox 的下拉箭头时还将引发其 DropDown 事件，下拉完毕之后则将引发其 DropDownClosed 事件。

ComboBox 是带下拉箭头的文本框，单击右侧的箭头可以展开一个下拉列表供用户选取。ComboBox 的 DropDownStyle 属性表示下拉的样式，其类型为 ComboBoxStyle 枚举，取值包括 DropDown（文本框可编辑，也可单击箭头显示下拉列表）、DropDownList（文本框不可编辑，只能单击箭头显示下拉列表）和 Simple（文本框可编辑，且显示整个列表）。

 对于组合框控件，只有单击其文本框部分才会引发 Click 和 MouseClick 事件；在单击下拉箭头时只会引发 DropDown 事件，而单击事件会被屏蔽。

ListBox 则将所有列表项都显示在一个矩形区域中（项数超出范围时还可显示滚动条），选中的项会被加亮显示。和 ComboBox 不同，ListBox 允许一次性选取多项内容，这是通过其 SelectionMode 属性来实现的，属性类型为 SelectionMode 枚举，取值包括 One（只能选择一项）、MultiSimple（可以选择多项）、MultiExtended（可以选择多项，而且可使用 Shift 键、Ctrl 键和方向键来进行选择）和 None（不能进行选择）。当此属性值为 MultiSimple 或 MultiExtended 时，ListBox 的 SelectedItems 属性表示被选中的项的集合，而 SelectedIndices 属性表示这些项的索引集合。

下面的程序模拟了一个学生选课的 Windows 窗体：左侧的 ComboBox 控件中包含了一组待选的课程集合，单击"加入"按钮可将其选取的课程加入到右侧的 ListBox 控件中，单击"删除"按钮则可删除 ListBox 控件中已有的课程。程序的运行效果如图 7-11 所示。

```
//程序 P7_6(Form1.cs)
using System;
using System.Windows.Forms;
```

```csharp
namespace P7_6
{
    public partial class Form1 : Form
    {
        private int totalHours = 0;

        public Form1()
        {
            InitializeComponent();
            this.Load += Form1_Load;
            button1.Click += button1_Click;
            button2.Click += button2_Click;
        }

        void Form1_Load(object sender, EventArgs e)
        {
            Course[] courses = new Course[7] { new Course("英语", 50), new Course("高等数学", 60), new Course("数理统计", 35), new Course("大学物理", 40), new Course("电子电工", 45), new Course("计算机应用基础", 40), new Course("计算机语言程序设计", 45) };
            for (int i = 0; i < 7; i++)
                comboBox1.Items.Add(courses[i]);
            textBox1.Text = "0";
        }

        void button1_Click(object sender, EventArgs e)
        {
            if (comboBox1.SelectedIndex != -1){
                Course c1 = (Course)comboBox1.SelectedItem;
                if (!listBox1.Items.Contains(c1)){
                    listBox1.Items.Add(c1);
                    totalHours += c1.hours;
                    textBox1.Text = totalHours.ToString();
                }
            }
        }

        void button2_Click(object sender, EventArgs e)
        {
            if (listBox1.SelectedIndex != -1){
                Course c1 = (Course)listBox1.SelectedItem;
                listBox1.Items.Remove(c1);
                totalHours -= c1.hours;
                textBox1.Text = totalHours.ToString();
            }
        }
    }

    public class Course
    {
        public string name;
        public int hours;

        public Course(string name, int hours)
        {
            this.name = name;
            this.hours = hours;
        }

        public override string ToString()
```

```
        return name;
    }
}
```

在程序 P7_6 中，列表框和组合框的 Items 集合都不是简单的一组文本，而是一组 Course 对象，这样就能够在选课的同时计算已选课程的总课时。这种通过 Items 属性处理特定对象集合的技术也是应当掌握的。

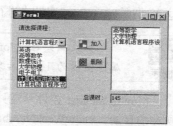

图 7-11　程序 P7_6 的示例输出结果

7.4.5　日历控件

为了方便用户查看和设置时间信息，.NET Framework 中提供了两个美观实用的 DateTimePicker 控件和 MonthCalendar 控件。前者的外观很像一个组合框，只不过单击下拉箭头将显示一个日历，用户在其中可以选择一个日期；后者则直接显示一个日历，用户在其中可以选择一组日期。图 7-12 给出了这两个控件的显示示例。

　　　　（a）DateTimePicker　　　　　　　　　　（b）MonthCalendar

图 7-12　DateTimePicker 控件和 MonthCalendar 控件

DateTimePicker 的 Value 属性表示当前选择的时间值，而属性 MinDate 和 MaxDate 则分别表示允许选取的最小时间和最大时间，这 3 个属性的都是 DateTime 类型。控件的显示格式则可通过 Format 属性来进行设置，其类型为 DateTimePickerFormat 枚举，取值包括 Long（默认日期）、Short（短日期）、Time（时间）和 Custom（自定义）。注意前 3 种格式只能显示日期或时间部分；要想同时显示日期和时间，那么应设置 Format 属性值为 Custom，并通过 CustomFormat 属性来设置字符串格式（和 DateTime 结构的 ToString 重载格式化方法类似，请参看本书第 4.5.5 小节），例如：

```
dateTimePicker1.Format = DateTimePickerFormat.Custom;
dateTimePicker1.CustomFormat = "yy年MM月dd日 hh时mm分ss秒";
```

图 7-13 给出了 DateTimePicker 控件 4 种显示格式示例。

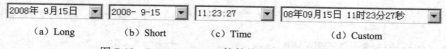

　　（a）Long　　　　（b）Short　　　　（c）Time　　　　　　（d）Custom

图 7-13　DateTimePicker 控件的时间显示格式示例

当 DateTimePicker 控件中的时间值改变时将引发其 ValueChanged 事件；单击控件右侧的下拉按钮，将引发控件的 DropDown 事件，选择完毕之后则会引发 CloseUp 事件，而整个下拉过程同样不会引发单击事件（和组合框类似）。

MonthCalendar 控件则只能显示时间的日期部分，它也通过 MinDate 和 MaxDate 两个属性来限制日期的选取范围。不过在该控件上可以选取一段时间（通过鼠标或方向键加 Shift 键选取），其 SelectionStart 属性表示开始时间，SelectionEnd 属性表示结束时间，MaxSelectionCount 属性则表示一次最多可选取的天数（默认为 7）。MonthCalendar 控件中还会突出显示日历中的"今天"，这可通过 TodayDate 属性来进行设置。

 如果只想在 MonthCalendar 控件中使用一个日期,那么可将其 MaxSelectionCount 属性值设为 1,而后只访问其 SelectionStart 属性即可。

MonthCalendar 控件外观具有很强的可配置性,如 ShowTodayCircle 表示是否突出显示"今天"所在的日期,ShowWeekNumbers 属性表示是否显示每周在当年的编号,BoldedDates 属性表示要粗体显示的日期集合,等等。

当用户在 MonthCalendar 上改变所选日期,那么控件将先后引发 DateSelected 和 DateChanged 事件,不过在程序代码中修改 SelectionStart/SelectionEnd 属性值将都只引发 DateChanged 事件。

7.4.6 滑块、进度条和滚动条

1. TrackBar 控件

TrackBar 控件又叫滑块控件,它由一个滑块和一栏刻度组成,用户可以通过鼠标或键盘来"拖动"滑块,从而改变滑块当前所在的刻度值。这个控件还支持鼠标滚轮的操作。

TrackBar 控件的 Value 属性表示当前刻度值,Minimum 和 Maximum 属性分别表示刻度的最小和最大值,TickFrequency 属性则表示刻度分布的密度。例如设置 Minimum 和 Maximum 属性值分别为 0 和 100,TickFrequency 属性值为 5,那么控件上一共会显示 20 条刻度线。当用户通过左右方向键来移动滑块时,每次增加或减少的刻度值可通过 SmallChange 属性来进行设置;而当用户通过上下翻页键来移动滑块,或是使用鼠标在滑块两侧单击,滑块每次将移动 LargeChange 属性所指定的刻度值。注意上述属性的类型都是 int。

当用户通过鼠标或键盘移动滑块时,将先后引发 TrackBar 的 Scroll 事件和 ValueChanged 事件;而在程序代码中改变 Value 属性值,将只引发 ValueChanged 事件。

下面的程序在窗体上放置了三个刻度范围均在 0~255 之间的 TrackBar 控件,每个滑块的刻度值分别代表当前窗体背景色的 R、G、B 分量。当用户拖动滑块时,窗体的背景色将发生相应的变化,并将颜色信息显示在窗体标题栏上(程序输出如图 7-14 所示)。

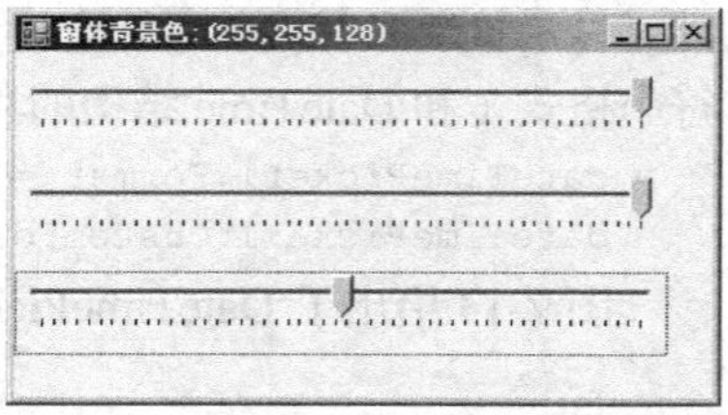

```
//程序 P7_7(Form1.cs)
using System;
using System.Drawing;
using System.Windows.Forms;
namespace P7_7
{
  public partial class Form1 : Form
  {
    public Form1()
    {
      InitializeComponent();
      trackBar1.ValueChanged += new EventHandler(trackBar_ValueChanged);
      trackBar2.ValueChanged += new EventHandler(trackBar_ValueChanged);
      trackBar3.ValueChanged += new EventHandler(trackBar_ValueChanged);
    }
    private void Form1_Load(object sender, EventArgs e)
    {
      this.BackColor = Color.FromArgb(0, 0, 0);
      this.Text = "窗体背景色:(0,0,0)";
    }

    void trackBar_ValueChanged(object sender, EventArgs e)
```

图 7-14 程序 P7_7 的输出结果示例

```
            {
                this.BackColor = Color.FromArgb(trackBar1.Value, trackBar2.Value, trackBar3.Value);
                this.Text = string.Format("窗体背景色:({0},{1},{2})", this.BackColor.R, this.BackColor.G, this.BackColor.B);
            }
        }
    }
```

2. ProgressBar 控件

ProgressBar 控件就是通常所说的进度条，主要用于显示某项任务完成的进度。它使用一系列实心矩形来逐步填充进度条，进度条填满表示任务完成。比如在 Windows 中复制大量文件或安装程序时都可以看到进度条的使用。

和 TrackBar 类似，ProgressBar 通过 Value 属性来表示当前进度值，通过 Minimum 和 Maximum 属性来表示最小和最大进度。例如当 Minimum 和 Maximum 属性值分别为 0 和 100，而 Value 属性值为 50 时，进度条将有一半被填充。此外，ProgressBar 的 Step 属性表示进度每次前进的步长。调用 ProgressBar 的 PerformStep 方法可使进度条前进一个步长单位，调用 Increment 方法则能使进度条前进指定的数值。

3. ScrollBar 控件

ScrollBar 本身是一个抽象类，它的派生类 HScrollBar 和 VScrollBar 分别表示了水平滚动条和垂直滚动条控件。滚动条在 Word、Internet Explorer 等程序中的应用非常普遍，如通过滚动条来改变当前界面的显示范围。类似的，ScrollBar 的 Value 属性表示滚动条当前所处的位置，Minimum 和 Maximum 属性分别表示滚动范围的最小值和最大值，SmallChange 和 LargeChange 属性则分别表示滚动的最小步长和最大步长，它们的类型均为整数 int。

不过，绝大多数需要显示大量信息的 Windows 控件，如 RichTextBox、ListBox、Panel 等，在其包含内容超出自身显示范围时都会自动显示滚动条，因此在程序中直接使用 ScrollBar 控件的机会并不是很多。

7.4.7 图片框控件

尽管 Label、Button 这些控件上都可以显示图像，但.NET 类库中还是提供了一个专门用于在 Windows 窗体上显示图像的控件 PictureBox。该控件的用法非常简单，只要设置其 Image 属性就能在控件上显示指定的图像。该属性即可以在 Visual Studio 的属性窗口中设置（此时图像在设计器中可见），也可以通过代码进行赋值，例如：

```
pictureBox1.Image = Image.FromFile("C:\\plane.bmp");
```

PictureBox 的 Image 属性类型为 System.Drawing 命名空间中的抽象类 Image，其派生类包括表示位图文件的 Bitmap 类和表示图元文件的 Metafile 类。上面代码使用了 Image 类的抽象方法 FromFile，即通过指定的图像文件来获得 Image 对象。此外，通过 Image 的 Width、Height 和 Size 属性可获得图像的尺寸；如果图像内容被修改，那么可通过 Save 方法将其保存到指定文件中。

另一种为 PictureBox 设置图像的方法是在其 ImageLocation 属性中直接指定图像的路径名，例如：

```
pictureBox1.ImageLocation = "C:\\plane.bmp";
```

PictureBox 的 SizeMode 属性表示图像的显示方式，其类型为 PictureBoxSizeMode 枚举，取值包括：

- Normal——默认值，图像在 PictureBox 控件左上角按原始尺寸显示，超出控件尺寸的部分被裁剪掉。
- AutoSize——PictureBox 控件的大小将被自动调整为图像的大小。

- StretchImage——图像的大小将被自动调整为 PictureBox 控件的大小。
- CenterImage——图像按原始尺寸居中显示，超出控件尺寸的部分被裁剪掉。
- Zoom——图像大小按原比例缩放。

PictureBox 控件本身没有提供滚动条的功能。为了在显示大图像时保持尺寸并支持滚动，程序员可增加 HScrollBar 和 VScrollBar 控件并编写相应的代码。但另一种简便的方式是将 PictureBox 放在一个 Panel 控件中，设置 PictureBox 的 SizeMode 属性值为 AutoSize，并设置 Panel 的 AutoScroll 属性值为 True，这样 PictureBox 控件大小超出范围时 Panel 控件就会自动显示滚动条，从而间接地实现了滚动查看图像的功能，如图 7-15 所示。

图 7-15　在 PictureBox 中滚动查看大图像

7.4.8　容器控件

窗体是一个最基本的容器控件，在它上面还可以放置选项卡控件 TabControl、面板控件 Panel、分组框控件 GroupBox、可分容器 SplitContainer 等其他容器控件，其中 TabControl 可以进行多页控制，GroupBox 控件可以显示标题，Panel 控件可以有滚动条，SplitContainer 控件可看成是两个 Panel 的复合控件，面板之间有一个拆分器，通过它可以动态调整两个面板的大小比例。

除了对单选框进行分组之外，容器控件的另一个功能是为其子控件设置 Tab 键导航顺序，这样用户可以方便地通过 Tab 键将焦点从一个控件转移到另一个控件，从而避免频繁地在鼠标和键盘之间进行切换。Visual Studio 的窗体设计视图中，选择"视图"菜单中的"Tab 键顺序"命令，那么各个控件的 Tab 键顺序就显示在其左上角，依次单击各个控件就能设置指定 Tab 键顺序，如图 7-16 所示。

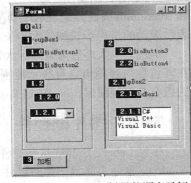

从图 7-16 中可以看到，每一个容器控件都对其子控件进行嵌套编号。容器控件还可以嵌套新的容器控件，那么在程序中可通过它们的 Controls 属性来依次访问子控件集合。下面的程序就在按钮 button1 的单击事件中遍历了窗体的所有控件，将控件的字体全部加粗加斜，并计算出总的控件数量（注意 EmphAllCtrls 方法的递归调用）：

图 7-16　设置 Tab 键导航顺序示例

```csharp
//程序 P7_8(Form1.cs)
using System;
using System.Collections.Generic;
using System.Drawing;
using System.Windows.Forms;
namespace P7_8
{
    public partial class Form1 : Form
    {
        public Form1()
        {
            InitializeComponent();
            button1.Click += button1_Click;
        }
        void button1_Click(object sender, EventArgs e)
        {
            int i = EmphAllCtrls(this);
```

```
            MessageBox.Show(string.Format("共处理{0}个控件", i));
        }
        public int EmphAllCtrls(Control ctrl)
        {
            int i = 0;
            foreach (Control c in ctrl.Controls){
                c.Font = new Font(c.Font, FontStyle.Bold | FontStyle.Italic);
                i++;
                if (c.HasChildren)
                    i += EmphAllCtrls(c);
            }
            return i;
        }
    }
}
```

7.4.9 列表视图和树型视图

1. ListView 控件

ListView 控件又叫列表视图，可以把它看作是一个增强版的列表框控件。ListView 可以显示带图标的列表项，还能进行分列显示。ListView 的显示格式是通过 View 属性来设置的，其类型就是 System.Windows.Forms 命名空间中的 View 枚举，取值包括 LargeIcon（大图标）、SmallIcon（小图标）、List（列表）、Details（细节）和 Tile（平铺）。

ListView 的 Items 属性表示列表项的集合，集合类型为 ListViewItemCollection，其中每个列表项的类型为 ListViewItem。构造 ListViewItem 对象时可以指定列表项的显示文本，例如下面的代码就创建了一个列表项，并将其加入到列表视图 listView1 中：

```
ListViewItem item1 = new ListViewItem("王小红");
listView1.Items.Add(item1);
```

如果不需要对 ListViewItem 对象进行单独操作，那么上面两行代码也可以合并为：

```
listView1.Items.Add("王小红");
```

如果要在 ListView 中进行多列显示，那么首先应将其 View 属性设置为 Details，而后向 ListView 的 Columns 属性中加入各列，最后通过 ListViewItem 对象的 SubItems 属性加入子列表项。下面的代码示例就在列表视图 listView1 中显示了"姓名"、"性别"、"年龄"三列内容：

```
listView1.View = View.Details;
listView1.Columns.Add("姓名");
listView1.Columns.Add("性别");
listView1.Columns.Add("年龄");
ListViewItem item1 = new ListViewItem("王小红");
item1.SubItems.Add("女");
item1.SubItems.Add("27");
listView1.Items.Add(item1);
```

ListView 的 SelectedItems 和 SelectedIndices 属性分别表示所选项的集合以及它们的索引集合。用户改变选项时将引发 ListView 的 SelectedIndexChanged 事件，而双击某列表项将导致该项被"激活"，从而引发 ItemActivate 事件。当 ListView 的 MultiSelect 属性值为 true 时，用户可在列表视图中进行多项选择。

2. TreeView 控件

TreeView 控件又叫树型视图。顾名思义，它显示一个树型结构，每个树节点可以有一组子节点，带有子节点的父节点可以通过左侧的加号或减号进行展开或折叠。例如 Windows 资源管理器就是一个典型的树型视图加列表视图的组合界面。

TreeView 的 Nodes 属性表示树的顶层节点的集合，集合类型为 TreeNodeCollection，其中每个节点的类型为 TreeNode，而 TreeNode 的 Nodes 属性又表示该节点的子节点集合，通过其 Parent 属性则可获得节点的父节点（根节点的该属性值为 null）。例如下面的代码就在列表视图 treeView1 中加入了"我的电脑"根节点及两个子节点：

```
TreeNode root = treeView1.Nodes.Add("我的电脑");
root.Nodes.Add("C");
root.Nodes.Add("D");
```

类似的，TreeView 的 SelectedNode 属性表示树型视图中所选择的节点，选择某一节点前后会分别引发 BeforeSelect 和 AfterSelect 事件，而展开或折叠某个节点的子节点前后将分别引发 BeforeExpand、AfterExpand、BeforeCollapse 和 AfterCollapse 事件。这几个事件都可通过事件参数类型来获得所操作的节点，例如：

```
private void treeView1_AfterSelect(object sender, TreeViewEventArgs e)
{
    MessageBox.Show("您所选中的节点为:" + e.Node.Text);
}
```

对于树节点 TreeNode 而言，其 Level 属性表示节点在树中的层次（顶层节点的的层次为 0，其余依次递增），Index 属性则表示节点在当前集合中的索引号。把它们和 TreeView 的事件参数结合起来，就可以准确定位树型视图中所操作的节点位置。

程序 P7_9 定义了一组依次包含的对象类型：一个 School 中包含一组 Department 集合，每个 Department 包含一组 Office 集合，每个 Office 又包含一组 Teacher 集合。程序主窗体 Form1 中创建了一个 School 对象，并按照集合包含的层次结构创建了一个树型视图；选中某个树节点后，列表视图中将显示当前对象所包含的集合内容。程序的运行效果如图 7-17 所示。

```csharp
//程序 P7_9(Form1.cs)
using System;
using System.Collections.Generic;
using System.Windows.Forms;
namespace P7_9
{
    public partial class Form1 : Form
    {
        private School school;

        public Form1()
        {
            InitializeComponent();
            Office o1 = new Office("计算机基础", "王军", "杨小勇", "何平", "姜涛");
            Office o2 = new Office("软件工程", "马建国", "陈君", "刘小燕");
            Office o3 = new Office("信息安全", "冯尧", "李建军", "张涛");
            Department d1 = new Department("计算机", o1, o2, o3);
            Office o4 = new Office("自动控制", "吴自力", "陈峰", "薛小龙");
            Office o5 = new Office("工业设计", "吴淑华", "方坤", "何力", "蔡聪");
            Department d2 = new Department("机电工程", o4, o5);
            Office o6 = new Office("信息管理", "赵民", "盛小楠", "徐小平");
            Office o7 = new Office("工商管理", "张敏", "李玲", "吕倩", "高剑");
            Department d3 = new Department("经济管理", o6, o7);
            school = new School("交通大学", d1, d2, d3);
        }
```

```csharp
        private void Form1_Load(object sender, EventArgs e)
        {
            TreeNode root = treeView1.Nodes.Add(school.ToString());
            foreach (Department d in school.Departments){
                TreeNode node1 = root.Nodes.Add(d.ToString());
                foreach (Office o in d.Offices)
                    node1.Nodes.Add(o.ToString());
            }
            listView1.Columns.Add("姓名");
            listView1.Columns.Add("电话");
            listView1.Columns.Add("电子邮件");
        }

        private void treeView1_AfterSelect(object sender, TreeViewEventArgs e)
        {
            listView1.Items.Clear();
            if (e.Node.Level == 0)
                foreach (Department d in school.Departments)
                    listView1.Items.Add(d.ToString(), 0);
            else if (e.Node.Level == 1)
                foreach(Office o in school.Departments[e.Node.Index].Offices)
                    listView1.Items.Add(o.ToString(), 1);
            else if (e.Node.Level == 2)
                foreach (string s in school.Departments[e.Node.Parent.Index]
.Offices[e.Node.Index].Teachers)
                    listView1.Items.Add(s, 2);
        }
    }

    public class School
    {
        private string name;
        public string Name
        {
            get { return name; }
        }

        private List<Department> departments;
        public List<Department> Departments
        {
            get { return departments; }
        }

        public School(string name, params Department[] departments)
        {
            this.name = name;
            this.departments = new List<Department>(departments);
        }

        public override string ToString()
        {
            return name;
        }
    }

    public class Department
    {
        private string name;
        public string Name
```

```csharp
        get { return name; }
    }

    private List<Office> offices;
    public List<Office> Offices
    {
        get { return offices; }
    }

    public Department(string name, params Office[] offices)
    {
        this.name = name;
        this.offices = new List<Office>(offices);
    }

    public override string ToString()
    {
        return name + "系";
    }
}

public class Office
{
    private string name;
    public string Name
    V{
        get { return name; }
    }

    private List<String> teachers;
    public List<String> Teachers
    {
        get { return teachers; }
    }

    public Office(string name, params string[] teachers)
    {
        this.name = name;
        this.teachers = new List<string>(teachers);
    }

    public override string ToString()
    {
        return name + "教研室";
    }
}
```

图 7-17　程序 P7_9 的输出结果

7.5 菜单栏、工具栏和状态栏

7.5.1 菜单栏

1. 主菜单

菜单是 Windows 应用程序中常见的一项基本元素。在 Visual Studio 工具箱的"菜单和工具栏"选项卡下，拖放一个 MenuStrip 控件到窗体的设计视图，那么就向窗体中加入了一个菜单栏。MenuStrip 类的 Items 属性表示菜单栏中的主菜单项集合，每个菜单项的类型为 ToolStripMenuItem；通过菜单项的 DropDownItems 属性还可以向其加入子菜单项。

下面的代码就向一个 menuStrip1 菜单栏中加入了一个"文件"主菜单项，并向该菜单项中又加入了 4 个子菜单项，其中"打印"菜单项还包含二级子菜单，那么得到的菜单栏内容将如图 7-18 所示。

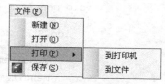

图 7-18 菜单栏示例

```
ToolStripMenuItem menuItem1 = new ToolStripMenuItem("文件(&F)");
menuItem1.DropDownItems.Add("新建(&N)");
menuItem1.DropDownItems.Add(new ToolStripMenuItem("打开(&O)"));
ToolStripMenuItem menuItem2 = new ToolStripMenuItem("打印(&P)");
menuItem2.DropDownItems.Add("到打印机");
menuItem2.DropDownItems.Add("到文件");
menuItem1.DropDownItems.Add(menuItem2);
menuItem1.DropDownItems.Add("保存(&S)", Image.FromFile("C:\\Save.bmp"));
menuStrip1.Items.Add(menuItem1);
```

从上述代码中可以看到，向菜单项集合中加入新项有多种方法：可以加入一个 ToolStripMenuItem 对象，也可以直接指定新菜单项的文本，还可以同时指定菜单项的文本和图像。菜单项文本中的"&"字符用于指定快捷键，例如"文件(&F)"将在字母 F 下面加上下划线，那么运行程序时通过快捷键 Ctrl+F 即可打开该菜单。

如果要对同级菜单项进行分组，那么只需将 ToolStripSeparator 对象加入到菜单项集合中的指定位置即可，例如。

```
menuItem1.DropDownItems.Add(new ToolStripSeparator());
```

在 Visual Studio 的窗体设计视图中设置菜单栏也是非常方便的。在设计视图中选择 MenuStrip 控件，那么控件就会自动转入设计模式，在每个菜单项的提示文本框中可输入其文本内容，还可以通过下拉箭头设置菜单项的风格，完成后就可以在属性窗口中进行图像、字体和颜色等更为详细的设置。如果不需要使用动态菜单的话，一般都是在设计时完成菜单栏的内容设置。

菜单项主要通过对 Click 事件的处理来执行有关操作。有时菜单项还可用来标记某些状态，此时可将 ToolStripMenuItem 的 Checked 属性值设为 true，那么在菜单项左侧就会显示一个"√"号；而设置其 CheckOnClick 属性值为 true，那么单击菜单项就会改变其选项标记。

在程序 P7_10 中，主窗体的菜单栏中包含一个"窗口"菜单项，其下又包含"大"（menuItemWindowBig）、"中"（menuItemWindowMiddle）、"小"（menuItemWindowSmall）3 个子菜单项，单击不同的菜单项会在其左侧显示勾号标记，并相应地调整窗体大小：

```
//程序 P7_10(Form1.cs)
using System;
using System.Drawing;
```

```
using System.Windows.Forms;
namespace P7_10
{
    public partial class Form1 : Form
    {
        public Form1()
        {
            InitializeComponent();
            menuItemWindowMiddle.Checked = true;
            menuItemWindowBig.Click += menuItem_Click;
            menuItemWindowMiddle.Click += menuItem_Click;
            menuItemWindowSmall.Click += menuItem_Click;
        }
        void menuItem_Click(object sender, EventArgs e)
        {
            ToolStripMenuItem item = (ToolStripMenuItem)sender;
            if (item == menuItemWindowBig){
                menuItemWindowBig.Checked = true;
                menuItemWindowMiddle.Checked = menuItemWindowSmall.Checked = false;
                this.Size = new Size(800, 450);
            }
            else if (item == menuItemWindowMiddle){
                menuItemWindowMiddle.Checked = true;
                menuItemWindowBig.Checked = menuItemWindowSmall.Checked = false;
                this.Size = new Size(480, 270);
            }
            else if (item == menuItemWindowSmall){
                menuItemWindowSmall.Checked = true;
                menuItemWindowBig.Checked = menuItemWindowMiddle.Checked = false;
                this.Size = new Size(240, 135);
            }
        }
    }
}
```

对于普通的程序菜单而言，我们只需要完成其菜单栏的设置，而后处理各个菜单项的 Click 事件即可。MenuStrip 和 ToolStripMenuItem 还提供了其他许多成员，可以实现组合菜单、悬浮菜单等更为强大的功能。

2. 快捷菜单

快捷菜单又叫上下文菜单，是指在窗体或控件上单击鼠标右键所弹出的菜单，以方便用户进行与当前上下文相关的操作。例如通过文本框的快捷菜单，就可以直接进行文本剪切、复制和粘贴等操作。

要使用快捷菜单，只要从 Visual Studio 工具箱中拖放 ContextMenuStrip 控件到窗体的设计视图即可。一个窗体通常只有一个主菜单，但可以有多个快捷菜单；通过 Control 类的 ContextMenuStrip 属性，可将不同快捷菜单分配给不同的控件。例如将窗体的 ContextMenuStrip 属性值设为某个快捷菜单对象，那么在窗体中没有其他控件的任何位置单击鼠标右键，就会弹出该菜单。

ContextMenuStrip 的设计和使用方式和 MenuStrip 类似，此处不再赘述。

7.5.2 工具栏

工具栏主要用于显示一组工具按钮，用鼠标单击其中的按钮就可快速执行相关的命令操作。通常情况下，应用程序中的绝大部分命令都应该能够通过菜单完成，工具栏只为其中最常用的部分提供快捷操作方式。

在.NET Framework 早期的版本中，菜单栏、工具栏和状态栏分别属于独立的类型。.NET Framework 2.0 以后的版本则采用了新的设计：ToolStrip 表示工具栏，菜单栏 MenuStrip 和状态栏 StatusStrip 都是 ToolStrip 的派生类型；ToolStrip 的 Items 属性表示其中的所有项集，集合的元素类型为 ToolStripItem，其派生类包括可下拉项 ToolStripDropDownItem、按钮项 ToolStripButton、文本项 ToolStripLabel、容器项 ToolStripControlHost，以及分隔项 ToolStripSeparator；下拉工具按钮和菜单项又属于 ToolStripDropDownItem 的派生类。这些类型之间的关系如图 7-19 所示。

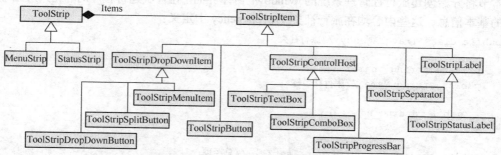

图 7-19　ToolStrip 及其 Items 集合项的类型继承结构

实际上，.NET Framework 2.0 之后版本中的菜单栏和工具栏没有本质的区别，菜单栏中可以有工具按钮，工具栏中也可以包含菜单项，它们都还能够嵌入文本框、组合框和进度条（Visual Studio 开发环境的工具栏就是一个典型的例子）。每一项都可通过 ToolStripItem 的 Text 和 Image 属性来设置文本和图像，通过 TextAlign 和 ImageAlign 属性来设置文本和图像的位置，通过 AutoToolTip 和 ToolTipText 属性来设置提示文字。只不过菜单项 ToolStripMenuItem 以显示文本为主，而工具按钮 ToolStripButton 以显示图像为主。

下面的代码就向一个 toolStrip1 工具栏中先后加入了工具按钮、带菜单项的下拉式工具按钮、分隔符和嵌入式文本框，得到的工具栏内容将如图 7-20 所示。

```
toolStrip1.Items.Add(new ToolStripButton(Image.FromFile("NewDocument.bmp")));
toolStrip1.Items.Add(new ToolStripButton(Image.FromFile("Open.bmp")));
ToolStripDropDownButton tsDrButton = new ToolStripDropDownButton(Image.FromFile("Print.bmp"));
tsDrButton.DropDownItems.Add(new ToolStripMenuItem("到打印机"));
tsDrButton.DropDownItems.Add(new ToolStripMenuItem("到文件"));
toolStrip1.Items.Add(tsDrButton);
toolStrip1.Items.Add(new ToolStripButton(Image.FromFile("Save.bmp")));
toolStrip1.Items.Add(new ToolStripSeparator());
toolStrip1.Items.Add(new ToolStripTextBox());
toolStrip1.Items.Add(new ToolStripButton(Image.FromFile("Find.bmp")));
```

图 7-20　工具栏示例

和菜单栏类似，如果不使用复杂的动态工具栏，那么一般都是在设计时完成工具栏的内容设置，然后通过处理工具栏按钮的 Click 事件来执行有关操作。

7.5.3　状态栏

同样，状态栏 StatusStrip 也是 ToolStrip 的派生类，因此它也可以包含菜单项和工具按钮。不过在一般情况下，状态栏主要通过 ToolStripStatusLabel 文本来显示一些状态信息，例如在

Windows 写字板的状态栏中可以看到键盘上的 CapsLock 键和 Num 键是否被按下。

7.6 案例研究——旅行社信息窗体和登录窗体

7.6.1 旅行社对象及其信息窗体

本节将开始创建旅行社管理系统的 Windows 窗体界面。在首次运行程序时，用户应当输入旅行社的基本信息，这些内容均在旅行社类 TravelAgency 中定义：

```
public class TravelAgency  //旅行社类
{
    private string _name;
    public string Name  //旅行社名称
    {
        get { return _name; }
    }
    public string Manager { get; set; }  //总经理
    public string Represtive { get; set; }  //法人代表
    public string Address { get; set; }  //通信地址
    public string PostCode { get; set; }  //邮编
    public string Website { get; set; }  //网址
    public string Phone { get; set; }  //电话
    public string Fax { get; set; }  //传真
    public string Bank { get; set; }  //开户行
    public string AccountName { get; set; }  //账户名
    public string AccountNumber { get; set; }  //账号
    public string License{ get; set; }  //经营许可证
    public TravelAgency(string name)
    {
        _name = name;
    }
}
```

接下来新建 Windows 应用程序项目 TravelWin，并为其添加对类库项目 TravelLib 的引用。默认主窗体 Form1 保留暂不处理，并向项目中添加一个新窗体 BasicForm。按照图 7-21 所示的内容向窗体添加各个控件。

切换到 BasicForm 的代码视图，为该类定义一个 TravelAgency 类型的字段，以及封装的只读属性：

```
private TravelAgency _agency;
public TravelAgency Agency
{
    get { return _agency; }
}
```

接下来修改 BasicForm 的构造函数，使其将指定的对象赋值给 _agency 字段：

```
public BasicForm(TravelAgency agency)
{
    _agency = agency;
    InitializeComponent();
}
```

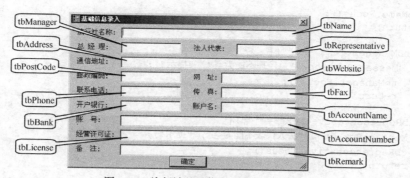

图 7-21 旅行社基础信息窗体 BasicForm

然后再为 BasicForm 添加两个成员方法，其中 LoadInfo 用于将_agency 对象的信息显示在窗体各个控件中，SaveInfo 则将各控件中的内容保存到_agency 对象：

```
public void LoadInfo()
{
    tbName.Text = _agency.Name;
    tbManager.Text = _agency.Manager;
    tbRepresentative.Text = _agency.Representive;
    tbAddress.Text = _agency.Address;
    tbPostCode.Text = _agency.PostCode;
    tbWebsite.Text = _agency.Website;
    tbPhone.Text = _agency.Phone;
    tbFax.Text = _agency.Fax;
    tbBank.Text = _agency.Bank;
    tbAccountName.Text = _agency.AccountName;
    tbAccountNumber.Text = _agency.AccountNumber;
    tbLicense.Text = _agency.License;
}
public void SaveInfo()
{
    _agency = new TravelAgency(tbName.Text);
    _agency.Manager = tbManager.Text;
    _agency.Representive = tbRepresentative.Text;
    _agency.Address = tbAddress.Text;
    _agency.PostCode = tbPostCode.Text;
    _agency.Website = tbWebsite.Text;
    _agency.Phone = tbPhone.Text;
    _agency.Fax = tbFax.Text;
    _agency.Bank = tbBank.Text;
    _agency.AccountName = tbAccountName.Text;
    _agency.AccountNumber = tbAccountNumber.Text;
    _agency.License = tbLicense.Text;
}
```

下面为 BasicForm 添加 Load 事件处理方法来载入_agency 对象信息；再为按钮 Button1 添加单击事件处理方法，它在检查完一些必需的输入之后保存_agency 对象信息：

```
private void BasicForm_Load(object sender, EventArgs e)
{
    if(_agency!= null)
        this.LoadInfo();
}

private void button1_Click(object sender, EventArgs e)
{
    if (tbName.Text == ""){
        MessageBox.Show("名称不能为空", "提示");
```

```
            tbName.Focus();
            return;
        }
        if (tbManager.Text == ""){
            MessageBox.Show("总经理不能为空", "提示");
            tbManager.Focus();
            return;
        }
        if (tbRepresentative.Text == ""){
            MessageBox.Show("法人代表不能为空", "提示");
            tbRepresentative.Focus();
            return;
        }
        this.SaveInfo();
        this.DialogResult = DialogResult.OK;
    }
```

完成 BasicForm 的设计和编码工作后，回到主窗体 Form1，为其也添加一个 TravelAgency 类型的字段：

```
private TravelAgency _agency;
```

然后再切换到主窗体的设计视图，为窗体添加 MenuStrip 菜单栏，在其中加入一个"系统管理(&S)"菜单 menuSystem 和一个"旅行社基本信息(&B)"菜单项 menuSystemBasic，为菜单项添加单击事件处理方法，通过 BasicForm 窗体对象来修改旅行社信息：

```
private void menuSystemBasic_Click(object sender, EventArgs e)
{
    BasicForm form = new BasicForm(_agency);
    if (form.ShowDialog() == DialogResult.OK)
        _agency = form.Agency;
}
```

之后可以编译运行程序，在主窗体中通过菜单命令"系统管理→单位基本信息"打开 BasicForm 窗体，在其中输入信息后单击"确定"按钮保存。然后再次打开该窗体可看到上次输入的内容。

7.6.2　系统用户及登录窗体

旅行团内部管理子系统需要登录才能使用。系统用户通过 User 类定义，它不仅包含用户名和密码，还设定了用户所对应的旅行社职员编号：

```
public class User// 系统用户类
{
    public string Username{ get; set; }//用户名
    public string Password{ get; set; } //密码
    public int StaffId{ get; set; }//对应职员编号

    public User(string username, string password, int staffId)
    {
        Username = username;
        Password = password;
        StaffId = staffId;
    }

    public static User[] GetAll() //获取所有用户（模拟）
    {
        User u1 = new User("ZJL", "8888", 105001);
        User u2 = new User("ZWQ", "zwq1980", 105003);
        User u3 = new User("MQP", "mqp1979", 105007);
```

```
        return new User[] { u1, u2, u3 };
    }
}
```

接下来再新建一个 Windows 窗体 Login-Form，按照图 7-22 所示的内容添加控件，设置文本框 tbPassword 的 PasswordChar 属性为"*"，并设置提示文本 Label3 的 Visible 属性为 false，其字体为红色粗体。

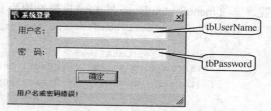

图 7-22　旅行社管理系统登录窗体 LoginForm

LoginForm 在 _users 字段中维护用户组，并通过一个 _count 字段来记录登录失败的次数。当用户单击"确定"按钮时，程序检查输入的用户名和密码是否与 _users 中的某个用户匹配，如是则登录成功（DialogResult 值为 OK），此时 _current 字段被赋值为该用户所对应的职员；如登录失败次数达到 3，那么将放弃登录（DialogResult 值为 Cancel）：

```
public partial class LoginForm : Form //用户登录窗体
{
    private int _count = 0;
    private User[] _users;
    private Staff _current;
    public Staff Current //成功登录用户所对应的职员
    {
        get { return _current; }
    }
    public LoginForm(User[] users)
    {
        InitializeComponent();
        _users = users();
    }
    private void button1_Click(object sender, EventArgs e)
    {
        foreach (User u in _users){
            if (tbUserName.Text == u.Username && tbPassword.Text == u.Password){
                _current = Staff.GetStaff(u.StaffId);
                this.DialogResult = DialogResult.OK;
                return;
            }
        }
        if (_count++ == 3){
            MessageBox.Show("您已连续输入错误超过 3 次", "错误", MessageBoxButtons.OK, MessageBoxIcon.Stop);
            this.DialogResult = DialogResult.Cancel;
            return;
        }
        label3.Visible = true;
    }
}
```

登录窗体应该在程序主窗体之前显示，登录成功之后主窗体也需要知道当前登录的是哪个职员。那么回到 Form1 的代码中，为其也添加一个 _users 字段来记录所有用户，并定义一个 _current 字段来记录登录职员：

```
private User[] _users;
private Staff _current;
```

然后为 Form1 添加 Load 事件处理方法，在其中载入用户信息并显示登录窗体，如登录失败则关闭窗体，否则将 LoginForm 对象的 Current 属性赋值给 Form1 的 _current 字段。此外，不同的系统用户有着不同的权限，比如希望旅行社的基本信息只有经理能够修改，那么可以先在设计视图中设置菜单项 menuSystemBasic 的 Enable 属性值为 false，而只有登录用户为经理时才使菜单项可用：

```csharp
private void Form1_Load(object sender, EventArgs e)
{
    _users = User.GetAll();
    LoginForm form = new LoginForm(_users);
    if (form.ShowDialog() != DialogResult.OK) //登录失败
    {
        this.Close();
        return;
    }
    _current = form.Current;
    InitializeComponent();
    menuSystemBasic.Enabled = (_current is Manager);
}
```

接下来再次编译运行程序，可以看到程序启动时将首先显示登录窗体，只有输入正确的用户名和密码后才能进入系统。

7.7 小　　结

.NET 类库提供了丰富的 Windows 窗体和控件类型，它们大都在 System.Windows.Forms 命名空间中定义。其中 Form 类是所有 Windows 窗体的基类，而 Control 则是 Form 及其他所有控件类的共同基类，这些常用控件包括用于显示文本的 Label，用于输入文本的 TextBox 和 RichTextBox，用于多选的 CheckBox 和单选的 RadioButton，用于包含其他控件的 TabControl、Panel 和 GroupBox，以及用于菜单命令的 MenuStrip 和用于工具栏命令的 ToolStrip 等，它们主要通过鼠标和键盘等事件来响应用户操作。在 Visual Studio 开发环境中能够方便地设置控件的各种属性和事件处理方法。

7.8 习　　题

1. 一个查询公交线路的 Windows 应用程序，在其界面设计上应注意哪些问题？
2. 基于不常用的字体族创建 Font 对象时要注意什么问题？在不同 Windows 系统的控制面板中查看现有的字体族集合，说明哪些是常用的字体族，哪些是不常用的。
3. 如何控制一个 Windows 窗体只能被横向拉伸，而纵向尺寸不会被改变？
4. 如何控制 TextBox 控件当中只能输入数字？写出相应的程序代码。
5. 扩展程序 P7_6 中的例子，向窗体中再加入一个组合框，用户可以在其中选择数学、计算机、电子工程等不同的专业，此时另一个组合框中的课程列表项会随着所选专业的不同而自动更换，用户单击"增加"或"删除"按钮无执行效果时则通过消息框给出提示。最后再说说组合框和列表框各有什么特点，它们分别适合在哪些场合中使用。
6. 对于一个 NumericUpDown 控件，下面两行语句的效果是否相同，为什么？
```csharp
numericUpDown1.UpButton();
numericUpDown1.Value += numericUpDown1.Increment;
```
7. 将程序 P7_10 的功能改由快捷菜单来实现。
8. 绘制一个 Windows 控件类继承示意图，要求包含 7.3 节中介绍过的所有控件。
9. 在 7.5 节案例的基础上，说说在 Windows 窗体上显示对象信息以及在不同的窗体间传递对象要注意哪些问题。

第 8 章
对象持久性——文件管理

应用程序运行在内存中,为了不使其中各种对象的信息随着程序的关闭而丢失,就必须要把信息保存在硬盘等持久性媒质上。本章将介绍.NET 类库中用于文件 I/O(输入/输出)操作的相关类型,通过这些文件功能就可以为对象添加持久性数据存取的功能。

8.1 文件和流

8.1.1 File 类

.NET 类库中用于文件 I/O 操作的类型大都在 System.IO 命名空间中定义,其中一个重要的类型就是 File 类。这是一个静态类,它通过一系列静态方法来提供对磁盘文件的操作功能,例如我们可以在 C 盘根目录下创建一个名为 "a.txt" 的文件,然后通过 File 类的 Copy、Move、Replace 和 Delete 方法来进行文件的复制、移动、替换和删除,还可通过其 Exists 方法来判断文件是否存在:

```
File.Copy("C:\\a.txt", "C:\\b.txt"); //将 a.txt 文件复制到 b.txt 文件
if (File.Exists("D:\\a.txt"))
    File.Replace("C:\\a.txt", "D:\\a.txt"); //用 C 盘的 a.txt 替换 D 盘的 a.txt
else
    File.Move("C:\\a.txt", "D:\\a.txt"); //将 a.txt 移动到 D 盘
File.Delete("C:\\b.txt"); //删除 b.txt 文件
```

在表示文件路径名的字符串中,要注意使用转义符 "\\" 来表示目录分隔符。而如果只写出文件名而不写出完整的路径名,那么默认的目录位置就是当前程序可执行文件所在的目录。对于 Windows 应用程序,可通过 Application 类的静态属性 StartupPath 来获得程序可执行文件所在的路径。

每个文件都有自己的属性信息,其中文件的创建时间、最近一次访问时间和最近一次修改时间可通过如下方法进行读写:

```
DateTime dt1 = File.GetCreationTime("C:\\a.txt"); //读取文件创建时间
File.SetCreationTime("C:\\a.txt", dt1.AddDays(10)); //设置文件创建时间
Console.WriteLine(File.GetLastAccessTime("C:\\a.txt")); //读取文件访问时间
Console.WriteLine(File.GetLastWriteTime("C:\\a.txt")); //读取文件修改时间
File.SetLastAccessTime("C:\\a.txt", dt1); //设置文件访问时间
File.SetLastWriteTime("C:\\a.txt", dt1); //设置文件修改时间
```

而文件的其他属性信息则可通过 File 类的 GetAttributes 方法来获得,该方法返回一个 FileAttributes 枚举值,其枚举值包括 Normal(普通文件)、Archive(存档文件)、ReadOnly(只读

文件）、Hidden（隐藏文件）、Compressed（压缩文件）等。注意一个文件的 FileAttributes 值可以是多个枚举值的"或"，例如对于一个只读隐藏文件，GetAttributes 方法返回的值将是 FileAttributes.ReadOnly | FileAttributes.Hidden。下面的代码演示了如何判断一个文件是只读的或隐藏的：

```
FileAttributes fa1 = File.GetAttributes("C:\\a.txt");
if ((fa1 & FileAttributes.ReadOnly) == FileAttributes.ReadOnly)
    Console.WriteLine("只读文件");
if ((fa1 & FileAttributes.Hidden) == FileAttributes.Hidden)
    Console.WriteLine("隐藏文件");
```

File 类还可用于读写文件内容，其中 ReadAllText 方法用于将文件所有内容读取为一个字符串，而 WriteAllText 方法则将字符串写入文件，例如：

```
File.WriteAllText("C:\\a.txt", "示例文本");
Console.WriteLine(File.ReadAllText("C:\\a.txt"));  //将输出"示例文本"
```

为了进一步方便文本读写，File 类还提供了将文件各行依次读取到一个字符串数组的 ReadAllLines 方法，以及将字符串数组逐行写入文件的 WriteAllLines 方法。此外，File 类的 ReadAllBytes 方法将文件的所有字节读取到一个 byte[]数组中，WriteAllBytes 方法则将一个 byte[]数组写入到文件中。下面的程序 P8_1 就定义了一个在控制台输出文件所有内容的 TypeFile 方法，以及将文件所有字节和 7 进行异或并写入另一个文件的 EncryptFile 方法（这是信息加密的一种基本技术）；程序主方法创建了一个文本文件 "a.txt"，并调用上面两个方法来处理文件内容：

```
//程序 P8_1
using System;
using System.IO;
namespace P8_1
{
    class Program
    {
        static void Main()
        {
            if (!File.Exists("a.txt"))
                File.Create("a.txt");
            string[] ss = { "2008-9-30 8:30am 中山路32号", "2009-2-9 2:00pm 华山路1号"};
            File.WriteAllLines("a.txt", ss);
            TypeFile("a.txt");
            EncryptFile("a.txt", "b.txt");
            TypeFile("b.txt");
            EncryptFile("b.txt", "c.txt");
            TypeFile("c.txt");
            Console.ReadLine();
        }

        static void TypeFile(string filename)
        {
            Console.WriteLine(filename);
            foreach (string s in File.ReadAllLines(filename))
                Console.WriteLine(s);
            Console.WriteLine();
        }

        static void EncryptFile(string sourceFile, string targetFile)
        {
            byte[] bs = File.ReadAllBytes(sourceFile);
            for (int i = 0; i < bs.Length; i++)
                bs[i] ^= 7;
```

```
            File.WriteAllBytes(targetFile, bs);
        }
    }
}
```
从图 8-1 可以看到，第一次调用 EncryptFile 方法得到的文件内容已被加密，而对加密文件再调用 EncryptFile 方法又得到了原始的文件内容。

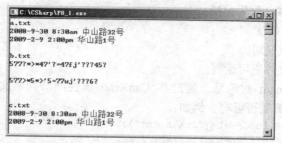

图 8-1　程序 P8_1 的输出结果

8.1.2　使用文件流

File 类的 ReadAllText 和 WriteAllText 等方法都只能一次性处理全部文件内容。要想读写文件的指定部分，那就要用到文件流对象 FileStream。文件流是对物理文件的封装，比如一个文件的内容可能分散在磁盘的多个地方，但通过 FileStream 仍可以进行连续的字节流读写，从字节流到物理存储的映射细节由 FileStream 隐藏。

例如下面的代码通过 File 的 Create 方法创建了一个文件，该方法返回一个 FileStream 对象，通过该对象的 WriteByte 方法可向文件中依次写入一组字节：

```
FileStream fs1 = File.Create("C:\\a.txt");
for (byte x = 1; x <= 10; x++)
    fs1.WriteByte(x);
```

使用完 FileStream 对象后，一定要记住使用其 Close 方法来关闭文件流，这样才能保持目标文件的完整性：

```
fs1.Close();
```

如果要向文件中写入字符串，那么就应当将各个字符先转换为 byte 类型，例如：

```
string s1 = "2008-9-30 8:30am";
for (int i = 0; i < s1.Length; i++)
    fs1.WriteByte((byte)s1[i]);
```

不过这种方式只适合 ASCII 字符（对应编码在 0～255 之间），而不适合汉字等 Unicode 字符。类似的，通过 File 的 Open 方法可打开文件并返回一个 FileStream 对象，通过该对象的 ReadByte 方法可依次从文件中读取字节，例如：

```
FileStream fs1 = File.Open("C:\\a.txt", FileMode.Open);
char[] chs = new char[fs1.Length];
for (int i = 0; i < fs1.Length; i++)
    chs[i] = (char)fs1.ReadByte();
Console.WriteLine(new string(chs));
fs1.Close();
```

上面使用到了 FileStream 的 Length 属性，它表示文件流的长度。而 File.Open 方法的第二个参数表示文件的打开模式，其类型为 FileMode 枚举，取值包括：

- Append——以追加方式打开文件，如果文件存在从文件尾开始操作，否则创建一个新文件。
- Create——创建并打开一个新文件，如果文件已存在，则覆盖旧文件。
- CreateNew——创建并打开一个新文件，如果文件已存在，则会发生异常。

- Open——打开现有文件，如果文件不存在，则会发生异常。
- OpenOrCreate——打开或新建一个文件，如果文件已经存在则打开它，否则创建并打开一个新文件。
- Truncate——打开现有文件并清空其内容。

打开文件时还可以指定文件流的读写方式，这可以通过 File.Open 方法的第三个参数来设置，其类型为 FileAccess 枚举，取值包括：

- Read——文件流只读。
- Write——文件流只写。
- ReadWrite——文件流可读写。

创建了一个 FileStream 对象后，通过其 CanRead 属性能够判断文件流是否可读，通过其 CanWrite 能够判断文件流是否可写。例如：

```
FileStream fs1 = File.Create("C:\\a.txt");
Console.WriteLine(fs1.CanRead); //输出 True
Console.WriteLine(fs1.CanWrite); //输出 True
fs1.Close();
```

如果不指定 FileAccess 枚举类型的参数，那么创建的 FileStream 对象默认都是可读写的。如果只需要对文件执行读操作，以只读方式打开文件能够显著提高其读取效率。

除了 ReadByte 和 WriteByte 方法外，FileStream 还可以通过 Read 方法进行一次性批量读入，以及通过 Write 方法进行批量读出，这两个方法都将读/写的内容放在一个 byte[]数组型的参数中，并通过两个参数来分别指定数组中读/写的开始位置以及要读/写的字节数。例如下面的代码就先将 20～39 这 20 个字节值写入文件中，而后将前 10 个值读入到数组 bs2 中：

```
byte[] bs1 = new byte[100];
for (byte i = 0; i < 100; i++)
    bs1[i] = i;
FileStream fs1 = File.Create("C:\\a.txt");
fs1.Write(bs1, 20, 20); //写入 20~39
fs1.Close();
byte[] bs2 = new byte[10];
fs1 = File.Open("C:\\a.txt", FileMode.Open, FileAccess.Read);
fs1.Read(bs2, 0, 10); //读出 20~29
fs1.Close();
```

无论是读还是写，FileStream 都是流的"当前"位置开始操作：新建或新打开一个文件时，流的当前位置位于文件的开头；每调用一次 ReadByte 或 WriteByte 方法，当前位置就向前移动一个字节；而每调用一次 Read 或 Write 方法，当前位置就向前移动所读/写的字节数量，直至达到文件末尾。FileStream 的 Position 属性用于指示流的当前位置，其值为 0 时代表文件开头，其值等于 Length 属性值时代表文件末尾。例如：

```
byte[] bs1 = new byte[100];
for (byte i = 0; i < 100; i++)
    bs1[i] = i;
FileStream fs1 =
    File.Open("C:\\a.txt", FileMode.OpenOrCreate, FileAccess.ReadWrite);
fs1.Write(bs1, 0, 100); //写入并到达文件末尾
Console.WriteLine(fs1.Position == fs1.Length); //输出 True
fs1.Position = 0; //重新移到开头位置
Console.WriteLine(fs1.ReadByte()); //读取并输出 0
fs1.Position = 50;
```

```
Console.WriteLine(fs1.ReadByte());   //读取并输出50
Console.WriteLine(fs1.ReadByte());   //读取并输出51
fs1.Close();
```
除了直接设置 Position 属性外，FileStream 的 Seek 方法也可修改流的当前位置，传递给方法的移动量为正数时表示向前（文件尾）移动，为负数时则表示向后（文件头）移动。

FileStream 的构造函数有多个重载形式也与 File 的静态方法效果相同，其中形如 FileStream(string path,FileMode mode)的构造函数得到的对象与 File.Open(string path,FileMode mode)方法返回的对象是相同的，例如：

```
FileStream fs1 = new FileStream("C:\\a.txt", FileMode.Open);
```

考虑数列 $a_1 = 1$, $a_2 = 1$, $a_n = a_{n-1} + a_{n-2}$ ($n>2$)，下面的程序 P8_2 就将该数列中小于 10000 的各项依次写入文件中，而后通过改变流的当前位置来读取数列的指定项。由于大部分数列项的值超出了 byte 的范围，程序将每个项值分为前 8 位和后 8 位分别写入，这样就能够有效处理 ushort 类型的数值：

```
//程序 P8_2
using System;
using System.IO;
namespace P8_2
{
    class Program
    {
        static void Main()
        {
            FileStream fs1 = new FileStream("num.txt", FileMode.Create);
            for (ushort a = 1, b = 1; a <= 10000; a += b){
                ushort t = b;
                b = a;
                a = t;
                fs1.WriteByte((byte)(a / 256));   //写入前8位
                fs1.WriteByte((byte)(a % 256));   //写入后8位
            }
            fs1.Position = 0;
            Console.Write("请输入要读取的数列项：");
            int i = int.Parse(Console.ReadLine());
            fs1.Position = 2 * i;
            int x = 256 * fs1.ReadByte() + fs1.ReadByte();
            Console.WriteLine("数列项为：" + x);
            fs1.Close();
        }
    }
}
```

8.1.3 FileInfo 类

静态类 File 的所有方法都是静态的，而其使用这些方法都需要指定目标文件的路径名。而 FileInfo 类是 File 类的非静态版本，可以指定文件的路径名来创建一个 FileInfo 对象，然后通过其他各种实例方法来实现与 File 类静态方法类似的功能。例如：

```
FileInfo fi = new FileInfo("C:\\a.txt");
fi.CopyTo("C:\\b.txt");
fi.MoveTo("D:\\a.txt");
fi.Delete();
```

FileInfo 类的其他许多成员也实现了和 File 类中提供的功能，比如用于判断文件是否存在的 Exists 方法，用于打开文件并返回 FileStream 对象的 Open 方法，包含文件基本信息的 Attributes

属性等，其具体用法此处不再赘述。

 如果只对文件执行一次操作，那么推荐使用 File 类的静态方法。而如果要反复对同一文件进行多次操作，那么创建并使用一个 FileInfo 对象能够提高程序的效率。

8.2 流的读写器

直接使用文件流对象在文件中读写字节的效率较高，但要处理其他各种类型就会比较麻烦。System.IO 命名空间中还提供了与文件流相配套的读写器类型，其中最主要的就是二进制读写器 BinaryReader 和 BinaryWriter，以及文本读写器 StreamReader 和 StreamWriter。

8.2.1 二进制读写器

BinaryReader 和 BinaryWriter 以二进制方式对流进行读写操作，它们都是基于流对象来创建相应的读写器对象，例如：

```
FileStream fs1 = File.Create("C:\\a.txt");
BinaryWriter bw1 = new BinaryWriter(fs1);
```

BinaryWriter 的 Write 方法用于向流中写入数据，但该方法有多种重载形式，接受的参数类型包括 bool、byte、short、int、long、float、double、string、char[]、byte[]等，因此可以直接写入各种类型的数值，例如：

```
bw1.Write(1000); //写入整数
bw1.Write(3.14); //写入实数
bw1.Write(new char[] { 'A', 'p', 'p' }); //写入字符数组
bw1.Write(DateTime.Now.ToString()); //写入字符串
```

对于从文件流构造的读写器，应先关闭读写器对象，再关闭文件流对象，例如：

```
bw1.Close();
fs1.Close();
```

向文件流中写入的数据将存放在系统缓冲区，关闭 BinaryWriter 对象时这些缓冲数据将一次性写入到物理文件中，这能够显著提高文件操作的效率。但如果希望在写文件的过程中及时保存数据，那么可通过 BinaryWriter 对象的 Flush 方法将缓冲数据强制写入文件，例如：

```
bw1.Write(1000);
bw1.Write(3.14);
bw1.Flush(); //强制写入缓冲数据
bw1.Write(new char[] { 'A', 'p', 'p' });
bw1.Write(DateTime.Now.ToString());
bw1.Close(); //一次性写入剩余数据
```

对应的，BinaryReader 提供了多个从流中读取数据的方法，如读取布尔值的 ReadBoolean 方法、读取字节的 ReadByte 方法、读取整数的 ReadInt32 方法、读取字符串的 ReadString 方法等。看下面的代码示例：

```
FileStream fs1 = File.Create("C:\\a.txt");
BinaryReader br1 = new BinaryReader(fs1);
int x = br1.ReadInt32(); //读取整数
double y = br1.ReadDouble(); //读取实数
string s1 = new string(br1.ReadChars(3)); //读取字符数组
DateTime dt1 = DateTime.Parse(br1.ReadString()); //读取字符串
br1.Close();
```

```
fs1.Close();
```
不论是 BinaryReader 还是 BinaryWriter，其读/写的位置总是流对象的当前位置，而且该位置会随着读/写的进行而前移相应的字节数量。看下面的代码示例：
```
FileStream fs1 = File.Create("C:\\a.txt");
BinaryWriter bw1 = new BinaryWriter(fs1);
bw1.Write(true);
Console.WriteLine(fs1.Position);    //写入1个字节，输出1
bw1.Write(1000);
Console.WriteLine(fs1.Position);    //写入4个字节，输出5
bw1.Write(3.14);
Console.WriteLine(fs1.Position);    //写入8个字节，输出13
bw1.Write(3.14M);
Console.WriteLine(fs1.Position);    //写入16个字节，输出29
bw1.Close();
fs1.Close();
```

8.2.2 文本读写器

StreamReader 和 StreamWriter 以文本方式对流进行读/写操作，它们也可通过指定的文件流来创建读写器对象。其中 StreamWriter 的 Write 方法也提供了一系列重载形式，用于写入不同类型的数值，例如：
```
FileStream fs1 = File.Create("C:\\b.txt");
StreamWriter sw1 = new StreamWriter(fs1);
sw1.Write(true);    //写入布尔值
sw1.Write(123.45M);    //写入小数
sw1.Write("Apple");    //写入字符串
int x = 2009, y = 10;
sw1.Write("{0}年{1}月{2}日", x, y, DateTime.Now.Day);    //写入格式化字符串
sw1.Close();
fs1.Close();
```

注意

StreamWriter 和 BinaryWriter 的 Write 方法的重载形式并不完全相同，比如 StreamWriter 在写入字符串时支持参数格式化。此外 StreamWriter 还提供了一系列重载的 WriteLine 方法，它在写入指定内容后会自动添加换行符（这类似于 Console 类的 Write 和 WriteLine 方法）。

提示

StreamWriter 的 Write 和 WriteLine 方法写入的是数值的文本表示，而 BinaryWriter 的 Write 方法写入的是数值的二进制内容。例如对于整数 100，采用二进制方式写入文件的内容是数值本身，只占 1 个字节（对应于字符 d）；而采用文本方式写入文件的内容是字符串 "100"，共占 3 个字节。

StreamWriter 的构造函数也有多种重载形式。在基于 FileStream 创建 StreamWriter 对象时，还可以通过构造函数的第二个参数来指定写入文本的编码格式（不指定时则采用默认的 UTF8 编码），例如：
```
StreamWriter w1 = new StreamWriter(fs1, Encoding.ASCII);    //使用 ASCII 编码
StreamWriter w2 = new StreamWriter(fs1, Encoding.Unicode);    //Unicode 编码
```

System.Text 命名空间下的 Encoding 类表示字符编码，它通过静态属性 ASCII、UTF7、UTF8、UTF32、Unicode、BigEndianUnicode 来返回相应的基本编码格式。此外，Encoding 的静态属性 Default 返回当前系统的默认字符编码，其静态方法 GetEncodings 则返回当前系统中的所有编码格式。

下面的 Windows 应用程序用于将 richTextBox1 中的内容写入指定文件，用户可通过一组单选框来选择写入方式，包括二进制格式和各种不同的字符编码格式，而写入的内容将以默认格式显示在 richTextBox2 中。程序的示例输出如图 8-2 所示，从中可看到 ASCII 编码不能有效处理汉字等 Unicode 字符。

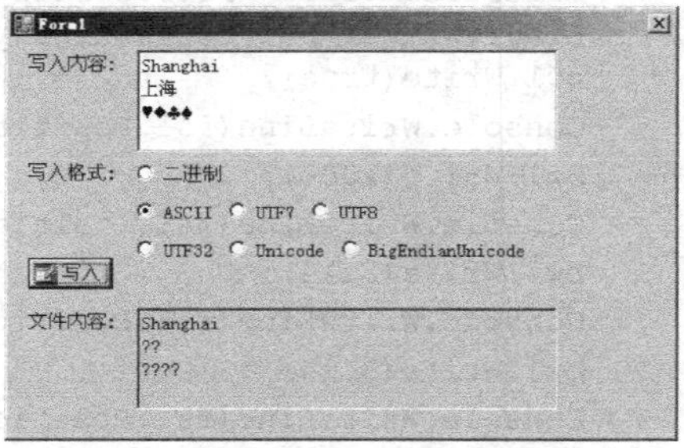

```csharp
//程序 P8_3(Form1.cs)
using System;
using System.Text;
using System.Windows.Forms;
using System.IO;
namespace P8_3
{
    public partial class Form1 : Form
    {
        public Form1()
        {
            InitializeComponent();
        }

        private void Form1_Load(object sender, EventArgs e)
        {
            string[] ss = { "Shanghai", "上海", "♥♦♣♠" };
            richTextBox1.Lines = ss;
        }

        private void button1_Click(object sender, EventArgs e)
        {
            FileStream fs1 = File.Create("demo.txt");
            if (radioButton1.Checked){
              BinaryWriter bw1 = new BinaryWriter(fs1);
              foreach (string s in richTextBox1.Lines)
                  bw1.Write(s);
              bw1.Close();
            }
            else{
              Encoding encoding = Encoding.ASCII;
              if (radioButton3.Checked)
                 encoding = Encoding.UTF7;
              else if (radioButton4.Checked)
                 encoding = Encoding.UTF8;
              else if (radioButton5.Checked)
                 encoding = Encoding.UTF32;
              else if (radioButton6.Checked)
                 encoding = Encoding.Unicode;
              else if (radioButton7.Checked)
                 encoding = Encoding.BigEndianUnicode;
              StreamWriter sw1 = new StreamWriter(fs1, encoding);
              foreach (string s in richTextBox1.Lines)
                 sw1.WriteLine(s);
              sw1.Close();
            }
            fs1.Close();
            richTextBox2.Lines = File.ReadAllLines("demo.txt");
        }
    }
}
```

图 8-2　程序 P8_3 的示例输出结果

类似地，基于 FileStream 创建 StreamReader 对象也可以指定编码格式，而通过 StreamReader 的 CurrentEncoding 属性以及 StreamWriter 的 Encoding 属性可获得当前读写器对象所使用的编码格式。例如：

```
Filetream fs1 = File.Create("C:\\b.txt");
StreamWriter sw1 = new StreamWriter(fs1);
if (sw1.Encoding == Encoding.ASCII)
    sw1.Write("Beijing");
else
    sw1.Write("北京");
StreamReadersr1 = new StreamReader(fs1, sw1.Encoding);
Console.WriteLine(sr1.CurrentEncoding);
```

StreamReader 也提供了不同的读操作方法。其中 Read 方法有两种重载形式，可分别用于读取单个字符或一组字符；另外 ReadLine 方法每次读取一行字符串，而 ReadToEnd 方法则从当前位置一直读取到流的末尾。下面的程序就先将一组字节写入到文件中，而后通过 StreamReader 分三次读出文件中的字符（字符 A～N 的 ASCII 编码对应为 65～78，而换行符的 ASCII 编码为 10）：

```
//程序 P8_4
using System;
using System.IO;
namespace P8_4
{
    class Program
    {
        static void Main()
        {
            FileStream fs1 = new FileStream("Test.txt", FileMode.Create, FileAccess.ReadWrite);
            BinaryWriter bw1 = new BinaryWriter(fs1);
            for (byte i = 65; i < 72; i++)
                bw1.Write(i);
            bw1.Write(10);
            for (byte i = 72; i < 79; i++)
                bw1.Write(i);
            bw1.Flush();
            fs1.Position = 0;
            char[] chs = new char[3];
            StreamReader sr1 = new StreamReader(fs1);
            sr1.Read(chs, 0, 3);
            Console.WriteLine(new string(chs)); //输出"ABC"
            Console.WriteLine(sr1.ReadLine()); //输出"DEFG"
            Console.WriteLine(sr1.ReadToEnd()); //输出" HIJKLMN"
            bw1.Close();
            sr1.Close();
            fs1.Close();
        }
    }
}
```

提示

上述代码只是演示 StreamReader 的读取效果。在通常情况下，二进制读写器和文本读写器不应混用，而使用文本读写器时 StreamReader 和 StreamWriter 所采用的编码格式也应一致，否则程序将很容易出错。

和二进制读写器不同，StreamReader 和 StreamWriter 还支持从文件名直接构造读写器对象，不过此时程序会自动生成隐含的文件流，读写器对文件的读写还是通过流对象进行的，而读写器关闭时会自动关闭流对象。通过读写器的 BaseStream 属性可获得隐含的文件流对象。看下面的代码示例：

```
StreamWriter sw1 = new StreamWriter("C:\\b.txt");
Console.WriteLine(sw1.BaseStream.CanRead); //输出 False
Console.WriteLine(sw1.BaseStream.CanWrite); //输出 True
```

```
sw1.Write("abc");
sw1.Flush();
Console.WriteLine(sw1.BaseStream.Position);   //输出3
sw1.Close();
StreamReader sr1 = new StreamReader("C:\\a.txt");
Console.WriteLine(sr1.BaseStream.CanRead);    //输出 True
Console.WriteLine(sr1.BaseStream.CanWrite);   //输出 False
Console.WriteLine(sr1.BaseStream.Position);   //输出 0
sr1.Close();
```

StreamReader 还提供了一个布尔类型的属性 EndOfStream，其值为 true 时表示流的当前位置已到达末尾，此时再进行读操作就无法再读取任何内容。例如下面的方法实现了与程序 P8_1 中 TypeFile 方法类似的功能：

```
static void TypeFile(string filename)
{
    Console.WriteLine(filename);
    StreamReader sr1 = new StreamReader(filename);
    while (!sr1.EndOfStream)
        Console.WriteLine(sr1.ReadLine());
    sr1.Close();
}
```

8.3　文件对话框

在 Windows 应用程序中操作文件，经常需要显示如图 8-3 所示的"打开"对话框和如图 8-4 所示的"另存为"对话框，以供用户选择要打开和保存的文件。在 System.Windows.Forms 命名空间中提供了 OpenFileDialog 和 SaveFileDialog 这两个对话框类以实现上述功能，它们都是抽象类 FileDialog 的派生类。

图 8-3　"打开"对话框

图 8-4　"另存为"对话框

FileDialog 的又是 CommonDialog 的派生类，因此可通过 ShowDialog 方法来模态显示对话框。此外，FileDialog 的 FileName 属性表示在对话框中选取的文件名，Filter 属性则表示可选文件名的筛选器。例如设置 Filter 属性值为"文本文件(*.txt)|*.txt"，那么该字符串将显示在"打开"对话框的"文件类型"或"另存为"对话框的"保存类型"选项框中，而在对话框中可选取后缀名为 txt 的文本文件：

```
OpenFileDialog dlg1 = new OpenFileDialog();
dlg1.Filter = "文本文件(*.txt)|*.txt";
```

如果要在文件对话框中支持多种文件类型的选择，那么可在相应的后缀名之间通过分号进行分割，例如：

```
dlg1.Filter = "网页文件(*.htm;*.html)|*.htm;*.html";
```

FileDialog 的 Filter 属性还支持多个筛选器的组合，即将各个筛选字符串通过符号"|"进行分割，例如下面的语句将使对话框的默认文件类型为文本文件，而通过"文件类型"的组合框可以选择网页文件类型或所有文件类型：

```
dlg1.Filter = "文本文件(*.txt)|*.txt|网页文件(*.htm;*.html)|*.htm;*.html|所有文件(*.*)|*.*";
```

如果使用了多个筛选器，那么通过 FileDialog 的 FilterIndex 可获得用户所选的筛选器索引，例如：

```
SaveFileDialog dlg2 = new SaveFileDialog();
dlg2.Filter = "文本文件(*.txt)|*.txt|二进制文件(*.bin)|*.bin";
if (dlg2.ShowDialog() == DialogResult.OK)
   if (dlg2.FilterIndex == 0)
      File.WriteAllLines(richTextBox1.Lines);
   else {
      byte[] bs1 = new byte[richTextBox1.Text.Length];
      richTextBox1.Text.ToCharArray().CopyTo(bs1, 0);
      File.WriteAllBytes(bs1);
}
```

FileDialog 还提供了 InitialDirectory 属性来设置对话框中初始显示的目录，例如下面的代码将使对话框 dlg1 的初始目录为 C 盘根目录，而 dlg2 的初始目录为程序当前目录：

```
dlg1.InitialDirectory = "C:\\";
dlg2.InitialDirectory = Application.StartupPath;
```

如果要记住用户在程序中上次使用文件对话框所选的目录，那么可设置 FileDialog 的 RestoreDirectory 属性值为 true，那么下次打开对话框还将显示相同的目录。当用户经常在同一个目录下操作文件时，这能够有效地简化用户操作。

使用打开文件对话框时，如果希望用户能一次性选取多个文件，那么应设置 OpenFileDialog 的 Multiselect 属性值为 true，此时通过对话框的 FileNames 属性可获得所选文件名的字符串数组，例如：

```
OpenFileDialog dlg1 = new OpenFileDialog();
dlg1.Filter = "文本文件(*.txt)|*.txt";
dlg1.Multiselect = true;
if (dlg1.ShowDialog() == DialogResult.OK)
   foreach (string file in dlg1.FileNames)
         richTextBox1 += File.ReadAllText(file);
```

而在使用保存文件对话框时，如果设置 SaveFileDialog 的 OverwritePrompt 属性值为 true，那么当用户指定的文件已存在，程序将显示如图 8-5（a）所示的警告消息框；如果设置其 CreatePrompt 属性值为 true，那么当用户指定的文件不存在，程序将显示如图 8-5（b）所示的警告消息框。此时只有选择"是"才能保存文件，否则将回到"另存为"对话框的界面。

（a）
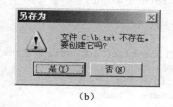
（b）

图 8-5 保存文件对话框的警告消息框

8.4 基于文件的对象持久性

8.4.1 实现对象持久性

在面向对象的程序中,各个对象的信息经常会发生变化。如果把对象信息(主要是成员字段值)保存到文件等永久性介质中,那么就可以在再次运行程序时载入对象信息,这称之为对象的持久性。

例如下面定义的 Student 类就能够通过 Save 方法将对象信息保存到指定文件,还能通过静态方法 Load 来从指定文件读取信息并创建对象:

```
Public class Student
{
  private int id;
  public int ID //学号
  {
    get { return id; }
    set { id = value; }
  }

  private string name;
  public string Name //姓名
  {
    get { return name; }
    set { name = value; }
  }

  private bool gender;
  public bool Gender //性别
  {
    get { return gender; }
    set { gender = value; }
  }

  private string department;
  public string Department //院系
  {
    get { return department; }
    set { department = value; }
  }

  private int grade;
  public int Grade //年级
  {
    get { return grade; }
    set { grade = value; }
  }

  public Student(int id, string name)
  {
    this.id = id;
    this.name = name;
  }
```

```csharp
public override string ToString()
{
    return string.Format("{0} {1} {2} {3}系{4}级", id, name, gender ? '男' : '女',
department, grade);
}

public static Student Parse(string s)
{
    string[] ss = s.Split(' ', '系', '级');
    Student s1 = new Student(int.Parse(ss[0]), ss[1]);
    s1.gender = (ss[2] == "男" ? true : false);
    s1.department = ss[3];
    s1.grade = byte.Parse(ss[4]);
    return s1;
}

public void Save(string filename)
{
    File.WriteAllText(filename, this.ToString());
}

public static Student Load(string filename)
{
    return Student.Parse(File.ReadAllText(filename));
}
```

上述读写方式在单个对象存储到单个文件时比较方便。在存储多个对象时，如果仍是每个对象对应一个文件，那么通常要求通过唯一的对象标识来区分文件，如学号为 80001 的 Student 对象存储在 80001.sti 文件中，学号为 80002 的对象存储在 80002.sti 文件中……当然，更常见的情况是将同一类型的多个对象存储在一个文件中，此时将对象关联到读写器对象（而非文件或文件流）更为方便，例如上面 Student 类的 Save 和 Load 方法可改写为如下内容：

```csharp
public void Save(BinaryWriter writer)
{
    writer.Write(id);
    writer.Write(name);
    writer.Write(gender);
    writer.Write(department);
    writer.Write(grade);
}

public static Student Load(BinaryReader reader)
{
    int id = reader.ReadInt32();
    string name = reader.ReadString();
    Student s1 = new Student(id, name);
    s1.gender = reader.ReadBoolean();
    s1.department = reader.ReadString();
    s1.grade = reader.ReadByte();
    return s1;
}
```

接下来就可以基于某个文件流创建读写器对象，很方便地对一组对象进行批量读写，例如：

```csharp
//从文件读取学生对象信息
Student[] students = new Student[5];
FileStream fs1 = new FileStream("students.bin", FileMode.Open);
```

```
BinaryReader br1 = new BinaryReader(fs1);
for (int i = 0; i < 5; i++)
    students[i] = Student.Load(br1);
br1.Close();
fs1.Close();
//修改学生对象信息
//……
//向文件写入学生对象信息
fs1 = new FileStream("students.bin", FileMode.OpenOrCreate);
BinaryWriter bw1 = new BinaryWriter(fs1);
foreach (Student s in students)
    s.Save(bw1);
bw1.Close();
fs1.Close();
```
如果改用文本读写器的话，上面 Student 类的 Save 和 Load 方法可改写为如下内容：
```
public void Save(StreamWriter writer)
{
    writer.WriteLine(id);
    writer.WriteLine(name);
    writer.WriteLine(gender);
    writer.WriteLine(department);
    writer.WriteLine(grade);
}
public static Student Load(StreamReader reader)
{
    int id = int.Parse(reader.ReadLine());
    string name = reader.ReadLine();
    Student s1 = new Student(id, name);
    s1.gender = bool.Parse(reader.ReadLine());
    s1.department = reader.ReadLine();
    s1.grade = byte.Parse(reader.ReadLine());
    return s1;
}
```

使用二进制读写器时，存储的文件内容较为紧凑，读写效率一般也较高。而使用文本读写器时，存储的文件内容可读性较强。具体选用哪类读写器应根据程序的实际需要来确定。

8.4.2 .NET 中的自动持久性支持

通常对象需要存取的都是其数据成员即字段值，这为我们提供了一种自动实现对象持久性的思路：定义一个格式化工具 Formatter，通过它自动获取对象所有字段的名称和值，并将其写入到文件流中；或是从文件流中解析出各字段的名称和值，并以此为依据来创建对象。这种技术也叫做串行化或序列化（Serialization）。

.NET Framework 提供了对串行化的良好支持。如果某个类型在定义前使用了 Serializable 标记，那么其对象就是可串行化的，例如：

```
[Serializable()]
public class Student
{
    ……
}
```

默认情况下，可串行化对象的所有字段值都会被保存到文件中。如果不希望某个存取字段，那么可将该字段标记为 NonSerializable，例如下面定义的 Student 类中，字段 id 和 name 可被存取，

而字段 department 则不会被处理:

```csharp
[Serializable()]
public class Student
{
    private int id;
    public int ID //学号
    {
        get { return id; }
        set { id = value; }
    }

    private string name;
    public string Name //姓名
    {
        get { return name; }
        set { name = value; }
    }

    [NonSerialized()]
    private string department;
    public string Department //院系
    {
        get { return department; }set { department = value; }
    }
}
```

对于可串行化的对象，就可以使用.NET 中提供的格式化工具来进行文件存取。其中 System.Runtime.Serialization.Formatters.Binary 命名空间中定义的 BinaryFormatter 类就是一个二进制格式化工具，其 Serialize 方法用于将对象的字段值写入文件流，Deserialize 方法则用于从文件中读取字段值并创建对象，例如：

```csharp
//串行化
Student s1 = new Student(800001, "王小红");
s1.Grade = 2;
FileStream fs1 = new FileStream("students.bin", FileMode.Create);
BinaryFormatter bf1 = new BinaryFormatter();
bf1.Serialize(fs1, s1);
fs1.Close();
//反串行化
fs1 = new FileStream("students.bin", FileMode.Open);
s1 = bf1.Deserialize(fs1);
fs1.Close();
```

当然，只要保证流的当前位置正确，格式化工具可以将多个对象依次保存到同一文件流中，或是从同一文件流中依次读取多个对象，例如：

```csharp
//串行化
Student s1 = new Student(800001, "王小红");
Student s2 = new Student(800002, "李强");
FileStream fs1 = new FileStream("students.bin", FileMode.Create);
BinaryFormatter bf1 = new BinaryFormatter();
bf1.Serialize(fs1, s1);
bf1.Serialize(fs1, s2);
fs1.Close();
//反串行化
fs1 = new FileStream("students.bin", FileMode.Open);
```

```
s1 = bf1.Deserialize(fs1);
s2 = bf1.Deserialize(fs1);
fs1.Close();
```

BinaryFormatter 以二进制格式来存取对象信息。.NET 类库中的另一个格式化工具 SoapFormatter 使用的则是 XML 格式。该类型在 System.Runtime.Serialization.Formatters.Soap 命名空间中定义（使用时需要添加对 System.Runtime.Serialization.Formatters.Soap.dll 程序集文件的引用），不过其使用方式和 BinaryFormatter 基本类似，例如：

```
//串行化
Student s1 = new Student(800001, "王小红");
s1.Grade = 2;
FileStream fs1 = new FileStream("students.xml", FileMode.Create);
SoapFormatter sf1 = new BinaryFormatter();
sf1.Serialize(fs1, s1);
fs1.Close();
//反串行化
fs1 = new FileStream("students.xml", FileMode.Open);
s1 = sf1.Deserialize(fs1);
fs1.Close();
```

在下面的 Windows 示例程序中，Student 类及其派生类 Graduate 都被标记为可串行化的。程序主窗体在启动时从文件中读取一组 Student 对象信息，用户可通过学号选择来显示具体的学生信息，并在窗体中编辑和保存学生信息（窗体的设计视图如图 8-6 所示）：

```
//程序 P8_5(Form1.cs)
using System;
using System.Text;
using System.Windows.Forms;
using System.IO;
using System.Runtime.Serialization.Formatters.Binary;
namespace P8_5
{
    public partial class Form1 : Form
    {
        Student[] students;

        public Form1()
        {
            InitializeComponent();
        }

        private void Form1_Load(object sender, EventArgs e)
        {
            students = new Student[5];
            FileStream fs1 = new FileStream("students.bin", FileMode.Open);
            BinaryFormatter bf1 = new BinaryFormatter();
            for (int i = 0; i < students.Length; i++){
                students[i] = (Student)bf1.Deserialize(fs1);
                cmbID.Items.Add(students[i].ID);
            }
            fs1.Close();
            cmbID.SelectedIndex = 0;
        }

        private void cmbID_SelectedIndexChanged(object sender, EventArgs e)
        {
            Student s1 = students[cmbID.SelectedIndex];
            tbName.Text = s1.Name;
```

```csharp
        if (s1.Gender)
            rdbMan.Checked = true;
        else
            rdbWoman.Checked = true;
        tbDepartment.Text = s1.Department;
        nudGrade.Value = s1.Grade;
        if (s1 is Graduate){
            tbSupervisor.Enabled = true;
            tbSupervisor.Text = ((Graduate)s1).Supervisor;
        }
        else {
            tbSupervisor.Text = "";
            tbSupervisor.Enabled = false;
        }
    }

    private void btnModify_Click(object sender, EventArgs e)
    {
        Student s1 = students[cmbID.SelectedIndex];
        s1.Name= tbName.Text;
        s1.Gender = rdbMan.Checked;
        s1.Department = tbDepartment.Text;
        s1.Grade = (byte)nudGrade.Value;
        if (s1 is Graduate)
            ((Graduate)s1).Supervisor = tbSupervisor.Text;
    }

    private void btnSave_Click(object sender, EventArgs e)
    {
        FileStream fs1 = new FileStream("students.bin", FileMode.OpenOrCreate);
        BinaryFormatter bf1 = new BinaryFormatter();
        foreach (Student s in students)
            bf1.Serialize(fs1, s);
        fs1.Close();
        MessageBox.Show("保存成功!");
    }
}

[Serializable()]
public class Student
{
    private int id;
    public int ID //学号
    {
        get { return id; }
        set { id = value; }
    }

    private string name;
    public string Name //姓名
    {
        get { return name; }
        set { name = value; }
    }

    private bool gender;
    public bool Gender //性别
    {
        get { return gender; }
```

```csharp
      set { gender = value; }
   }

   private string department;
   public string Department  //院系
   {
      get { return department; }
      set { department = value; }
   }

   private byte grade;
   public byte Grade  //年级
   {
      get { return grade; }
      set { grade = value; }
   }

   public Student(int id, string name)
   {
      this.id = id;
      this.name = name;
   }
}

[Serializable()]
public class Graduate : Student
{
   private string supervisor;
   public string Supervisor  //导师
   {
      get { return supervisor; }
      set { supervisor = value; }
   }

   public Graduate(int id, string name, string supervisor)
      : base(id, name)
   {
      this.supervisor = supervisor;
   }
}
```

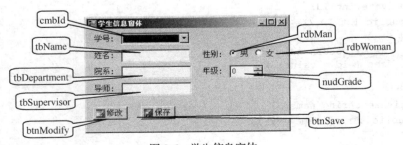

图 8-6 学生信息窗体

.NET 实现自动持久性的关键在于反射（Reflection）技术，它能够自动解析程序类型的元数据，如类型的名称、成员类型、继承结构等，从而支持动态的对象创建和对象数据存取。当然，比起在类型中手动实现持久性代码，自动持久性的类型解析需要进行很多额外的分析工作，因此数据存取的效率也会有所下降。

8.5 案例研究——旅行社信息和系统用户的持久性

8.5.1 旅行社对象的持久性

本节将为旅行社管理系统中的一些类型实现对象持久性。首先考虑旅行社类 TravelAgency，它的对象信息可存放在一个文件当中，那么只需要在该类的定义中增加如下两个成员方法：

```
public void Save(string filename)
{
    StreamWriter sw1 = new StreamWriter(filename);
    sw1.WriteLine(_name);
    sw1.WriteLine(Manager);
    sw1.WriteLine(Representive);
    sw1.WriteLine(Address);
    sw1.WriteLine(PostCode);
    sw1.WriteLine(Website);
    sw1.WriteLine(Phone);
    sw1.WriteLine(Fax);
    sw1.WriteLine(Bank);
    sw1.WriteLine(AccountName);
    sw1.WriteLine(AccountNumber);
    sw1.WriteLine(License);
    sw1.Close();
}

public static TravelAgency Load(string filename)
{
    StreamReader sr1 = new StreamReader(filename);
    TravelAgency a1 = new TravelAgency(sr1.ReadLine());
    a1.Manager = sr1.ReadLine();
    a1.Representive = sr1.ReadLine();
    a1.Address = sr1.ReadLine();
    a1.PostCode = sr1.ReadLine();
    a1.Website = sr1.ReadLine();
    a1.Phone = sr1.ReadLine();
    a1.Fax = sr1.ReadLine();
    a1.Bank = sr1.ReadLine();
    a1.AccountName = sr1.ReadLine();
    a1.AccountNumber = sr1.ReadLine();
    a1.License = sr1.ReadLine();
    sr1.Close();
    return a1;
}
```

设系统中的当前旅行社信息保存在 agency.txt 文件中，那么可以在程序主窗体 Form1 的 Load 事件处理方法中加入下列代码，从而在程序启动时读取旅行社信息：

```
if (File.Exists("agency.txt"))
    _agency = TravelAgency.Load("agency.txt");
```

BasicForm 的 SaveInfo 方法最后也应添加下面这行代码，那么当用户在该窗体中单击"确定"按钮后就可将旅行社信息保存到文件中：

```
_agency.Save("agency.txt");
```

8.5.2 系统用户对象的持久性

系统需要在一个文件中存取一组用户对象，那么可先定义单个 User 对象的 Save 和 Load 方法，

然后在整个文件的存取操作中调用这些方法。应加入到 User 类定义中的相应方法代码如下:

```csharp
public void Save(BinaryWriter writer)
{
    writer.Write(_username);
    writer.Write(_password);
    writer.Write(_staffId);
}

public static User Load(BinaryReader reader)
{
    stringusername = reader.ReadString();
    stringpassword = reader.ReadString();
    stringstaffId = reader.ReadInt32();
    return new User(username, password, staffId);
}

public static void SaveAll(User[] users, string filename)
{
    FileStream fs1 = new FileStream(filename, FileMode.Create);
    BinaryWriter bw1 = new BinaryWriter(fs1);
    bw1.Write(users.Length);
    foreach (User u in users)
        u.Save(bw1);
    bw1.Close();
    fs1.Close();
}

public static User[] GetAll(string filename)
{
    FileStream fs1 = new FileStream(filename, FileMode.Open);
    BinaryReader br1 = new BinaryReader(fs1);
    User[] users = new User[br1.ReadInt32()];
    for (int i = 0; i < users.Length; i++)
        users[i] = User.Load(br1);
    return users;
}
```

在上面的 SaveAll 和 GetAll 方法中，文件中最先被写入和读取的是用户的数量。其中 GetAll 方法替换了第 7 章中模拟生成一组用户的同名方法。设用户信息保存在 users.bak 文件中，那么在主窗体 Form 的 Load 事件处理方法中，_user 字段的赋值代码也应被替换为:

```csharp
_users = User.GetAll("users.bak");
```

接下来再实现用户信息的修改和保存。为 TravelWin 项目添加一个新的 Windows 窗体 UserForm，参照如图 8-7 所示的内容向其加入控件。上方的组合框将列出旅行社的所有职员，在其中选择某个职员后便可对其进行用户的创建、修改或删除操作。

和 LoginForm 类似，UserForm 需要维护所有的系统用户集合，此外它还分别通过 _staff 和 _current 来表示当前选择的职员和用户，那么为该窗体添加如下的字段和属性定义:

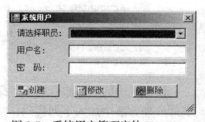

图 8-7 系统用户管理窗体 UserForm

```csharp
private Staff _staff = null;
privateUser _current
privateUser[] _users;;
public User[] Users
{
    get { return _users; }
}
```

UserForm 在构造函数中载入所有用户信息,并在 Load 事件处理方法中向组合框中加入所有职员的集合:

```csharp
public UserForm(users)
{
    InitializeComponent();
    _users = users;
}

private void UserForm_Load(object sender, EventArgs e)
{
    foreach (Staff s in Staff.GetAll())
        comboBox1.Items.Add(s);
    comboBox1.SelectedIndex = 0;
}
```

在组合框中选取了一组职员后,程序将在用户组中寻找与之对应的用户,找到后则在两个文本框中分别显示用户名和密码以供修改,否则允许针对该职员创建新的用户:

```csharp
private void comboBox1_SelectedIndexChanged(object sender, EventArgs e)
{
    _staff = (Staff)comboBox1.SelectedItem;
    foreach (User u in _users){
        if (u.StaffId == _staff.Id){
            _current = u;
            tbUserName.Text = u.Username;
            tbPassword.Text = u.Password;
            btnAdd.Enabled = false;
            btnModify.Enabled = true;
            btnDelete.Enabled = !(_staff is Manager); //经理用户不能删除
            return;
        }
    }
    tbUserName.Text = tbPassword.Text = "";
    btnAdd.Enabled = true;
    btnModify.Enabled = btnDelete.Enabled = false;
}
```

以"修改"按钮为例,其单击事件的处理方法将修改当前用户信息,并将结果保存到文件中("新建"和"删除"按钮的事件处理方法请读者自行完成):

```csharp
private void btnModify_Click(object sender, EventArgs e)
{
    if (tbUserName.Text == "" || tbPassword.Text == "")
    {
        MessageBox.Show("用户名和密码不能为空", "提示");
        return;
    }
    _current.Username = tbUserName.Text;
    _current.Password = tbPassword.Text;
    User.SaveAll(_users, "users.bak");
}
```

最后在主窗体的"系统管理"菜单中加入一个菜单项"系统用户管理(&U)",在其单击事件处理方法中添加如下代码来显示 UserForm,这就实现了系统的用户管理功能:

```csharp
private void menuSystemUsers_Click(object sender, EventArgs e)
{
    UserForm form = new UserForm(_users);
    if (form.ShowDialog() == DialogResult.OK)
        _users = form.Users;
}
```

8.6 小　　结

.NET 类库中定义了一系列用于文件操作的类型。其中文件流是对物理文件的封装，而使用读写器可方便地对文件流进行读写，读写的方式主要包括二进制方式和文本方式。通过这些文件存取操作，可以将程序对象信息保存到文件中，并在将来运行程序时再载入这些信息，从而实现对象的持久化特性。

8.7 习　　题

1. 扩展程序 P8_2，使其能够通过 FileStream 来读写 int 型和 long 型的数值。
2. 编写一个 Windows 窗体应用程序，在窗体的状态栏中显示该程序上一次运行的时间。
3. 写出下面程序的输出结果：

```csharp
class Program
{
    static void Main()
    {
        FileStream fs1 = File.Create("C:\\a.txt");
        StreamWriter sw1 = new StreamWriter(fs1);
        for (byte i = 0; i < 10; i++)
            sw1.Write(i);
        sw1.Flush();
        fs1.Position = 0;
        BinaryReader br1 = new BinaryReader(fs1);
        for (byte i = 0; i < 10; i++)
            Console.WriteLine(br1.Read());
        sw1.Close();
        br1.Close();
        fs1.Close();
    }
}
```

4. 采用二进制方式和文本方式来读写文件各有什么优缺点？
5. 除了本章介绍的基本功能外，File、FileStream、StreamWriter 和 StreamReader 等类型的许多成员方法还有多种重载形式。在 MSDN 中查看这些方法的重载形式，并进行编程测试。
6. 对于文件对话框，有哪些属性应当在显示对话框之前设置，又有哪些属性需在对话框返回 DialogResult.OK 结果后才能读取？
7. 如果要将程序中不同类型（不一定存在继承关系）的多个对象信息都存储在同一个文件中，那么在编写持久化代码时要注意哪些问题？
8. 8.5 节中的案例程序以二进制格式来保存用户信息，即使这样，用户名和密码的安全仍然存在较大隐患。试通过文件加密来进一步提高程序的安全性。
9. 为旅行社管理系统中的 Staff 及其派生类实现对象持久性。如果还要保存用户登录系统的记录，那么应如何来编程实现？

第9章
异常处理

程序在运行过程中可能遇到各种各样的"异常"情况,异常处理就是对这些情况进行处理,使得程序能够继续有效地运行。本章将讲解 C#语言中的异常处理功能,这一技术对于提高程序可靠性有着及其重要的作用。

9.1 异常的基本概念

程序中的错误包括语法错误和逻辑错误。语法错误是指代码不符合语言的语法规则,这类错误能够被编译器检查出来,只有改正之后才能生成应用程序。例如 C#不允许在整数除法中除以 0,那么下面的代码就不能通过编译:

```
int y = 100 / 0;
```

逻辑错误是指程序能够通过编译,但运行结果和预期结果不一致,比如开发人员将 x+y 不小心写成了 x*y。再如在整数除法中,C#编译器不会去检查每个变量的值,那么下面的代码可以通过编译,但在程序运行时就会发生错误:

```
int x = 0, y = 0;
y = 100 / x;
Console.WriteLine("y = {0}", y);
```

这种运行时错误会导致程序的非正常终止,这叫做程序异常。例如上面的代码段会在运行到第二行时终止,并给出如图 9-1 的异常信息,其中包括异常的类型和引发异常的代码位置,而第三行代码则不会被执行。

图 9-1 试图除以 0 的异常信息

上述错误能够被细心的开发人员检查出来,但有很多错误是不可能完全避免的。例如程序可能要求用户输入一个整数 x,然后再进行除法运算:

```
Console.Write("请输入 x 的值:");
int x, y = 0;
x = int.Parse(Console.ReadLine());
y = 100 / (10 - x) / (x - 5) / x;
Console.WriteLine("y = {0}", y);
```

此时如果用户输入的不是一个整数,或是输入了 10、5、0,程序都会出现异常。一种解决办法是在代码中增加检查语句,例如:

```
Console.Write("请输入 x 的值:");
```

```csharp
int x, y = 0;
if (!int.TryParse(Console.ReadLine(), out x))
    Console.WriteLine("输入格式不正确");
else{
    if (x == 10 || x == 5 || x == 0)
        Console.WriteLine("x的值不能为10、5或0");
    else
        y = 100 / (10 - x) / (x - 5) / x;
}
Console.WriteLine("y = {0}", y);
```

但这种方式会大大降低程序的开发效率和可读性。如果需要检查的条件很多，代码就会变得十分冗长，且包含大量的 if-else 嵌套语句。再比如文件操作，当遇到磁盘损坏、文件不存在、文件内容不正确、文件已被其他程序占用、用户没有操作权限等情况都会导致程序异常，其中很多异常情况是难以检测的。如果不对其进行有效处理，我们就无法保证程序的可靠性。

异常处理则提供了一种新的解决途径，它假设程序一直按照预期的方式运行，如果在某个时候假设不成立，那么程序就转入异常处理状态，等到异常被成功处理之后再回到正常运行状态。

下面的程序 P9_1 示范了简单的异常处理技术：try 关键字之后的大括号对中是假设能够正常运行的代码；如果其中发生了错误，那么程序跳过其中剩余的代码，并转而执行 catch 关键字之后的大括号对中的代码，也就是说程序进入异常处理状态。

```csharp
//程序 P9_1
using System;
namespace P9_1
{
    class Program
    {
        static void Main()
        {
            Console.Write("请输入x的值:");
            int x, y = 0;
            try
            {
                x = int.Parse(Console.ReadLine());
                y = 100 / (10 - x) / (x - 5) / x;
            }
            catch
            {
                Console.WriteLine("输入不正确");
            }
            Console.WriteLine("y = {0}", y);
        }
    }
}
```

运行该程序时，如果用户输入正确，那么 catch 代码段中的内容不会被执行。但无论用户输入正确与否，catch 代码段之后的内容总是会被执行，如图 9-2 所示。

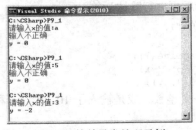

图 9-2 简单异常处理示例

9.2 异常处理结构

9.2.1 try-catch 结构

程序 P9_1 示范的是 try-catch 异常处理结构，其 catch 语句只能捕获前面与之配套的 try 代码段中的异常。当程序在 try 代码段的某条语句上发生错误后，程序控制权就转入 catch 代码段，而 try 代码段中剩余的代码会被忽略。

如果要给程序增加异常处理的功能，可以使用一个 try 语句将可能引发异常的代码段包围起来，并在其后增加一个 catch 代码段来处理异常。处理的方式包括记录异常、向用户报告异常、寻找异常原因并予以解决等。

C#采用面向对象的方式来处理程序异常，每一个异常都是一个 Exception 对象，在 catch 语句中可以访问该对象，例如：

```
int x, y = 0;
try
{
    x = int.Parse(Console.ReadLine());
    y = 100 / (10 - x) / (x - 5) / x;
}
catch(Exception ex)
{
    Console.WriteLine("输入错误:" + ex.Message);
}
```

其中 catch 关键字后面括号中的 Exception 对象 ex 就表示所捕获的异常对象，其 Message 属性表示 CLR 给出的异常消息。那么在运行这段代码时，如果输入了 10、5 或 0，程序将提示"输入错误：试图除以 0"；而如果输入的内容不是一个整数（比如 abc），程序将提示"输入错误：输入的字符串格式不正确"。

Exception 类型有一系列派生类，它们表示更为具体的异常类型，比如 DivideByZeroException 就表示除以 0 所引发的异常。如果在 catch 语句中指定了某个具体的异常类型，那么它只能捕获这一类异常。例如下面的 try-catch 语句只能捕获除以 0 的错误，而当用户输入一个非整数时仍会引发程序异常：

```
int x, y = 0;
try
{
    x = int.Parse(Console.ReadLine());
    y = 100 / (10 - x) / (x - 5) / x;
}
catch(DivideByZeroException ex)
{
    Console.WriteLine("错误：除数为 0");
}
```

catch 关键字后指定的异常对象类似于方法的参数，它是 catch 代码段中的局部变量，在该代码段之外无效。如果 catch 代码段中没有使用到异常对象，那么在 catch 关键字后的括号里可以只写出异常的类型，而不必写出异常对象的名称；如果只写出 catch 关键字，那么默认捕获的异常类型是 Exception。

在 try-catch 异常处理结构中，一个 try 语句后面可以有多个并列的 catch 语句，每个 catch 语句用于捕获和处理不同类型的异常。那么在发生了某个特定类型的异常后，程序就转入相应的 catch 代码段，并在执行完后退出整个 try-catch 语句；如果列出的所有 catch 语句都不能处理异常，当前程序将非正常中止。程序 P9_2 演示了多 catch 语句的 try-catch 结构，它能够针对不同的异常类型给出不同的提示：

```
//程序 P9_2
using System;
namespace P9_2
{
    class Program
    {
        static void Main()
        {
            Console.Write("请输入 x 的值:");
            int x, y = 0;
            try
            {
                x = int.Parse(Console.ReadLine());
                y = 100 / (10 - x) / (x - 5) / x;
            }
            catch (FormatException)
            {
                Console.WriteLine("输入的格式不正确，应输入一个整数");
            }
            catch (DivideByZeroException)
            {
                Console.WriteLine("错误：除数为 0");
            }
            catch (Exception)
            {
                Console.WriteLine("程序发生意外错误");
            }
            Console.WriteLine("y = {0}", y);
        }
    }
}
```

上面的 FormatException 表示方法的参数格式不正确所引发的异常，因此第一个 catch 语句能够处理 int.Parse 方法的输入不是整数的情况。不过程序 P9_2 永远不会非正常中止，因为最后一个 catch 语句捕获的异常类型是 Exception，它是所有异常的基类。

在使用多个 catch 语句时，应当先捕获更"具体"的异常，再捕获更"一般"的异常。换句话说，如果两个异常类存在继承关系，那么应当使捕获派生异常的 catch 语句在前、捕获基础异常的 catch 语句在后。例如在上面的程序中，捕获 Exception 的 catch 语句必须放在最后。

9.2.2 try-catch-finally 结构

在 try-catch 结构之后再加上一个 finally 代码段，这就形成了 try-catch-finally 结构。它对异常的捕获和处理方式与 try-catch 结构相同；但不论程序在执行过程中是否发生异常，finally 代码段总是会被执行。

下面的程序 P9_3 由用户依次输入一组数值，并输出这些数的总和及平均值：

```
//程序 P9_3
using System;
namespace P9_3
```

```csharp
{
    class Program
    {
        static void Main()
        {
            int m = -1, n = 0;
            double sum = 0;
            Console.WriteLine("请依次输入一组数值,输入 END 结束:");
            while (true)
            {
                try
                {
                    string s = Console.ReadLine();
                    if (s.ToUpper() == "END")
                        break;
                    sum += double.Parse(s);
                    n++;
                }
                catch (FormatException)
                {
                    Console.Write("输入的格式不正确,请重新输入:");
                }
                finally
                {
                    m++;
                }
            }
            Console.WriteLine("您总共输入了{0}次,其中正确输入{1}次", m, n);
            Console.WriteLine("数组之和为{0},平均值为{1}", sum, sum / n);
        }
    }
}
```

该程序通过 try-catch-finally 结构来处理输入格式错误所引发的异常,其中变量 m 记录了总的输入次数,而 n 记录了输入的数值个数。也就是说,只有输入格式正确,语句 n++ 才会被执行;而无论输入正确与否,finally 代码段中的 m++ 都会被执行。程序 P9_3 的输出结果示例如图 9-3 所示。

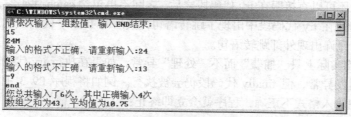

图 9-3 程序 P9_3 的运行结果示例

注意上面程序中 m 的初始值是-1,这是因为结束标记 END 不记入总的输入次数。而当用户输入 END 时,程序将执行其中的 break 语句,但此时并不是立即跳出外围的 while 循环,而是执行完 finally 代码段后才跳出循环。如果不使用 finally 代码段,而是将语句 m++直接放在 catch 代码段之后,用户输入 END 后就不会执行该语句,那么 m 的初始值就应当设为 0。

这说明 finally 语句具有强制执行的特性,即使是在前面的 try 代码段或 catch 代码段中遇到 break、continue、goto 等转移语句,finally 代码段也不会被跳过。通常可将清理资源、保存数据等工作放在 finally 代码段中完成。

程序 P9_3 中，即使直接输入 END（此时 n=0）也不会引发异常，这是因为 C#的实数除法中除以 0 不会引发异常：0 除以 0 得到的结果是非数字（NaN），而其他数除以 0 的结果是无穷大（NegativeInfinity 或 PositiveInfinity）。

C#还规定：在 finally 代码段中不允许使用 return 语句；如果在 finally 代码段中使用了 break、continue 或 goto 语句，那么该转移语句的目的地也必须在此代码段中。也就是说，finally 代码段中的跳转语句不能跳出代码段之外。例如不能在程序 P9_3 中使用如下的 finally 代码段来控制总的输入次数，因为其 break 语句将跳出外围的 while 循环：

```
while (true)
{
    try
    {
        string s = Console.ReadLine();
        if (s.ToUpper() == "END")
            break;
        sum += double.Parse(s);
        n++;
    }
    catch (FormatException)
    {
        Console.Write("输入的格式不正确，请重新输入:");
    }
    finally
    {
        if (m++ == 50)
        {
            Console.WriteLine("输入次数超过限制");
            break;  //错误：控制不能离开 finally 代码段！
        }
    }
}
```

9.2.3 try-finally 结构

finally 代码段还可以直接跟在 try 代码段之后，之间不再包含 catch 代码段，这就是 try-finally 结构。类似的，即使在 try 代码段中出现了跳转语句，其后的 finally 代码段也总是会被执行，而 finally 代码段中不允许出现外部跳转语句。

try-finally 结构实际上只"捕获"而不"处理"异常，如果在执行 try 代码段的过程中发生了错误，程序就会引发异常，但 finally 代码段仍会被执行。例如将程序 P9_3 中的循环代码改为如下内容，只要用户输入格式不正确，程序就会立即终止：

```
try
{
    string s = Console.ReadLine();
    if (s.ToUpper() == "END")
        break;
    sum += double.Parse(s);
    n++;
}
finally
{
    m++;
}
```

9.3 异常的捕获和传播

9.3.1 传播过程

程序引发异常后，程序的控制权将在异常处理结构中转移，直至找到一个能够处理该异常的 catch 语句，否则程序终止运行，这个过程叫做异常传播，其步骤为：

（1）如果错误代码所在的异常处理结构中包含能够处理该异常的 catch 语句，那么程序控制权就转移给第一个这样的 catch 语句，异常传播结束。

（2）如果没有找到能够处理该异常的 catch 语句，则程序通过当前的异常处理结构（如果存在 finally 代码段则执行它）。

（3）如果程序到达更外层的一个异常处理结构，则转到第（1）步。

（4）如果异常在当前方法中没有得到处理，则当前方法的执行被终止；若当前方法是程序主方法，那么整个程序结束运行。

（5）否则，程序控制权转移给调用当前方法的代码，重复第（1）步。

例如在程序 P9_4 中，Main 方法调用了 ReadRecip 方法，ReadRecip 方法又调用了 ReadInt 方法：

```
//程序 P9_4
using System;
namespace P9_4
{
    class Program
    {
        static void Main()
        {
            try
            {
                int y = ReadRecip();
                Console.WriteLine(100 / (10 - y));
            }
            catch(Exception ex)
            {
                Console.WriteLine("Main 方法中发生异常:{0}", ex.GetType());
            }
        }

        static int ReadRecip()
        {
            try
            {
                int x = ReadInt();
                return 100 / x;
            }
            catch(DivideByZeroException ex)
            {
                Console.WriteLine("ReadRecip 方法中发生异常:{0}", ex.GetType());
                return 0;
            }
        }

        static int ReadInt()
```

```
        {
            Console.Write("请输入一个整数:");
            return int.Parse(Console.ReadLine());
        }
    }
}
```

在运行程序 P9_4 时，如果用户输入的格式不正确，那么 ReadInt 方法的第二行代码将引发 FormatException 异常；但由于 ReadInt 方法没有进行异常处理，那么异常就传播到 ReadRecip 方法中，而 ReadRecip 方法无法捕获 FormatException 异常，异常将会传播到 Main 方法中并被捕获，那么程序的输出为 "Main 方法中发生异常: System.FormatException"。

如果用户的输入为 0，ReadInt 方法没有错误，ReadRecip 方法的除法语句将引发异常，被其后的 catch 语句捕获并处理，该异常将不会继续传播，所以此时程序的第一行将输出 "ReadRecip 方法中发生异常: System.DivideByZeroException"，而后第二行直接输出 10。

当用户输入的数字为 10，那么 ReadInt 方法和 ReadRecip 方法都没有错误，程序的输出将是 "Main 方法中发生异常: System.DivideByZeroException"。

图 9-4 给出了该程序进行不同的异常传播后得到的不同结果:

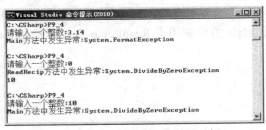

图 9-4　程序 P9_4 的运行结果示例

9.3.2　Exception 和异常信息

在程序 P9_4 中，Main 方法中的 catch 语句指定的是 Exception 异常类型，但在格式错误时异常对象 ex 的 GetType 方法将得到 FormatException，而在除以 0 时 ex 的 GetType 方法将得到 DivideByZeroException。也就是说，实际捕获到的异常对象类型总是最具体的异常类型。

Exception 类是其他所有异常的基类，通过该类型的下列属性能够访问异常对象的基本信息:
- Message——描述异常信息的字符串。
- Source——引发异常的程序或对象的名称。
- StackTrace——对异常传播的方法调用堆栈的描述。
- TargetSite——引发异常的方法。

在程序 P9_5 中，只有 Main 方法提供了异常处理结构，并通过 WriteExceptionInfo 方法来输出异常的详细信息:

```
//程序 P9_5
using System;
namespace P9_5
{
    class Program
    {
        static void Main()
        {
            try
            {
```

```
            Console.WriteLine(100 / (10 - ReadRecip()));
        }
        catch (Exception ex)
        {
            WriteExceptionInfo(ex);
        }
    }

    static int ReadRecip()
    {
        return 100 / ReadInt();
    }
    static int ReadInt()
    {
        Console.Write("请输入一个整数:");
        return int.Parse(Console.ReadLine());
    }

    static void WriteExceptionInfo(Exception ex)
    {
        Console.WriteLine("异常类型:{0}", ex.GetType());
        Console.WriteLine("引发程序:{0}", ex.Source);
        Console.WriteLine("引发方法:{0}", ex.TargetSite);
        Console.WriteLine("异常信息:{0}", ex.Message);
        Console.WriteLine("调用堆栈:\n{0}", ex.StackTrace);
    }
}
```

运行程序 P9_5，当用户输入 0 后输出结果如下（它表示异常由当前程序的 ReadRecip 方法引发，并传播到主方法 Main 中）：

异常类型:System.DivideByZeroException

引发程序:P9_5

引发方法:Int32 ReadRecip()

异常信息:试图除以零。

调用堆栈:

在 P9_5.Program.ReadRecip() 位置 D:\Prjs\CSOO\Ch7\P9_5\P9_5.cs:行号 21

在 P9_5.Program.Main() 位置 D:\Prjs\CSOO\Ch7\P9_5\P9_5.cs:行号 11

而当用户输入格式不正确时，程序的输出结果如下（这是因为异常来源于系统程序集 mscorlib 中定义的 StringToNumber 方法（int.Parse 方法正是调用该方法来解析字符串的），并且经过 ReadInt 和 ReadRecip 方法一直传播到主方法 Main 中）：

异常类型:System.FormatException

引发程序:mscorlib

引发方法:Void StringToNumber(System.String, System.Globalization.NumberStyles, NumberBuffer ByRef, System.Globalization.NumberFormatInfo, Boolean)

异常信息:输入字符串的格式不正确。

调用堆栈:

在 System.Number.StringToNumber(String str, NumberStyles options, NumberBuffer& number, NumberFormatInfo info, Boolean parseDecimal)

在 System.Number.ParseInt32(String s, NumberStyles style, NumberFormatInfo info)

在 System.Int32.Parse(String s)

在 P9_5.Program.ReadInt() 位置 D:\Prjs\CSOO\Ch7\P9_5\P9_5.cs:行号 27

在 P9_5.Program.ReadRecip() 位置 D:\Prjs\CSOO\Ch7\P9_5\P9_5.cs:行号 21

在 P9_5.Program.Main() 位置 D:\Prjs\CSOO\Ch7\P9_5\P9_5.cs:行号 11

9.3.3 异常层次结构

Exception 有两个的常用的派生类：ApplicationException 和 SystemException。前者表示开发人员在程序中所引发的异常，后者表示系统所引发的的系统异常。.NET 类库中又定义了 SystemException 的一系列派生类，图 9-5 给出了其中一些常用异常的继承层次结构。

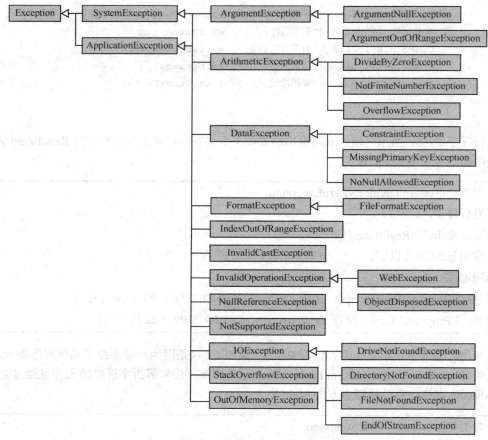

图 9-5 .NET 类库中常用异常的继承结构

1. 参数和格式异常

ArgumentException 类和 FormatException 类都是由于方法参数错误所引发的异常。其中 FormatException 表示参数格式错误；ArgumentException 表示参数无效，其 ParamName 属性表示出错的参数名称，它的派生类 ArgumentNullException 表示将空值 null 传递给了方法参数，而

ArgumentOutOfRangeException 表示参数值超出了预定范围。

例如使用 File.Open 方法打开一个文件时，如果表示文件名的字符串参数值为 null，那么就会引发 ArgumentNullException 异常；而在创建一个 DateTime 变量时，如果指定的月份超过了 12，就会引发 ArgumentOutOfRangeException 异常：

```
string s1 = null;
File.Open(s1, FileMode.Open);; //引发 ArgumentNullException 异常
DateTime dt = new DateTime(2008,13,1); //ArgumentOutOfRangeException 异常
```

2. 与算术运算有关的异常

ArithmeticException 类表示与算术运算有关的异常，其派生类 DivideByZeroException 表示除以零所引发的异常，NotFiniteNumberException 表示浮点数运算中出现无穷大或非数值时所引发的异常，OverflowException 则表示运算溢出时所引发的异常。

数值类型的变量在超出取值范围都会发生溢出，例如 byte 类型的取值范围为 0～255，那么下面代码中的后两句就会导致溢出：

```
byte b1 = 0, b2 = 255;
b1--;
b2++;
```

如果采用默认选项来编译程序，上述语句并不会引发异常，溢出之后的结果是 b1 的值为 255，而 b2 的值为 0；而如果编译时指定 "/check+" 选项，程序就会在溢出时引发异常。

3. 与对象操作有关的异常

InvalidOperationException 是指无效操作所引发的异常，它主要表示对象的当前状态不支持某项操作。特别的，如果试图操作一个已被释放（析构）的对象，那么将引发其派生异常 ObjectDisposedException。例如试图对一个已关闭的文件流进行读写就会引发该异常：

```
FileStream fs1 = File.Create("C:\\a.txt");
BinaryWriter bw1 = new BinaryWriter(fs1);
bw1.Write('a');
fs1.Close();
bw1.Write('b'); //引发 ObjectDisposedException 异常
```

4. 与文件操作有关的异常

在进行文件读写时可能发生各种异常情况，如文件不存在、磁盘物理故障等。.NET 类库中 System.IO 命名空间下的 IOException 表示文件 I/O 操作所引发的基本异常，其派生类 DirveNotFoundException、DirectoryNotFoundException 和 FileNotFoundException 分别表示目标磁盘、目录和文件不存在而引发的异常，EndOfStreamException 则表示读写位置超过了文件流的末尾而引发的异常。合理捕获和处理这些异常能够有效地提高文件 I/O 操作的安全性。

5. 与数据访问有关的异常

数据库访问也是异常可能性较多的一类操作。System.Data 命名空间下的 DataException 表示与数据访问组件相关的异常，比如其派生类 ConstraintException 就表示违反数据库中的约束所引发的异常。不过 DataException 并不代表所有的数据访问异常，更多的数据异常是和特定的数据库平台而非.NET 类库相关，比如表示 SQL Server 数据库异常的 SqlException 和表示 Oracle 数据库异常的 OracleException。在第 15 章中将会看到这些异常的有关用法。

6. 其他常见异常

下列异常在程序中也可能经常会遇到：

- IndexOutOfRangeException——数组越界所引发的异常。
- InvalidCastException——类型转换失败所引发的异常，比如将一个基类对象强制转换为派生类对象。

- NullReferenceException——空对象引用所引发的异常,比如调用空对象的成员方法。
- NotSupportedException——调用的方法不受支持所引发的异常,比如对只读的文件流进行写入操作。
- OutOfMemoryException——内存不足所引发的异常。
- StackOverflowException——操作系统堆栈溢出所引发的异常,比如方法的递归调用发生死循环。

提示

对于一般的引用类型,如果只是声明了其变量而未赋值,那么访问对象成员的代码将不能通过编译;而如果将 null 值赋予其变量,那么访问对象成员的代码能够通过编译,直到运行时才会发生异常。

9.4 自定义异常

9.4.1 主动引发异常

try-catch 和 try-catch-finally 这样的异常处理结构是为了避免在发生异常后程序终止。而在另一些情况下,程序代码需要主动引发异常,以便向用户报告错误,或是阻止对象和方法的不合理使用。此时异常代码的调用者就应当对异常进行捕获和处理。

考虑下面的 Student 类的定义,它对构造函数的参数没有什么限制,那么其他开发人员就可以在创建对象时指定任意的姓名和年级,这通常不是 Student 类的开发者所期望的。

```csharp
public class Student
{
    private string name;
    public string Name
    {
        get { return name; }
    }
    private int grade;
    public int Grade
    {
        get { return grade; }set { grade = value; }
    }
    public Student(string name, int grade)
    {
        this.name = name;
        this.grade = grade;
    }
}
```

一种有效的解决途径就是在试图创建不合理的对象时引发异常,例如可将 Student 的构造函数改为如下内容:

```csharp
public Student(string name, int grade)
{
    if (name == null || name.Length == 0 ||grade <= 0 || grade > 6)
        throw new ArgumentException();
    this.name = name;
    this.grade = grade;
}
```

其中的 throw 语句用于引发一个异常，异常对象应跟在 throw 关键字之后。程序执行到这里就会引发并传播异常，之后的语句不再被执行；如果异常没有被捕获，整个程序就会终止。这就能防止非法的 Student 对象被创建。

在自定义类型的方法代码中使用 throw 语句，其目的主要是为了提醒方法的调用者要遵循有关约定，以便更好地排除错误或处理异常，从而提高程序的可靠性。显然，throw 语句通常不应出现在程序的主方法里。

在创建 Exception 或其派生对象时，通常还可以在构造函数中指定专门的异常消息（即 Exception 的 Message 属性值），以便向类型或方法的使用者进行更详细的说明。例如 Student 构造函数中引发异常的语句可作进一步的细化：

```
public Student(string name, int grade)
{
   if (name == null)
      throw new ArgumentNullException();
   else if (name.Length == 0)
      throw new ArgumentException("姓名不能为空");
   if (grade <= 0 || grade > 6)
      throw new ArgumentOutOfRangeException("年龄必须在 1～6 之间");
   this.name = name;
   this.grade = grade;
}
```

9.4.2 自定义异常类型

在程序中引发异常时，其原则是首先在.NET 类库已定义的异常类中选择最合适的一个。例如对于素数类 Prime，如果用户在创建对象是指定了一个非素数，那么可以引发 ArgumentOutOfRangeException 异常；而对于自定义类型的 Parse 方法，在字符串格式不正确时通常都应引发 FormatException 异常（请参见程序 P4_7 中的示例代码）。

如果已有的异常类型都不合适，这时可以考虑自定义一个异常类，并在其中维护特定的异常信息。例如在学校的学籍管理系统中，当学生使用不正确的学号进行选课、办理图书证、申请奖学金等操作时，ArgumentException 可能不足以表达所需的信息，那么可以自定义一个 InvalidIDException 类，并在其属性中指出学号格式错误、无此学生，或是学生已毕业等原因。

为了明确用途，自定义的异常类通常应从 Exception 或 ApplicationException 派生（除非是开发.NET Framework 的扩展类库时才应考虑从 SystemException 派生），并以 "Exception" 作为类名的结尾。

下面的类库程序 P9_6 定义了一个整数数列类 NumberSequence，其整数字段 _a0 表示数列的首项，委托字段 _recur 则表示相邻项之间的递推关系；其方法 GetNumber 和 GetNumbers 分别用于计算数列的第 n 项和前 n 项。为了描述数列计算中可能发生的错误，程序还自定义了一个异常类型 NumberSequenceException，其属性 Item 表示错误发生在数列的第几项，它还重载了基类的常用构造函数以方便使用。

```
//程序 P9_6
using System;
namespace P9_6
{
    public class NumberSequence
    {
        public delegate int Recur(int a);

        protected int _a0 = -1;
```

```csharp
        public int a0
        {
            get { return _a0; }
        }
        protected Recur _recur;
        public NumberSequence(int a0, Recur recur)
        {
            _a0 = a0;
            _recur = recur;
        }

        public virtual int GetNumber(int n)
        {
            int a = _a0;
            for (int i = 1; i < n; i++)
                a = _recur(a);
            return a;
        }

        public virtual int[] GetNumbers(int n)
        {
            int[] numbers = new int[n];
            numbers[0] = _a0;
            for (int i = 1; i < n; i++)
                numbers[i] = _recur(numbers[i - 1]);
            return numbers;
        }
    }
    public class NumberSequenceException : ApplicationException
    {
        private int _item = -1;
        public int Item
        {
            get { return _item; }
        }

        public NumberSequenceException(int item)
            : base(string.Format("数列第{0}项发生异常", item))
        {
            _item = item;
        }

        public NumberSequenceException(string msg, int item)
            : base(msg)
        {
            _item = item;
        }
    }
}
```

程序 P9_7 则定义了 NumberSequence 的两个派生类：递增数列 IncrementalNumberSequence 和递减数列 DecrementalNumberSequence。在进行数列项计算的过程中，如果发现递增或递减关系没有得到满足，程序就会引发异常（请注意在程序中添加对 P9_6 的引用）。

```csharp
//程序 P9_7
using System;
using P9_6;
namespace P9_7
```

```csharp
    {
        class Program
        {
            static void Main()
            {
                try
                {
                    NumberSequence ns1 = new NumberSequence(1, delegate(int a) { return a + 3; });
                    Console.Write("数列 ns1 前 10 项为: ");
                    foreach (int i in ns1.GetNumbers(10)){
                        Console.Write(i);
                        Console.Write(' ');
                    }
                    NumberSequence ns2 = new IncrementalNumberSequence(1, delegate(int a) { return a + 3; });
                    Console.Write("\n数列 ns2 前 10 项为: ");
                    foreach (int i in ns2.GetNumbers(10)){
                        Console.Write(i);
                        Console.Write(' ');
                    }
                    NumberSequence ns3 = new DecrementalNumberSequence(1, delegate(int a) { return a + 3; });
                    Console.Write("\n数列 ns3 前 10 项为: ");
                    foreach (int i in ns3.GetNumbers(10)){
                        Console.Write(i);
                        Console.Write(' ');
                    }
                }
                catch (NumberSequenceException ex)
                {
                    Console.WriteLine("发生数列异常: {0}", ex.Message);
                }
                catch (Exception ex)
                {
                    Console.WriteLine("发生异常: {0}", ex.Message);
                }
            }
        }

        public class IncrementalNumberSequence : NumberSequence
        {
            public IncrementalNumberSequence(int a0, Recur recur)
                : base(a0, recur)
            { }

            public override int GetNumber(int n)
            {
                int a = _a0;
                for (int i = 1, b; i < n; i++){
                    b = _recur(a);
                    if (b < a)
                        throw new NumberSequenceException("数列项非递增", i);
                    a = b;
                }
                return a;
            }

            public override int[] GetNumbers(int n)
```

```csharp
        {
            int[] numbers = new int[n];
            numbers[0] = _a0;
            for (int i = 1; i < n; i++){
                numbers[i] = _recur(numbers[i - 1]);
                if (numbers[i] < numbers[i - 1])
                    throw new NumberSequenceException("数列项非递增", i);
            }
            return numbers;
        }
    }

    public class DecrementalNumberSequence : NumberSequence
    {
        public DecrementalNumberSequence(int a0, Recur recur)
         : base(a0, recur)
        { }

        public override int GetNumber(int n)
        {
            int a = _a0;
            for (int i = 1, b; i < n; i++){
                b = _recur(a);
                if (b > a)
                    throw new NumberSequenceException("数列项非递减", i);
                a = b;
            }
            return a;
        }

        public override int[] GetNumbers(int n)
        {
            int[] numbers = new int[n];
            numbers[0] = _a0;
            for (int i = 1; i < n; i++){
            numbers[i] = _recur(numbers[i - 1]);
            if (numbers[i] > numbers[i - 1])
                throw new NumberSequenceException("数列项非递减", i);
            }
            return numbers;
        }
    }
}
```

程序主方法中创建了 3 个不同类型的数列对象，并对数列的使用进行了异常捕获和处理。由于第 3 个数列对象不符合递减数列类型的要求，程序将引发异常，最后等到的输出结果如下：

> 数列 ns1 前 10 项为: 1 4 7 10 13 16 19 22 25 28
> 数列 ns2 前 10 项为: 1 4 7 10 13 16 19 22 25 28
> 数列 ns3 前 10 项为: 发生数列异常: 数列项非递减

9.5 使用异常的指导原则

异常处理是提高程序可靠性的重要手段，但它也会增加编程的工作量，同时造成程序性能下降。例如在程序不出错的情况下，常规代码段的运行效率明显要高于将其放在 try 代码段中的运行效率。因此，开发人员需要根据具体情况来确定异常的使用时机和范围。下面总结了异常处理的一些实践指导原则。

- **不要掩盖错误**：最不可取的做法是把所有可能出错的代码都放进异常处理结构，然后假装什么错误也没有发生。这虽然能够避免程序意外终止，但会使程序丧失应有的功能，因为发生异常的语句之后的程序代码都不会被执行。
- **事先避免异常**：首先应考虑防止错误的发生，而不是完全依赖于错误发生后的异常处理。例如在判断整数 b 是否为 a 的倍数时，逻辑表达式 "(a != 0) && (b / a == 0)" 就借助了 "短路效应" 来避免除以 0 的异常；再如在 Windows 等图形用户界面中，可通过控件设置来限制用户输入的范围，而不是等用户输入错误之后再进行处理。
- **限定异常范围**：如果需要，应仅将可能出错的语句放在 try 代码段中，以尽量减少性能损失。比如变量的定义和赋值语句通常不会出错。
- **精确异常类型**：在异常处理结构中，捕获的异常类型越具体，越有利于错误的判断和处理，而且对性能的损害也越小。例如在处理方法参数时捕获 ArgumentException 异常，在读写文件时捕获 IOException 异常，而不是不加思索地都使用基类异常 Exception。同样，主动引发异常时也应尽量使用具体的异常类型。
- **控制异常结构**：尽量不要在异常处理结构中嵌套异常处理结构，也尽量避免在 catch 代码段或 finally 代码段中又使用 throw 语句引发新的异常。
- **简化异常信息**：Exception 及其派生类型能够提供较为丰富的异常信息，但不应把这些信息的内容一股脑地显示给用户。特别是对于面向普通用户的应用程序，应尽量使用非专业性的、通俗易懂的文字来进行说明；更为详细的信息则可以根据需要保存到文件或系统日志中。
- **在高层次处理异常**：尽量在方法调用链的高层来捕获和处理异常，这能够有效地减少异常处理的代码量，并节约因异常传播而消耗的系统资源。例如，某图书馆程序的主方法调用了借书方法，借书方法又调用了读取图书信息的方法，那么可以只在主方法中捕获和处理读取图书信息失败的异常。
- **良性异常处理**：这里所说的 "良性" 是指在处理异常的过程中尽可能地解决问题、恢复程序应用的功能，而不仅仅是简单地报告错误。例如，在网络或数据库连接中断的异常处理中，程序可以尝试重新建立连接，或者到脱机缓存中寻找所需的内容，用户甚至没有察觉异常的发生，这才能真正体现程序的友好性。
- **限制引发异常**：某些方法可以进行功能扩展，从而减少或避免使用 throw 语句来抛出异常。TryParse 方法使用返回值来指示操作是否成功就是一个很好的范例。
- **自定义异常**：优先考虑引发已定义的异常类型，只在需要时才定义自己的异常类型，而且自定义异常的继承层次不宜过深（从 Exception 开始通常不超过 5 级）。
- **区分不同版本的程序异常**：软件的早期开发和测试需要暴露而不是屏蔽错误，因此通常是先编写常规代码，通过基本功能测试后再逐步增加异常处理语句。此外，测试过程中程序的异常信息应尽可能丰富，而发布的程序则需要加以控制，特别是要防止在异常中泄露某些安全信息（如用户名和密码）。

9.6 案例研究——旅行社管理系统中的异常处理

在旅行社管理系统的开发中，对异常处理的基本方式是：在业务类中对一些无效或非法的操作引发异常,在用户界面部分捕获并处理有关异常。

9.6.1 文件 I/O 异常处理

首先考虑与文件读取操作有关的异常处理。在主窗体启动时,从文件读入系统用户信息和旅行社基本信息都可能引发异常,不过处理的方式不同:如果读取用户信息失败,程序将直接退出;而如果读取旅行社信息失败,程序则允许输入这些信息。那么 Form1 的 Load 事件处理方法中可使用如下代码来捕获和处理异常:

```csharp
private void Form1_Load(object sender, EventArgs e)
{
    try //系统登录
    {
        _users = User.GetAll("users.bak");
    }
    catch (Exception ex)
    {
        MessageBox.Show("初始信息载入失败:" + ex.Message, "错误", MessageBoxButtons.OK, MessageBoxIcon.Error);
        this.Close();
        return;
    }
    LoginForm frm1 = new LoginForm(_users);
    if (frm1.ShowDialog() != DialogResult.OK) {
        this.Close();
        return;
    }
    _current = frm1.Current;
    menuSystemBasic.Enabled = menuSystemUsers.Enabled = (_current is Manager);
    try //载入基础信息
    {
        if (File.Exists("agency.txt"))
            _agency = TravelAgency.Load("agency.txt");
    }
    catch
    {
        _agency = null;
    }
    if (_agency == null && MessageBox.Show("旅行社基础信息为空,是否现在输入?", "提示", MessageBoxButtons.YesNo, MessageBoxIcon.Question) == DialogResult.Yes)
    {
        BasicForm frm2 = new BasicForm(null);
        if (frm2.ShowDialog() == DialogResult.OK)
            _agency = frm2.Agency;
    }
}
```

而在 UserForm 保存系统用户信息的代码中,以及在 BasicForm 保存旅行社基本信息的代码中,也都应加入类似的异常处理代码:

```csharp
try
{
    User.SaveAll(_users, "users.bak");
    MessageBox.Show("保存成功", "提示", MessageBoxButtons.OK, MessageBoxIcon.
```

```csharp
Information);
        }
        catch (Exception ex)
        {
            MessageBox.Show("保存失败:" + ex.Message, "提示", MessageBoxButtons.OK, MessageBoxIcon.Warning);
        }

        try
        {
            _agency.Save("agency.txt");
            MessageBox.Show("保存成功", "提示", MessageBoxButtons.OK, MessageBoxIcon.Information);
        }
        catch (Exception ex)
        {
            MessageBox.Show("保存失败:" + ex.Message, "提示", MessageBoxButtons.OK, MessageBoxIcon.Warning);
        }
```

9.6.2 旅行社业务异常

旅行社业务对象的很多操作都可能有无效或非法的情况，如游客向已经满员的旅行团报名、给旅行团安排不在岗的导游、休假的职工请假等，这类无效操作通常应抛出 InvalidOperationException 类型的异常。仔细分析可以发现，其中大部分异常都是由于对象所处的状态不满足操作要求，而相关状态都是通过枚举类型来定义的。那么只要记录下枚举类型和枚举值，就能够以一种统一的方式来描述异常信息。下面就自定义了一个异常类型来描述这类异常：

```csharp
// 旅行社对象处于无效状态所引发的异常
public class BadStateException : InvalidOperationException
{
    private Type _stateType;
    public Type StateType   //状态类型
    {
        get { return _stateType; }
    }

    private int _state;
    public int State   //状态值
    {
        get { return _state; }
    }

    private object _object;
    public object Object   //处于无效状态的对象
    {
        get { return _object; }
    }

    public BadStateException(Type stateType, int state, object o)
        : base(string.Format("处于{0}状态的{1}不支持该操作", Enum.GetName(stateType, state), o.GetType().Name))
    {
        _object = o;
        _stateType = stateType;
        _state = state;
    }
}
```

其中 Enum 所有枚举类型的基类，其 GetName 方法可以获得指定枚举值的字符串描述。考虑 Staff 类的 Apply 方法，其执行代码一开始可加入下面的语句：

```
if (State == StaffState.休假 || State == StaffState.离职)
    throw new BadStateException(typeof(StaffState), (int)State, this);
```

那么当某个处于休假状态的 Staff 对象执行 Apply 操作时，程序就会抛出异常，且异常信息为"处于休假状态的 Staff 不支持该操作"。

类似地，Director 类的 Assign 方法也应对所安排的职员状态进行检查，并在状态无效时引发异常：

```
public void Assign(Package p, Agent a)    //安排业务员
{
    if (a.State != StaffState.试用&& a.State != StaffState.在岗)
        throw new BadStateException(typeof(StaffState), (int)a.State, a);
    p.Agent = a;
}

public void Assign(Tour t, Guide g)    //安排导游
{
    if (g.State != StaffState.试用&& g.State != StaffState.在岗)
        throw new BadStateException(typeof(StaffState), (int)g.State, g);
    t.Guide = g;
}
```

再考虑与旅行团有关的情况，游客报名时旅行团应当处于报名状态，此外，系统还要求报名的游客不能是儿童，那么 Customer 类的 Enroll 方法开始处应加入下面的语句：

```
if (t.State != TourState.报名 )
    throw new BadStateException(typeof(TourState), (int)t.State, t);
if (_type == CustomerType.Child)
    throw new BadStateException(typeof(CustomerType), (int)_type, this);
```

旅行团只有在出发之前才能取消报名，因此 Customer.Cancel 方法开头的一行语句也应改为如下内容：

```
Tour t = enr.Tour;
if (t.State != TourState.报名 || t.State != TourState.满员)
    throw new BadStateException(typeof(TourState), (int)t.State, t);
if (enr.Applier != this)
    throw new InvalidOperationException("只有报名者本人才能取消");
```

类似地，业务员在接受和拒绝报名时，也要保证旅行团或报名的状态合法：

```
public void Accept(Enrollment enr)    //接受报名
{
    if (enr.State != EnrollmentState.Unconfirmed)
        throw new BadStateException(typeof(EnrollmentState), (int)enr.State, enr);
    Tour t = enr.Tour;
    if (t.State != TourState.报名)
        throw new BadStateException(typeof(TourState), (int)t.State, t);
    //…
}

public void Reject(Enrollment enr, string reason)    //拒绝报名
{
    if (enr.State != EnrollmentState.Unconfirmed)
        throw new BadStateException(typeof(EnrollmentState), (int)enr.State, enr);
    //…
}
```

9.7 小结

C#程序的异常处理模型有两方面：(1) 异常处理结构，包括 try-catch 语句、try-catch-finally 语句、try-finally 语句和 throw 语句；(2) 以 Exception 为基类的异常类型集合，在 catch 语句中使用不同的类型能够捕获指定的异常。基于该模型能够对异常进行高效的处理，从而提高程序的可靠性和安全性；但要注意滥用异常也会降低程序的性能和可维护性。

9.8 习题

1. 简述 try-catch、try-catch-finally 和 try-finally 语句有哪些不同之处。
2. 写出下面程序的输出结果：

```
class Program
{
    static void Main()
    {
        try{
            FA();
        }
        catch (ArithmeticException){
            Console.WriteLine("发生算术异常");
        }
        catch (ArgumentException){
            Console.WriteLine("发生参数异常");
        }
        Catch{
            Console.WriteLine("发生异常");
        }
    }
    static void FA()
    {
        try {
            FB();
        }
        Catch{
            throw new ArgumentNullException();
        }
        finally {
            throw new OverflowException();
        }
    }
    static void FB()
    {
        throw new FormatException();
    }
}
```

3. 在程序中进行数学运算时，如何在溢出时引发程序异常？在哪些情况下应该引发异常？
4. 使用 File.Open 方法打开一个文件时，可能发生哪些异常？在程序当中又应如何处理？
5. 在程序 P9_7 的基础上再定义一个正整数数列 PositiveNumberSequence，以及 NumberSequenceException 的派生类 NegativeNumberException，并在数列计算出现负数项时引发该异常。
6. 编写程序，模拟旅行社对象的业务操作，捕获并处理发生的 BadStateException 异常。

第10章 基于接口的程序设计

随着软件的规模、复杂度及版本数量的不断增长，软件对象之间的交互越来越困难，软件接口技术也越来越为人们所重视，目前已成为面向对象技术的一个重要组成部分。本章将介绍 C# 语言中接口的概念，并说明接口在对象系统设计，特别是继承设计中的用途。

10.1 接口的定义和使用

10.1.1 接口的定义

从第 5 章我们了解到：抽象类不能被实例化，但它仍可以有自己的成员字段和非抽象方法。接口则是比抽象类更为"抽象"的一种数据类型，它所描述的是功能契约，即"能够提供什么服务"，而不考虑与实现有关的任何因素。

C#中使用关键字 interface 来定义接口类型，其成员只能是一般方法、属性和索引函数，而不能有字段和构造函数。例如下面的代码就定义了一个"可取款"接口 IDrawable，其中包含了一个接口方法 Draw：

```
public interface IDrawable
{
   bool Draw(decimal money);
}
```

接口和抽象类都不能创建实例，接口方法和抽象方法也一样没有执行代码；此外，接口方法不能是静态的，也不能使用任何访问限制修饰符。从某种意义上说，可以把接口看成是只包含抽象方法的抽象类，其中每个接口方法都表示一项"服务契约"。

接口之间也可以继承。在下面代码中，"可存取"接口 IDepositDraw 就从 IDrawable 派生，那么它也就继承了 IDrawable 的 Draw 方法：

```
public interface IDepositDraw : IDrawable
{
    void Deposit(decimal money);
}
```

提示

> 子接口可以包含与父接口相同的方法，但此时子接口的方法必须使用关键字 new 来隐藏父接口的方法，而不能使用 override 修饰符来重载父接口的方法。

接口之间的继承同样存在传递性。例如接口 IB 从接口 IA 派生，而接口 IC 又从 IB 派生，那么它也就间接地从 IA 派生。不过接口不能从类派生。

10.1.2 接口的实现

如果一个类声明支持某个接口,它就必须履行该接口的契约,即支持该接口中定义的所有方法。这可以分为以下两种情况:

- 如果支持接口的类型是非抽象的,那么它必须支持接口中的所有方法,并为这些方法提供具体实现。
- 如果支持接口的类型是抽象类,那么它必须支持接口中的所有方法,且这些方法要么提供具体实现,要么是抽象的。

类对接口的支持声明和类的继承声明类似,有时也称之为类从接口继承。这种继承关系同样存在传递性:如果一个类声明支持某个接口,那么它的所有派生类也就自动支持该接口。例如下面定义的 BankAccount 类就声明支持 IDrawable 接口,并提供了对接口方法的实现:

```
public class BankAccount : IDrawable
{
    private decimal balance = 1000;

    public bool Draw(decimal money)
    {
        if (balance >= money){
            balance -= money;
            return true;
        }
        else
            return false;
    }
}
```

1. 隐式实现

在上面的代码中,BankAccount 是通过公有成员方法来实现所支持的接口方法,这叫做对接口方法的隐式实现。此时既可以通过类的实例来进行方法调用,也可以隐式转换为接口实例再进行方法调用,例如:

```
BankAccount account1 = new BankAccount();
account1.Draw(500);
IDrawable i1 = account1;
i1.Draw(300);
```

在进行隐式实现时,成员方法可以使用 abstract 或 virtual 修饰符(表示允许派生类重载),但不能使用 sealed 和 override 修饰符。

2. 显式实现

类对接口方法的实现还有另一种形式,即在方法名之前加上接口名,这叫做对接口方法的显式实现。以 BankAccount 类为例,其显式实现接口方法的代码如下:

```
bool IDrawable.Draw(decimal money)
{
    if (balance >= money){
        balance -= money;
        return true;
    }
    else
        return false;
}
```

此时方法不能使用任何修饰符,那么它实际上就是类的私有成员,因此不能通过类的实例来访问,而是需要转换为接口实例才能访问,例如:

```
IDrawable i1 = new BankAccount();
```

```
            i1.Draw(500);
```
　　显式实现明确指出了所实现的方法来自于哪一个接口，因此方法前的接口名必须是定义方法的原始接口。例如将 BankAccount 所支持的接口改为 IDepositDraw，那么其显式实现的两个接口方法就应当是 IDepositDraw.Deposit 和 IDrawable.Draw：

```
        public interface IDrawable
        {
            bool Draw(decimal money);
        }
        public interface IDepositDraw : IDrawable
        {
            void Deposit(decimal money);
        }
        public class BankAccount : IDepositDraw
        {
            private decimal balance = 0;
            void IDepositDraw.Deposit(decimal money)
            {
                balance += money;
            }
            bool IDrawable.Draw(decimal money) //不能写成 IDepositDraw.Draw
            {
                if (balance >= money){
                    balance -= money;
                    return true;
                }
                else
                    return false;
            }
        }
```

10.2　接口与多态

10.2.1　通过接口实现多态性

　　在本书 5.2 节中介绍了通过虚拟方法和重载方法来实现多态性。类似的，我们可以对声明为接口类型的变量调用接口方法，那么程序会根据实现变量的实际类型来实现其方法。

　　在下面的程序 P10_1 中，BankAccount、CreditCard 和 DebitCard 都支持"可支付"接口 IPayable，主程序 Program 的 Receive 方法用于接受付款：只要对象声明支持 IPayable 接口，就可通过 Pay 方法来要求其提供付款服务；至于付款对象的具体类型是什么、其内部的付款操作是如何实现的，客户方都不需要关心。

```
//程序 P10_1
using System;
namespace P10_1
{
    class Program
    {
        static void Main()
        {
            IPayable[] payers = new IPayable[4];
```

```csharp
            payers[0] = new BankAccount(3000);
            payers[1] = new BankAccount(5000);
            payers[2] = new CreditCard(5000);
            payers[3] = new DebitCard((BankAccount)payers[1]);
            foreach (IPayable payer in payers){
                Receive(payer, 1500);
                Receive(payer, 2000);
            }
        }

        static void Receive(IPayable payer, decimal money)
        {
            if (payer.Pay(money))
                Console.WriteLine("{0}成功付款{1}元", payer, money);
            else
                Console.WriteLine("{0}付款失败", payer);
        }
    }

    public interface IPayable
    {
        bool Pay(decimal money);
    }

    public class BankAccount : IPayable
    {
        protected decimal balance = 0;
        public decimal Balance
        {
            get { return balance; }
            set { balance = value; }
        }

        public BankAccount(decimal balance)
        {
            this.balance = balance;
        }

        public virtual bool Pay(decimal money)
        {
            if (balance >= money){
                balance -= money;
                return true;
            }
            else
                return false;
        }
    }

    public class CreditCard : BankAccount
    {
        private decimal credit;
        public decimal Credit
        {
            get { return credit; }
        }

        public CreditCard(decimal credit) : base(0)
        {
            this.credit = credit;
```

```csharp
        }
        public override bool Pay(decimal money)
        {
            if (money <= credit + balance){
                balance -= money;
                return true;
            }
            else
                return false;
        }
    }
    public class DebitCard : IPayable
    {
        private BankAccount account;
        public DebitCard(BankAccount account)
        {
            this.account = account;
        }
        public bool Pay(decimal money)
        {
            return account.Pay(money);
        }
    }
}
```

10.2.2 区分接口方法和对象方法

派生类可以隐藏基类中的方法，也可以重载基类中的方法，但二者不能同时存在。以第 5 章的程序 P5_4 为例，基类 Automobile 定义了虚拟的 Speak 方法，其派生类 Truck 又重载了该方法以实现多态性，那么无论是将一个 Truck 对象声明为 Automobile 类型还是 Truck 类型，调用其 Speak 方法的结果都是一样的；反之，如果希望根据声明类型来选择调用不同的方法，那么就只能隐藏基类成员，而牺牲多态性。

利用接口来实现多态性，就可以达到"鱼和熊掌兼得"的效果，其关键就在于接口方法的隐式实现和显式实现可以并存。例如 BankAccount 可以同时提供 IDrawable.Draw 方法和公有的 Draw 方法：

```csharp
public class BankAccount : IDrawable
{
    private decimal balance = 1000;
    bool IDrawable.Draw(decimal money)
    {
        if (balance >= money){
            balance -= money;
            return true;
        }
        else {
            Console.WriteLine("余额不足");
            return false;
        }
    }
    public bool Draw(decimal money)
    {
```

```
            if (balance - money>= 10){
                balance -= money;
                return true;
            }
            else {
                Console.WriteLine("余额不能少于10元");
                return false;
            }
        }
    }
```

在这种情况下，显式实现的方法才是对接口方法的真正实现，而隐式实现的方法可以看做是对原接口方法的覆盖。此时通过接口实例调用的是显式实现的方法，而通过对象实例调用的则是隐式实现的方法。

考虑这样一种情况：客户在使用账户进行消费时，如果出示VIP卡还可享受九折优惠；但使用信用卡消费时，无论是否为VIP客户都不享受优惠。下面的程序P10_2就定义了这样一个BankAccount类，它声明支持接口IPayable，并实现了两个版本的Pay方法，前者用于常规支付，后者能用于优惠支付：

```
//程序P10_2
using System;
namespace P10_2
{
    class Program
    {
        static void Main()
        {
            BankAccount a1 = new BankAccount(1000, true);
            IPayable p1 = a1;
            Console.WriteLine("p1付款1100: "+ (p1.Pay(1100)? "成功":"失败"));
            Console.WriteLine("a1付款1100: "+ (a1.Pay(1100)? "成功":"失败"));
            a1 = new CreditCard(1000);
            Console.WriteLine("a1付款1100: "+ (a1.Pay(1100)? "成功":"失败"));
        }
    }

    public interface IPayable
    {
        bool Pay(decimal money);
    }

    public class BankAccount : IPayable
    {
        private bool vip = false;
        protected decimal balance = 0;
        public BankAccount(decimal balance, bool vip)
        {
            this.balance = balance;
            this.vip = vip;
        }

        bool IPayable.Pay(decimal money)
        {
            if (balance >= money){
                balance -= money;
                return true;
            }
```

```csharp
        else
            return false;
    }
    public virtual bool Pay(decimal money)
    {
        IPayable p1 = (IPayable)this;
        if(!vip)
            return p1.Pay(money);
        else
            return p1.Pay(money * 0.9M);
    }
}
public class CreditCard : BankAccount
{
    private decimal credit;
    public CreditCard(decimal credit) : base(0, true)
    {
        this.credit = credit;
    }
    public override bool Pay(decimal money)
    {
        if (money <= credit + balance){
            balance -= money;
            return true;
        }
        else
            return false;
    }
}
```

该程序的输出结果如下:

p1 付款 1100: 失败
a1 付款 1100: 成功
a1 付款 1100: 失败

也就是说,在通过 IPayable 实例调用 Pay 方法时,实际执行的将是显式实现的接口方法;而通过 BankAccount 实例调用 Pay 方法时,实际执行的是对象的公有方法,这是根据对象的声明类型来选择要调用的方法。但 BankAccount 的虚拟方法和 CreditCard 的重载方法所实现的多态性不会受到影响。从中可以看到,引入了接口之后,就能够在程序中实现比类继承更为丰富的多态性。

不过对于程序 P10_2 中的 CreditCard 类,尽管它间接地从 BankAccount 继承了 IPayable 接口,其对象也可以显式转换为 IPayable 类型,但它不能显式实现 IPayable 的 Pay 方法,这是因为它的定义中并没有直接声明支持该接口。

10.3 接口和多继承

10.3.1 多继承概述

单继承是指一个类最多只能从一个基础类型直接派生,这样的继承层次可以用一棵树来描述。

例如在图 10-1（a）中，银行卡 BankCard 的派生类有一卡通 GeneralCard 和信用卡 CreditCard，而 VisaCard 又从 CreditCard 中派生。

单继承在表达能力上有一定的限制。例如并不是所有的银行卡都能在 POS 机上消费，那么消费功能的描述就需要在 BankCard 的派生类中重复多次。如果采用多继承技术，那么就可以将消费功能抽象在一个 Payer 类中，这样 GeneralCard/CreditCard 就能同时从 BankCard 和 Payer 中派生，如图 10-1（b）所示。

多继承的层次结构是一个有向图，这就会带来二义性的问题。考虑 Payer 还可以有一个支持外币消费的派生类 GlobalPayer，而 VisaCard 又同时从 CreditCard 和 GlobalPayer 中派生，这时两个基类中的同名方法就会发生冲突（VisaCard 中的 base.Pay 是 CreditCard.Pay 还是 GlobalPayer.Pay）。此外基类过多也会严重影响程序性能，因为派生类的对象所占的内存空间将是其所有基类对象的内存空间之和，派生类对象在创建时还会依次调用其所有基类的构造函数。不少专家认为，程序设计语言中多继承带来的问题远远大于其所带来的优越性。

C#语言规定一个类只能有一个直接基类，但可以同时支持多个接口，这能够有效弥补单继承在表达能力上的不足。而与类的多继承相比，C#基于接口的多继承是轻量级的：接口之间的继承只是"契约"继承，不需要提供实现；当一个类支持多个接口时，它也只是实现各接口的"契约"服务，而不需要处理多重构造函数和重载方法等复杂情况。例如可以把消费和外币消费的功能描述分别放在接口 IPayable 和其派生接口 IGlobalPayable 中，并由 CreditCard 和 VisaCard 各自实现所需的接口，如图 10-1（c）所示。

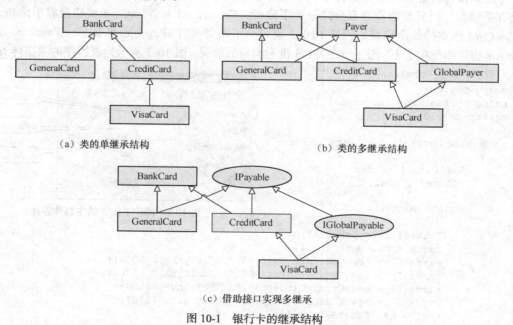

（a）类的单继承结构　　　　　　　　　（b）类的多继承结构

（c）借助接口实现多继承

图 10-1　银行卡的继承结构

10.3.2　基于接口的多继承

和类之间的单继承不同，一个接口可以有多个父接口，此时在继承声明中父接口之间通过逗号分隔（先后顺序不限）。例如下面定义的接口 IC 就有两个父接口 IA 和 IB，那么它也就继承了 IA 的 F 方法和 IB 的 G 方法：

```
public interface IA
{
```

```csharp
    void F();
}
public interface IB
{
    void G();
}
public interface IC : IA, IB
{
    void H();
}
```

一个类同样可以支持多个接口；如果同时存在类继承，那么继承声明中的基类应出现在所有接口之前。例如：

```csharp
public class CA : IA, IB
{
    void IA.F() { }
    void IB.G() { }
}
public class CC : CA, IC
{
    void IC.H() { }
}
```

考虑某银行可能发行多种银行卡，不同的银行卡有着不同的消费方式：GeneralCard 本地消费没有手续费，但异地消费有百分之一的手续费；CreditCard 和 VisaCard 消费没有手续费，其中 VisaCard 还支持外币消费。程序 P10_3 就定义了消费接口 IPayable 和 IGlobalPayable，并在 Windows 窗体中模拟了不同银行卡在 POS 机上消费的情况。图 10-2 所示为该程序的主窗体界面。

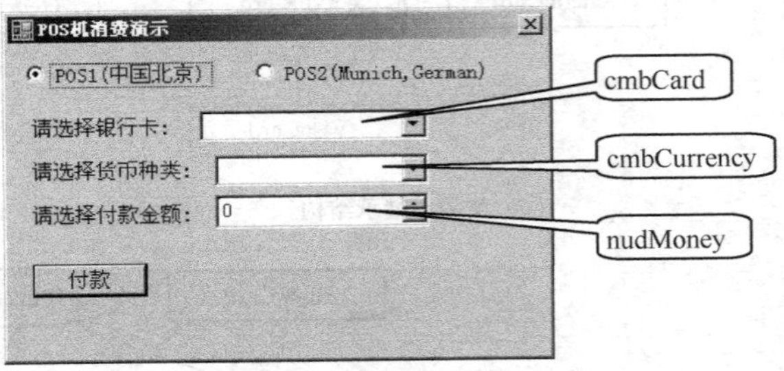

```csharp
//程序 P10_3(Form1.cs)
using System;
using System.Windows.Forms;
namespace P10_3
{
    public partial class Form1 : Form
    {
        BankCard[] cards;
        POS pos1, pos2;

        public Form1()
        {
            InitializeComponent();
            cards = new BankCard[4];
            cards[0] = new GeneralCard("g001", "Beijing", 1000);
            cards[1] = new GeneralCard("g002", "Shanghai", 1000);
            cards[2] = new CreditCard("c001", "Beijing", 1000);
            cards[3] = new VisaCard("v001", "Shanghai", 1000);
            pos1 = new POS("Beijing");
            pos2 = new POS("Shanghai");
        }

        private void Form1_Load(object sender, EventArgs e)
        {
            foreach (BankCard card in cards)
                cmbCard.Items.Add(card);
            for (int i = 0; i < 4; i++)
                cmbCurrency.Items.Add((Currency)i);
            cmbCard.SelectedIndex = cmbCurrency.SelectedIndex = 0;
        }
```

图 10-2　程序 P10_3 的主窗体界面

```csharp
        private void button1_Click(object sender, EventArgs e)
        {
            POS pos = radioButton1.Checked ?pos1 : pos2;
            BankCard card = (BankCard)cmbCard.SelectedItem;
            Currency cur = (Currency)cmbCurrency.SelectedIndex;
            try
            {
                if (cur == Currency.人民币)
                    ((IPayable)card).Pay(nudMoney.Value, pos);
                else
                    ((IGlobalPayable)card).Pay(nudMoney.Value, cur, pos);
                MessageBox.Show("消费成功, 余额" + card.Balance);
            }
            catch (InvalidCastException)
            {
                MessageBox.Show("该卡不支持此项消费");
            }
            catch (Exception exp)
            {
                MessageBox.Show("消费失败: " + exp.Message);
            }
        }
    }

    public enum Currency
    {
        人民币, 美元, 英镑,欧元
    }

    public class POS
    {
        private string region;
        public string Region
        {
            get { return region; }
        }

        public POS(string region)
        {
            this.region = region;
        }
    }

    public interface IPayable
    {
        void Pay(decimal money, POS pos);
    }

    public interface IGlobalPayable : IPayable
    {
        void Pay(decimal money, Currency currency, POS pos);
    }

    public class BankCard
    {
        protected string id, region;

        protected decimal balance = 0;
```

```csharp
        public decimal Balance
        {
            get { return balance; }
        }

        public BankCard(string id, string region)
        {
            this.id = id;
            this.region = region;
        }
    }

    public class GeneralCard : BankCard, IPayable
    {
        public GeneralCard(string id, string region, decimal money)
            : base(id, region)
        {
            balance = money;
        }

        public void Pay(decimal money, POS pos)
        {
            decimal cost = 0;
            if (this.region != pos.Region)
                cost = money * 0.01M;
            if (balance >= money + cost)
                balance -= (money + cost);
            else
                throw new InvalidOperationException("余额不足");
        }

        public override string ToString()
        {
            return string.Format("银联卡{0}", id);
        }
    }

    public class CreditCard : BankCard, IPayable
    {
        private decimal credit;

        public CreditCard(string id, string region, decimal credit)
            : base(id, region)
        {
            this.credit = credit;
        }

        public void Pay(decimal money, POS pos)
        {
            if (money <= credit + balance)
                balance -= money;
            else
                throw new InvalidOperationException("超出额度");
        }

        public override string ToString()
        {
            return string.Format("信用卡{0}", id);
        }
```

```csharp
    }

    public class VisaCard : CreditCard, IPayable, IGlobalPayable
    {
        public VisaCard(string id, string region, decimal credit)
            : base(id, region, credit)
        { }

        public void Pay(decimal money, Currency currency, POS pos)
        {
            decimal rate = 1;
            if (currency == Currency.美元)
                rate = 7;
            else if (currency == Currency.英镑)
                rate = 13;
            else if (currency == Currency.欧元)
                rate = 10;
            base.Pay(rate * money, pos);
        }

        public override string ToString()
        {
            return string.Format("Visa卡{0}", id);
        }
    }
}
```

10.3.3 解决二义性

C#语言解决多继承二义性的方式很简单，即通过显式实现来明确方法来自于哪一个接口。例如在下面的代码中，接口 IA 和 IB 以及类 CA 都定义了方法 F：

```csharp
public interface IA
{
    void F();
}
public interface IB
{
    void F();
}
public class CA
{
    public virtual void F()
    {
        Console.WriteLine("类CA的F方法");
    }
}
```

假定 CA 的派生类 CB 同时还支持接口 IA 和 IB，那么 CB 对基类方法的继承以及对接口方法的实现可以分为以下几种情况。

（1）如果 CB 没有定义方法 F，那么它从 CA 继承的方法 F 就是对接口 IA 和 IB 的方法的隐式实现：

```csharp
class Program
{
    static void Main()
    {
```

```csharp
        CB cb = new CB();
        cb.F();         //输出"类CA的F方法"
        ((CA)cb).F();   //输出"类CA的F方法"
        ((IA)cb).F();   //输出"类CA的F方法"
        ((IB)cb).F();   //输出"类CA的F方法"
    }
}

public class CB : CA, IA, IB
{ }
```

（2）如果CB只重载或隐藏了CA中的方法F，那么CB定义的成员方法F就是对接口方法的隐式实现：

```csharp
class Program
{
    static void Main()
    {
        CB cb = new CB();
        cb.F();         //输出"类CB的F方法"
        ((CA)cb).F();   //输出"类CB的F方法"
        ((IA)cb).F();   //输出"类CB的F方法"
        ((IB)cb).F();   //输出"类CB的F方法"
    }
}

public class CB : CA, IA, IB
{
    public override void F()
    {
        Console.WriteLine("类CB的F方法");
    }
}
```

（3）否则，CB可通过显式实现的方式来明确指定所实现的接口方法，但这不会影响到它对基类方法的继承或重载：

```csharp
class Program
{
    static void Main()
    {
        CB cb = new CB();
        cb.F();         //输出"类CB的F方法"
        ((CA)cb).F();   //输出"类CA的F方法"
        ((IA)cb).F();   //输出"接口IA的F方法"
        ((IB)cb).F();   //输出"接口IB的F方法"
    }
}

public class CB : CA, IA, IB
{
    public new void F()
    {
        Console.WriteLine("类CB的F方法");
    }

    void IA.F()
    {
```

```
            Console.WriteLine("接口 IA 的 F 方法");
        }

        void IB.F()
        {
            Console.WriteLine("接口 IB 的 F 方法");
        }
    }
```

下面的程序 P10_4 演示了一个非常典型的多继承例子：水上飞机类 Seaplane 同时支持接口 IFlyable 和 ISwimmable；两栖车辆类 Amphicar 不仅继承了汽车类 Automobile，还支持接口 ISwimmable。

```
//程序 P10_4
using System;
namespace P10_4
{
    class Program
    {
        static void Main()
        {
            Seaplane plane = new Seaplane();
            IFlyable flyer = plane;
            Console.WriteLine("{0}空中飞行 1000 公里需{1}小时", flyer, flyer.Run(1000));
            ISwimmable swimmer = plane;
            Console.WriteLine("{0}水上航行 1000 公里需{1}小时", swimmer, swimmer.Run(1000));
            Amphicar car = new Amphicar(80);
            Console.WriteLine("{0}地面行驶 1000 公里需{1}小时", car, car.Run(1000));
            swimmer = car;
            Console.WriteLine("{0}水上航行 1000 公里需{1}小时", swimmer, swimmer.Run(1000));
        }
    }

    public interface IFlyable
    {
        float Run(float distance);
    }

    public interface ISwimmable
    {
        float Run(float distance);
    }

    public class Seaplane : IFlyable, ISwimmable
    {
        float ISwimmable.Run(float distance)
        {
            return distance / 50;
        }

        float IFlyable.Run(float distance)
        {
            return distance / 400;
        }
    }

    public class Automobile
```

```
    {
        private float speed;
        public float Speed
        {
            get { return speed; }
        }
        public virtual float Run(float distance)
        {
            return distance / speed;
        }
        public Automobile(float speed)
        {
            this.speed = speed;
        }
    }
    public class Amphicar : Automobile, ISwimmable
    {
        public Amphicar(float speed) : base(speed)
        { }
        float ISwimmable.Run(float distance)
        {
            return base.Run(distance) * 3;
        }
    }
}
```

这样通过区分基类和接口中的方法，就能够在程序中实现真正所需要的多态性。程序 P10_4 的输出结果如下：

```
P10_4.Seaplane 空中飞行 1000 公里需 2.5 小时
P10_4.Seaplane 水上航行 1000 公里需 20 小时
P10_4.Amphicar 地面行驶 1000 公里需 12.5 小时
P10_4.Amphicar 水上航行 1000 公里需 37.5 小时
```

10.4 接口与集合

10.4.1 集合型接口及其实现

像列表、堆栈、队列等集合类型在计算机程序中有着广泛的应用。.NET 类库的 System.Collections 命名空间下定义了一组与集合有关的接口，并通过在预定义集合类型中支持不同的接口来提供相关服务。表 10-1 简要地描述了.NET 类库中与集合有关的接口类型，表 10-2 则介绍了.NET 类库中支持这些接口的主要集合类型。

从接口方法的原型可以看出，这些集合操作都以 object 为元素类型，那么实际上就可以在集合中存储任意类型的对象。表 10-2 中列出的类型还都支持 IEnumerable 和 ICollection 接口，因此可使用 foreach 语句来遍历集合元素，并通过 Count 属性来获取集合长度。此外，C#中的所有数组类型都默认从 Array 类继承，而该类也支持 IEnumerable 和 ICollection 接口，只不过它对 Count 属性是显式接口实现，并通过 Length 属性来返回该值。例如下面最后两行代码的输出内容实际上是相同的：
```
int[] x = new int[5];
ICollection iCol = x;
Console.WriteLine(x.Length);
Console.WriteLine(iCol.Count);
```

表 10-1　　　　　　　　　　　.NET 类库中与集合有关的接口类型

接口	支持的概念描述	继承的接口	主要方法
IEnumerable	支持 foreach 遍历的可枚举类型	无	IEnumerator GetEnumerator()
IEnumerator	枚举器，提供对可枚举类型的遍历操作	无	bool MoveNext() void Reset() object Current { get; }
ICollection	对象集合	IEnumerable	void CopyTo(Array, int) int Count { get; } bool IsSynchronized{ get; } object SyncRoot{ get; }
IList	对象列表，支持对象的插入、删除和索引访问	IEnumerable ICollection	int Add(object) void Clear() bool Contains(object) void Insert(int, object) int IndexOf(object) void Remove(object) void RemoveAt(int) object this[int index] { get; set; }
IDictionary	有字典序（键/值对）的集合	IEnumerable ICollection	int Add(object, object) void Clear() bool Contains(object) void Remove(object) ICollection Keys { get; } ICollection Values { get; } object this[objectkey] { get; set; }
IDictionaryEnumerator	有字典序的枚举器	IEnumerator	Object Key { get; } Object Value { get; }

表 10-2　　　　　　　　　　　.NET 类库中的主要集合类型

接口	类型描述	继承的接口
ArrayList	链表类，集合大小可动态改变	IEnumerable, ICollection, IList
Queue	先进先出的队列类	IEnumerable, ICollection
Stack	先进后出的堆栈类	IEnumerable, ICollection
SortedList	维护键/值对的列表类，元素按键进行排序	IEnumerable, ICollection, IDictionary
Hashtable	维护键/值对的哈希表	IEnumerable, ICollection, IDictionary

10.4.2　列表、队列和堆栈

ArrayList 类实现了 IList 接口的各个方法，并由此提供了动态数组的功能。例如它不仅可通过 Add 方法在集合末端加入元素，通过 Remove 方法删除指定元素，还可通过 Insert 和 RemoveAt 方法在任何位置插入和删除元素，而集合的大小会自动进行调整。ArrayList 自定义的成员方法 BinarySearch 和 Sort 还可用于集合元素的查找和排序。

下面的程序 P10_5 演示了 ArrayList 集合的简单用法。

```csharp
//程序 P10_5
using System;
using System.Collections;
namespaceP10_5
{
    class Program
    {
        static void Main()
        {
            ArrayList al1 = new ArrayList();
            al1.Add("王小红");
            al1.Add("周军");
            al1.Insert(0, "方小白");
            al1.Add("Smith");
            al1.Insert(1, "Jerry");
            Console.WriteLine("排序前:");
            foreach (object obj in al1){
                Console.Write(obj);
                Console.Write(' ');
            }
            al1.Sort();
            Console.WriteLine("\n排序后:");
            foreach (object obj in al1){
                Console.Write(obj);
                Console.Write(' ');
            }
        }
    }
}
```

BinarySearch 和 Sort 方法要求集合中的元素类型应能够相互比较，否则它们起不到搜索和排序的效果。根据 String 类型的比较规则，程序 P10_5 的输出结果如下：

```
排序前:
方小白 Jerry 王小红 周军 Smith
排序后:
Jerry Smith 方小白 王小红 周军
```

此外，Queue 类的 Enqueue 和 Dequeue 方法分别用于集合元素的入队和出队，而 Stack 类的 Push 和 Pop 方法分别用于集合元素的入栈和出栈。下面给出了这两个类型的使用示例：

```csharp
Queue queue1 = new Queue();
queue1.Enqueue(10);
queue1.Enqueue("中国");
queue1.Enqueue(DayOfWeek.Sunday);
while (queue1.Count > 0)
    Console.WriteLine(queue1.Dequeue());  //依次输出"10"、"中国"和"Sunday"
Stack stack1 = new Stack();
stack1.Push(10);
stack1.Push("中国");
stack1.Push(DayOfWeek.Sunday);
while (stack1.Count > 0)
    Console.WriteLine(stack1.Pop());  //依次输出"Sunday"、"中国"和"10"
```

除了默认的无参构造函数外，ArrayList、Queue 和 Stack 都提供了以 ICollection 为参数类型的构造函数，它表示在创建集合对象的同时从指定的集合复制元素，这样就能在这些集合类型以及数组之间方便地进行数据交换。例如：

```
int[] x = { 5, 9, 6, 3, -6, 12 };
Queue q1 = new Queue(x);
foreach (int i in q1)
    Console.WriteLine(i);  //输出顺序和数组 x 中的元素顺序相同
Stack s1 = new Stack(x);
foreach (int i in s1)
    Console.WriteLine(i);  //输出顺序和数组 x 中的元素顺序相反
ArrayList al1 = new ArrayList(s1);
foreach (int i in al1)
Console.WriteLine(i);  //输出顺序和数组 x 中的元素顺序相反
```

10.4.3 自定义集合类型

如果用户要在自定义类型中支持这些集合型接口，就必须实现有关集合操作的接口方法，如通过 Count 属性返回集合长度，通过 CopyTo 方法将集合元素复制到数组中等。

在下面的程序 P10_6 中，集合类型 Phones 用于维护一组电话号码的集合。它声明支持 ICollection 接口，并在内部维护一个长度为 5 的字符串数组。不过其 Count 属性只返回不为空的电话号码数量：如果新设置了一个号码，那么集合长度加 1；而如果将一个现有号码清空，那么集合长度减 1。接下来定义的联系人类 Contact 中又使用了 Phones 类型的字段和属性。

```
//程序 P10_6
using System;
using System.Collections;
namespace P10_6
{
    class Program
    {
        static void Main()
        {
            Contact c1 = new Contact("王小红");
            c1.Phones[PhoneType.手机] = "1331121122";
            c1.Phones[PhoneType.小灵通] = "1234567";
            Contact c2 = new Contact("方小白");
            c2.Phones[PhoneType.家庭电话] = "88338844";
            c2.Phones[PhoneType.办公电话] = "66116622";
            c2.Phones[PhoneType.传真] = "66116623";
            ICollection iCol = c1.Phones;
            Console.WriteLine("c1 电话数量为{0}", iCol.Count);
            iCol = c2.Phones;
            Console.WriteLine("c2 电话数量为{0}", iCol.Count);
            c1.Print();
            c2.Print();
        }
    }

    public enum PhoneType
    {
        家庭电话, 办公电话, 手机, 小灵通, 传真
    }

    public class Phones : ICollection
    {
        private string[] phones;
```

```csharp
        private int count = 0;
        public int Count
        {
            get { return count; }
        }
        public string this[PhoneType type]
        {
            get{ return phones[(int)type];}
            set{
                int index = (int)type;
                if (phones[index] == null && value != null)
                    count++;
                else if (phones[index] != null && value == null)
                    count--;
                phones[index] = value;
            }
        }
        public Phones()
        {
            phones = new string[5];
        }
        public void CopyTo(Array array, int index)
        {
            foreach (string s in phones)
                if (s != null)
                    array.SetValue(s, index++);
        }
        IEnumerator IEnumerable.GetEnumerator()
        {
            for (int i = 0; i < 5; i++)
            if (phones[i] != null)
                yield return string.Format("{0}: {1}", (PhoneType)i, phones[i]);
        }
        bool ICollection.IsSynchronized
        {
            get { return false; }
        }
        object ICollection.SyncRoot
        {
            get { return null; }
        }
    }
    public class Contact
    {
        private string name;
        public string Name
        {
            get { return name; }
        }
        private string address;
        public string Address
        {
            get { return address; }
            set { address = value; }
```

```
        }
        private Phones phones;
        public Phones Phones
        {
            get { return phones; }
        }

        public Contact(string name)
        {
            this.name = name;
            phones = new Phones();
        }

        public void Print()
        {
            Console.WriteLine("姓名:" + name);
            Console.WriteLine("地址:" + address);
            foreach (string s in phones)
                Console.WriteLine(s);
        }
    }
}
```

ICollection 接口继承了 IEnumerable 接口,因此需要提供 GetEnumerator 方法来支持 foreach 遍历,其实现机制请参看本书 12.2 节。此外 ICollection 接口还要求提供 IsSynchronized 和 SyncRoot 属性(支持异步操作),不过程序 P10_6 中仅仅定义了它们的原型,而没有进行完整的实现。

程序 P10_6 的输出结果如下:

c1 电话数量为 2
c2 电话数量为 3
姓名:王小红
地址:
手机: 1331121122
小灵通: 1234567
姓名:方小白
地址:
家庭电话: 88338844
办公电话: 66116622
传真: 66116623

10.5 案例研究——旅行社管理系统中的集合类型

10.5.1 职员列表与数据绑定

在本节的案例研究中,我们首先使用 ArrayList 来维护系统中所有职员的信息。为 TravelWin

项目添加一个新的 Windows 窗体 StaffForm，内容设计如图 10-3 所示，并在"类别"、"状态"和"学历"组合框的 Items 集合中分别加入对应的枚举值。

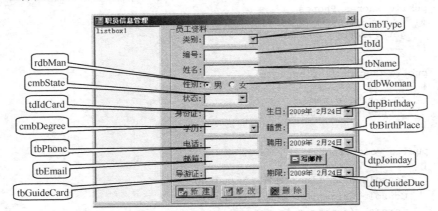

图 10-3　职员管理窗体 StaffForm

接下来为 StaffForm 类添加一个 Staff 类型的字段_staff 和一个 ArrayList 类型的字段_staffs，分别表示当前所选职员和所有职员集合；在窗体构造函数中对_staffs 进行初始化；再为窗体添加 Load 事件处理方法，在其中将所有职员的信息显示在左侧列表框中：

```
private Staff _staff;
private ArrayList _staffs;

public StaffForm(ArrayList staffs)
{
    InitializeComponent();
    _staffs = staffs;
}

private void StaffForm_Load(object sender, EventArgs e)
{
    foreach (Staff s in _staffs)
        listBox1.Items.Add(s);
    listBox1.SelectedIndex = 0;
}
```

切换到程序主窗体 Form1 的代码视图，为其也添加一个同样的_staffs 字段，并在其 Load 事件处理方法的最后一行调用 Staff.GetAll 方法来获取所有职员信息：

```
_staffs = new ArrayList(Staff.GetAll());
```

然后为主窗体添加一个菜单项"职员信息(&I)"，在其单击事件处理方法中显示 StaffForm 窗体：

```
private void menuStaffInfo_Click(object sender, EventArgs e)
{
    StaffForm frm1 = new StaffForm(_staffs);
    frm1.ShowDialog();
}
```

此时编译运行程序，通过菜单命令打开职员信息窗体，在列表框中已能看到列出的职员信息。下面继续为 StaffForm 添加新建、修改和删除职员的功能。当用户在列表框中选择一个职员时，窗体上应当显示该职员的详细信息，这可以通过列表框的选项改变事件来实现（注意下方的"导游证"和"期限"仅在所选职员为 Guide 对象时可用）：

```
private void listBox1_SelectedIndexChanged(object sender, EventArgs e)
{
    if (listBox1.SelectedItem == null)
        return;
```

```
    _staff = (Staff)listBox1.SelectedItem;
    tbId.Text = _staff.Id.ToString();
    tbName.Text = _staff.Name;
    if (_staff.Gender)
        rdbMan.IsChecked = true;
    else
        rdbWoman.IsChecked = true;
    dtpBirthday.SelectedDate = _staff.Birthday;
    cmbState.SelectedIndex = (int)_staff.State;
    cmbDegree.SelectedIndex = (int)_staff.Degree;
    tbBirthPlace.Text = _staff.Birthplace;
    tbIdCard.Text = _staff.IdCard;
    dtpJoinday.SelectedDate = _staff.Joinday;
    tbPhone.Text = _staff.Phone;
    if (_staff.Email != null)
        tbEmail.Text = _staff.Email;
    if (_staff is Guide)
    {
        Guide g = (Guide)_staff;
        cmbType.SelectedIndex = 3;
        tbGuideCard.Text = g.GuideCard;
        dtpGuideDue.SelectedDate = g.GuideDue;
        return;
    }
    tbGuideCard.Text = "";
    if (_staff is Manager)
        cmbType.SelectedIndex = 0;
    else if (_staff is Director)
        cmbType.SelectedIndex = 1;
    else if (_staff is Agent)
        cmbType.SelectedIndex = 2;
}
```

注意窗体下方的"导游证"和"期限"仅在所选职员为导游时可用：

```
private void cmbType_SelectionChanged(object sender, SelectionChangedEventArgs e)
{
    if(cmbType.SelectedIndex != 3)
        tbGuideCard.IsEnabled = dtpGuideDue.IsEnabled = false;
    else
        tbGuideCard.IsEnabled = dtpGuideDue.IsEnabled = true;
}
```

在单击"新建"按钮时，程序首先检查必填项是否都已输入，并确保_staffs中没有相同编号的职员，满足条件的话就根据所选的职员类别来创建相应的 Staff 对象，并将其加入列表中：

```
private void btnAdd_Click(object sender, EventArgs e)
{
    if (tbName.Text == "" || tbIdCard.Text == "")
    {
        MessageBox.Show("姓名和身份证不能为空", "错误");
        return;
    }
    int id;
    if (tbId.Text.Length < 6 || !int.TryParse(tbId.Text, out id))
    {
        MessageBox.Show("职员编号格式不正确", "错误");
        return;
    }
    for(int i=0; i<_staffs.Count; i++)
        if (_staffs[i].Id == id){
            MessageBox.Show("该编号已经存在", "错误");
```

```csharp
            return;
    }
    return;
    if (cmbType.SelectedIndex == 0)
        _staff = new Manager(tbName.Text, rdbMan.Checked, dtpBirthday.Value, dtpJoinday.Value);
    else if (cmbType.SelectedIndex == 1)
        _staff = new Director(tbName.Text, rdbMan.Checked, dtpBirthday.Value, dtpJoinday.Value);
    else if (cmbType.SelectedIndex == 2)
        _staff = new Agent(tbName.Text, rdbMan.Checked, dtpBirthday.Value, dtpJoinday.Value);
    else if (cmbType.SelectedIndex == 3)
    {
        Guide guide = new Guide(tbName.Text, rdbMan.Checked, dtpBirthday.Value, dtpJoinday.Value);
        guide.GuideCard = tbGuideCard.Text;
        guide.GuideDue = dtpGuideDue.Value;
        _staff = guide;
    }
    else
        _staff = new Staff(tbName.Text, rdbMan.Checked, dtpBirthday.Value, dtpJoinday.Value);
    _staff.Birthplace = tbBirthPlace.Text;
    _staff.IdCard = tbIdCard.Text;
    _staff.Degree = (Degree)cmbDegree.SelectedIndex;
    _staff.Phone = tbPhone.Text;
    _staff.Email = tbEmail.Text;
    _staffs.Add(_staff);
}
```

类似地,"修改"按钮用于修改现有的职员信息,"删除"按钮用于删除职员对象,详细代码此处不再列出。

10.5.2 使用自定义集合

1. 报名优先级队列

前面提到过,不同游客的报名具有不同的优先级,也就是说 Enrollment 对象之间可以进行比较。System 程序集中的 IComparable 接口就表示对象之间可比较的契约,它要求通过 CompareTo 方法来返回比较结果。那么可以声明 Enrollment 支持该接口,并通过比较两个 Enrollment 的优先级来实现其 CompareTo 方法:

```csharp
public class Enrollment : IComparable// 报名类
{
    //…
    public int CompareTo(object obj)
    {
        return _priority - ((Enrollment)obj)._priority;
    }
}
```

在处理旅行团的报名集合时,业务员应当先取出优先级高的报名进行处理。为此可考虑定义一个专门的报名队列类 EnrollmentQueue,它内部使用一个 ArrayList 对象来维护报名列表,其中的 Enrollment 对象总是按照优先级由低到高的顺序排列(相同优先级的对象则按加入队列的时间顺序排列),而每次只能取出优先级最高的一个 Enrollment。

下面给出了 EnrollmentQueue 的类型定义,其 Enqueue 和 Dequeue 方法分别用于向队列中加入和取出 Enrollment 对象,Peek 方法则访问下一个将要取出的对象。该类型声明支持 IEnumerable 和 ICollection 接口,而对接口方法的实现主要是依靠 ArrayList 对象来完成的:

```csharp
public class EnrollmentQueue : IEnumerable, ICollection// 报名队列类
{
    private ArrayList _queue;
```

```csharp
public int Count
{
    get { return _queue.Count; }
}

public EnrollmentQueue()
{
    _queue = new ArrayList();
}

public bool Contains(Enrollment enr)
{
    return _queue.Contains(enr);
}

public Enrollment Peek()
{
    return (Enrollment)_queue[_queue.Count - 1];
}

public void Enqueue(Enrollment enr)
{
    int i = 0;
    while (i < _queue.Count && enr.CompareTo(_queue[i]) > 0)
        i++;
    _queue.Insert(i, enr);
}

public Enrollment Dequeue()
{
    Enrollment enr = (Enrollment)_queue[_queue.Count - 1];
    _queue.RemoveAt(_queue.Count - 1);
    return enr;
}

public void Remove(Enrollment enr)
{
    _queue.Remove(enr);
}

public void Clear()
{
    _queue.Clear();
}

public void CopyTo(Array array, int index)
{
    _queue.CopyTo(array, index);
}

public IEnumerator GetEnumerator()
{
    return _queue.GetEnumerator();
}

bool ICollection.IsSynchronized
{
    get { return false; }
}

object ICollection.SyncRoot
{
```

```csharp
        get { return null; }
    }
}
```
那么在 Tour 类的成员定义中，报名集合的类型可改用 EnrollmentQueue 进行定义：
```csharp
private EnrollmentQueue _enrollments;
public EnrollmentQueue Enrollments //报名队列
{
    get { return _enrollments; }
}
```
当然接下来还要在 Tour 的构造函数中对队列对象进行初始化。使用了该队列类型后，Customer 类的 Enroll 和 Cancel 方法代码就能大大简化，而且能保证加入到 Tour 对象中的各个 Enrollment 都是按优先级顺序排列的：
```csharp
public virtual Enrollment Enroll(Tour t, Customer[] customers)
{
    if (t.State != TourState.报名)
        throw new BadStateException(typeof(TourState), (int)t.State, t);
    if (_type == CustomerType.Child)
        throw new BadStateException(typeof(CustomerType), (int)_type, this);
    Enrollment enr = new Enrollment(t, this, customers);
    t.Enrollments.Enqueue(enr);
    t.OnChange += new EventHandler(tour_OnChange);
    t.OnPreSend += new EventHandler(tour_OnPreSend);
    t.OnComplete += new EventHandler(tour_OnComplete);
    t.OnCancel += new EventHandler(tour_OnCancel);
    return enr;
}

public void Cancel(Enrollment enr) //取消报名
{
    Tour t = enr.Tour;
    if (t.State != TourState.报名 || t.State != TourState.满员)
        throw new BadStateException(typeof(TourState), (int)t.State, t);
    if (enr.Applier != this)
        throw new InvalidOperationException("只有报名者本人才能取消");
    if (t.Enrollments.Contains(enr)){
        t.Enrollments.Remove(enr);
        this.Unsubscribe(t);
    }
}
```

2. 游客集合

系统中对游客集合的操作也很频繁，使用静态数组很不方便。在旅行社业务操作中还需要统计游客中的学生、儿童和老人的数量，那么可考虑在 ArrayList 的基础上封装一个自定义集合类型 CustomerCollection，它维护一组游客集合，能够进行游客查询、增加和删除等操作，并自动更新各种游客的数量。下面给出了该类型的详细定义代码：
```csharp
// 游客集合类
public sealed class CustomerCollection : IEnumerable, ICollection, IList
{
    private ArrayList _collection;

    private int _childCount, _studentCount, _oldCount;
    public int ChildCount //儿童数量
    {
        get { return _childCount; }
```

```csharp
    }
    public int StudentCount  //学生数量
    {
        get { return _studentCount; }
    }
    public int OldCount  //老年人数量
    {
        get { return _oldCount; }
    }
    public int Count  //游客总数量
    {
        get { return _collection.Count; }
    }

    public Customer this[int index]
    {
        get { return (Customer)_collection[index]; }
        set { _collection[index] = value; }
    }

    public CustomerCollection()
    {
        _collection = new ArrayList();
    }

    public bool Contains(Customer c)  //是否包含指定的游客对象
    {
        return _collection.Contains(c);
    }

    public int IndexOf(Customer c)  //返回游客对象在集合中的索引
    {
        return _collection.IndexOf(c);
    }

    public int Add(Customer c)  //向集合中加入游客
    {
        if (c.Type == CustomerType.Child)
            _childCount++;
        else if (c.Type == CustomerType.Student)
            _studentCount++;
        else if (c.Type == CustomerType.Old)
            _oldCount++;
        return _collection.Add(c);
    }

    public void Insert(int index, Customer c)  //在集合的指定位置插入游客
    {
        _collection.Insert(index, c);
        if (c.Type == CustomerType.Child)
            _childCount++;
        else if (c.Type == CustomerType.Student)
            _studentCount++;
        else if (c.Type == CustomerType.Old)
            _oldCount++;
    }
```

```csharp
public void RemoveAt(int index)  //删除指定位置的游客
{
    CustomerType type = ((Customer)_collection[index]).Type;
    if (type == CustomerType.Child)
        _childCount--;
    else if (type == CustomerType.Student)
        _studentCount--;
    else if (type == CustomerType.Old)
        _oldCount--;
    _collection.RemoveAt(index);
}

public void Remove(Customer c)   //删除指定游客对象
{
    if (c.Type == CustomerType.Child)
        _childCount--;
    else if (c.Type == CustomerType.Student)
        _studentCount--;
    else if (c.Type == CustomerType.Old)
        _oldCount--;
    _collection.Remove(c);
}

public void Clear()  //清空集合
{
    _childCount = _studentCount = _oldCount = 0;
    _collection.Clear();
}
//……
}
public void CopyTo(Array array, int index)//将集合复制到数组
{
    _collection.CopyTo(array, index);
}

public IEnumerator GetEnumerator()
{
    return _collection.GetEnumerator();
}

object IList.this[int index]
{
    get { return _collection[index]; }
    set { _collection[index] = (Customer)value; }
}

bool IList.Contains(object o)
{
return (o is Customer && this.Contains((Customer)o));
}

int IList.IndexOf(object o)
{
    if (o is Customer) return this.IndexOf((Customer)o);
        elsereturn -1;
}

int IList.Add(object o)
{
```

```csharp
        if (o is Customer)return this.Add((Customer)o);
            elsethrow new InvalidOperationException("加入的对象必须是Customer");
    }

    void IList.Insert(int index, object o)
    {
        if (o is Customer)this.Insert(index, (Customer)o);
            elsethrow new InvalidOperationException("插入的对象必须是Customer");
    }

    void IList.Remove(object o)
    {
        if (o is Customer)this.Remove((Customer)o);
            elsethrow new InvalidOperationException("删除的对象必须是Customer");
    }

    bool IList.IsReadOnly{get { return false; }}
    bool IList.IsFixedSize{get { return false; }}
    bool ICollection.IsSynchronized{get { return false; }}
    object ICollection.SyncRoot{get { return null; }}
}
```

CustomerCollection 声明支持了 IEnumerable、ICollection 和 IList 三个集合型接口，其中许多接口方法与公有成员方法的功能相同，只不过操作的对象是 object 类型。

同样，Tour 类的成员 Customers 可改用 CustomerCollection 类型进行定义：

```csharp
private CustomerCollection _customers;
public CustomerCollection Customers //游客集合
{
    get { return _customers; }
}
```

相应的，Agent 类用于接受报名的 Accept 方法也可以大大简化：

```csharp
public void Accept(Enrollment enr) //接受报名
{
    Tour t = enr.Tour;
    if (enr.State != EnrollmentState.待确认 || t.State != TourState.报名)
        throw new BadStateException(typeof(TourState), (int)t.State, t);
    foreach (Customer c in enr.Customers)
        t.Customers.Add(c);
    enr.State = EnrollmentState.接受;
    if (t.Customers.Count == t.Package.Number)
        t.State = TourState.满员;
    //…
}
```

10.6 小　　结

接口是只提供声明、不提供实现的抽象数据结构，接口技术是对面向对象程序设计的重要扩展。一个类型要支持某个接口，它就要实现接口中声明的所有成员，实现的方式包括隐式实现和显示实现。在接口的基础上能够实现更为丰富的多态性，并且能够以清晰的程序结构来支持多继承。

10.7 习 题

1. 简述接口和抽象类的相同点和不同点。
2. 在实现接口方法时，哪些情况下更适合隐式实现？哪些情况下更适合显式实现？
3. 程序 P10_1 中定义的 DebitCard 类是和 BankAccount 类相关联的，那么其 Pay 方法的实现代码可否改为如下内容，为什么？

```
public override bool Pay(decimal money)
{
    return account.Pay(money);
}
```

4. 考虑一个客户可开设多个账户的情况，定义一个银行账户集合类 Accounts，它支持 ICollection 接口，并维护活期、定期、零存整取、定活两便 4 个账户；只有开通了某个账户，才可以设置账户的金额；类型的 Count 属性返回已开通的账户数量（可参考程序 P10_6 的实现方法）。
5. 如何对组合框和列表框中的元素项集合进行排序？
6. 将程序 P10_6 扩展成一个联系人管理程序。
7. 定义集合类型 BinaryTree，它支持 ICollection 接口并实现二叉树的数据结构。
8. 参照 10.5.1 小节中的内容，在系统的用户管理窗体 UserForm 中使用 ArrayList 类型来替换 User[]数组。
9. 扩充旅行社管理系统中游客报名的功能，使一名游客能够代表多名游客报名，并使用 CustomerCollection 类型来实现这一功能。

第 11 章 泛型程序设计

泛型的核心思想就是将算法从数据结构中抽象出来，使得预定义的操作能够作用于不同的类型，从而提高程序的效率、通用性和类型安全性，进而简化整个编程模型。本章将以泛型类为基础，讨论泛型使用中的一般性问题，逐步展现泛型程序设计的优越性。

11.1 为什么要使用泛型

本书10.4节介绍了列表、队列、堆栈等集合的用法，这些集合都是以 object 为元素类型，从而支持在集合中存储任意类型的对象，例如：

```
stack1.Push(10);
stack1.Push("中国");
stack1.Push(DayOfWeek.Sunday);
```

不过在大多数场合中，把不同类型的对象存放在一个集合中没有太多的意义，人们更感兴趣的是同一类型的对象集合，比如整数的集合、学生对象的集合，等等。例如下面的代码就将整数数组 x1 的元素依次放入一个堆栈，然后又将堆栈的元素依次读入另一个数组 x2 中，那么数组 x2 实际上就是 x1 的元素的反向排列：

```
int[] x1 = { 5, 9, 6, 3, -6, 12 };
Stack stack1 = new Stack();
for (int i = 0; i < x1.Length; i++)
    stack1.Push(x1[i]);   //装箱转换
int[] x2 = new int[x1.Length];
for (int i = 0; i < x2.Length; i++)
    x2[i] = (int)stack1.Pop();   //拆箱转换
```

在上面的代码段中，每次将整数放入堆栈时都要进行装箱转换，而在整数出栈时又要进行拆箱转换，这种频繁的转换会大大降低程序的性能。此外，如果开发人员不小心将一个其他类型的对象（如一个小数或一个字符串）放入堆栈中，程序也能够通过编译，但在将该元素其拆箱转换为整数时会发生 InvalidCastException 异常。

一种解决办法是定义单独的整数堆栈类 IntStack，它只允许存放整数类型的元素。不过这种方式的可复用性太差，今后程序可能还需要使用 DoubleStack、StringStack、StudentStack 等一系列堆栈，它们实现的功能基本类似，但需要为每一种元素类型都定义一个新的堆栈类型。

下面的程序则采用了泛型技术来"一劳永逸"地解决此问题：

```
//程序 P11_1
using System;
using System.Collections.Generic;
namespace P11_1
{
```

```
class Program
{
    static void Main()
    {
        int[] x1 = { 5, 9, 6, 3, -6, 12 };
        Stack<int> stack1 = new Stack<int>();
        for (int i = 0; i < x1.Length; i++)
            stack1.Push(x1[i]);
        int[] x2 = new int[x1.Length];
        for (int i = 0; i < x2.Length; i++)
            x2[i] = stack1.Pop();
        foreach (int i in x2)
            Console.WriteLine(i);
        string[] ss1 = { "one", "day", "when", "we", "were", "young" };
        Stack<string> stack2 = new Stack<string>();
        for (int i = 0; i < ss1.Length; i++)
            stack2.Push(ss1[i]);
        string[] ss2 = new string[ss1.Length];
        for (int i = 0; i < ss1.Length; i++)
            ss2[i] = stack2.Pop();
        foreach (string s in ss2)
            Console.WriteLine(s);
    }
}
```

该程序中使用的不是 System.Collections 命名空间下的普通类 Stack，而是 System.Collections.Generic 命名空间下的泛型类 Stack<T>，其中 T 表示一个抽象数据类型，在使用时可被不同的具体类型所替代。上面的 Stack<int>和 Stack<string>分别表示使用 int 和 string 类型来替换类型参数 T，它们都叫做 Stack<T>的构造类型：

```
Stack<int> stack1 = new Stack<int>();
Stack<string> stack2 = new Stack<string>();
```

那么堆栈 stack1 中的元素类型就必须是 int，而 stack2 中的元素类型必须是 string，入栈和出栈时无须进行拆箱和装箱转换，而试图将其他类型的对象放入堆栈中则是不允许的。比如下面的代码就根本不能通过编译：

```
stack1.Push("abc");    //错误：类型不匹配！
```

泛型类本身不能创建对象，而只能针对其构造类型来创建对象，例如下面的代码是不合法的，因为 C#编译器并不知道该堆栈中要存储什么类型的数据：

```
Stack<T> stack3 = new Stack<T>();
```

也就是说，类是对一组对象的抽象，而泛型类是对一组普通类的抽象。泛型技术允许我们一次性定义通用的算法，并使其能够作用于不同类型的数据结构，从而大大提高程序的灵活性和可重用性。

和泛型类一样，在结构中也可以使用类型参数，这就是泛型结构。考虑到类更具有一般性，本章的介绍都围绕泛型类展开，其中的许多内容也适用于泛型结构。

11.2 泛型类

11.2.1 泛型类的定义和使用

Stack<T>是.NET 类库提供的，下面我们定义一个自己的泛型类 LinkNode<T>：

```
//程序 P11_2
using System;
namespace P11_2
{
    public class LinkNode<T>
    {
        protectedT data;
        protectedLinkNode<T> next;

        public T Data  // 获取或设置节点值
        {
            get { return data; }
            set { data = value; }
        }

        public LinkNode<T> Next  // 获取或设置节点的下一个节点
        {
            get { return next; }
            set { next = value; }
        }

        public LinkNode()  // 创建链表节点
        { }

        public LinkNode(T t)  // 指定数据创建链表节点
        {
            data = t;
        }

        //移至 node 的第 n 个后续节点
        public static LinkNode<T> operator >>(LinkNode<T> node, int n)
        {
            LinkNode<T> node1 = node;
            for (int i = 0; i < n && node1 != null; i++)
                node1 = node1.next;
            return node1;
        }
    }
}
```

该类表示链表节点，即每个节点不仅包含自身的数据 Data，还包含指向下一个节点的链接指针 Next。不过其包含的数据类型不是具体类型，而是抽象类型 T，它在使用时可以替换为各种具体类型（但不能是静态类）。下面的示例代码就创建了依次链接的一系列 LinkNode<int>的节点：

```
int x = 0;
LinkNode<int> node1= new LinkNode<int>(x);
while(x++ < 1000)
{
    if (x % 3 == 0 && x % 4 == 1 && x % 5 == 2){
        node1.Next = new LinkNode<int>(x);
        node1 = node1.Next;
    }
}
```

在泛型类的定义中，类型参数可以作为字段、方法的参数和返回值，以及方法代码中局部变量的类型。而在创建泛型类的构造类型的对象时，这些类型参数都被替换为对应的具体类型。例如对于构造类型 LinkNode<int>，其_data 字段的类型就成为 int，_next 字段的类型成为 LinkNode<int>，而使用其带参构造函数时传递的参数类型也应是 int。

从上面的程序中可以看到，泛型类的构造函数名称与类名相同，但不包含类型参数：

```
publicLinkNode() { }
```
泛型类 LinkNode<T>还重载了右移位操作符 ">>"。对泛型类而言，本书 4.3.4 小节中介绍的操作符重载规则同样适用，此外还应注意以下几点：
- 重载一元操作符时，参数类型应为当前泛型类（而不能是其构造类型）。
- 重载二元操作符时，至少有一个参数类型应为当前泛型类。
- 重载类型转换操作符时，还可以使用类型参数来指代转换的目标类型。

11.2.2 使用"抽象型"变量

在定义泛型时，我们并不能确定类型参数会被替换为哪种具体类型。因此在泛型类的方法代码中，对使用类型参数定义的"抽象型"变量，不能随意进行赋值和成员调用。例如下面为 LinkNode<T>重载的相等和不等操作符都是错误的：

```
public static bool operator ==(LinkNode<T> node1, LinkNode<T> node2)
{
    return (node1._data == node2._data);
}

public static bool operator !=(LinkNode<T> node1, LinkNode<T> node2)
{
    return (node1._data != node2._data);
}
```

这是因为：node1 和 node2 的_data 字段都是"抽象型" T，我们不能保证操作符"=="和"!="能够作用于该类型上。假设某个自定义类 Student 没有重载相等和不等操作符，那么对于两个 LinkNode<Student>对象调用上述代码就是不合法的。

不过将操作符重载定义改为如下的代码就是正确：

```
public static bool operator ==(LinkNode<T> node1, LinkNode<T> node2)
{
    return (node1.data.Equals(node2.data));
}

public static bool operator !=(LinkNode<T> node1, LinkNode<T> node2)
{
    return !(node1.data.Equals(node2.data));
}
```

上面的代码中对 T 型_data 字段调用了 Equals 方法，这是没有问题的，因为 C#所有类型的基类 Object 定义了该方法。同理，对"抽象型"变量也可以放心地调用其 ToString 方法和 GetType 方法（但在变量值为 null 时会引发异常）。

再看对"抽象型"变量的赋值问题。在 LinkNode<T>的定义中使用如下代码也都是错误的：

```
private T data = null;  //错误

public LinkNode()
{
    data = 0;  //错误
}
```

这是因为在创建 LinkNode<T>的构造类型时，T 既可能被替换成一个值类型，也可能被替换成一个引用类型。对于前者，不能把 null 赋值给_data 字段；而对于非数值类型，则不能把 0 赋值给_data 字段。

为此，C#中允许通过关键字 default 来获得"抽象型"变量的默认值，例如：

```
public LinkNode()
{
    data = default(T);
```

}
```

那么在构造 LinkNode<int>、LinkNode<double>等对象时,其_data 字段的初始值就是 0;而在构造 LinkNode<object>、LinkNode<string>等对象时,_data 字段的初始值就是 null。

 这里的 default 实际上是一个操作符,它和 switch 选择结构中的 default 标签的含义是不同的。

### 11.2.3 使用多个类型参数

泛型类可以有多个类型参数,其中每个参数指代一个抽象数据类型。定义时这些类型参数之间通过逗号分隔,例如下面的代码就定义了一个泛型类 Pair<L, R>,它的两个字段 Left 和 Right 分别对应类型参数 L 和 R:

```
public class Pair<L, R>
{
    public L Left;
    public R Right;

    public Pair(L left, R right)
    {
        Left = left;
        Right = right;
    }

    public override string ToString()
    {
        return string.Format("Left: {0}, Right: {1}", Left, Right);
    }
}
```

下面的代码则使用了 Pair<L, R>的不同构造类型的对象:

```
Pair<int, int> p1 = new Pair<int, int>(1, 1);
Pair<char, string> p2 = new Pair<char, string>('S', "王小红");
Pair<int, string> p3 = new Pair<int, string>(p1.Left, p2.Right);
```

11.2.4 类型参数与标识

泛型类的标识由名称和类型参数共同组成,因此类型参数足以区分不同的类型,例如可以在程序中同时定义普通类 Pair、一元泛型类 Pair<T>和二元泛型类 Pair<L, R>,它们分别表示不同的类型,不会引起编译错误:

```
public class Pair
{
    Public int Left;
    Public string Right;
}

public class Pair<T>
{
    Public int Left;
    Public T Right;
}

public class Pair<L, R>
{
    public L Left;
```
```

```
    public R Right;
}
```

但类型参数的名称仅仅是一个指代作用,并不能区分不同的类型。例如不能在一个程序中同时定义泛型类 Pair<T>和 Pair<S>,也不能同时定义 Pair<L, R>和 Pair<R, L>。

同样,泛型类的类型参数也足以区分不同的方法成员。例如泛型类 Pair<L, R>中可以同时包含下面两个 SetValue 方法:

```
public class Pair<L, R>
{
    public L Left;
    public R Right;

    public Pair(L left, R right)
    {
        Left = left;
        Right = right;
    }

    public void SetValue(L l)
    {
        Left = l;
    }

    public void SetValue(R r)
    {
        Right = r;
    }
}
```

那么对于一个 Pair<int, string>类型的对象,下面的代码会将 100 赋值给其 Left 字段,将"王小红"赋值给其 Right 字段:

```
Pair<int, string> p1 = new Pair<int, string>(0, null);
p1.SetValue(100); //调用 SetValue(L l)
p1.SetValue("王小红"); //调用 SetValue(R r)
```

不过像 Pair<int, int>这样的构造类型就会导致两个 SetValue 方法的标识相同,那么下面的代码就是错误的,因为 C#编译器不能确定应该调用哪一个方法:

```
Pair<int, int> p1 = new Pair<int, int>(0, 0);
p1.SetValue(100); //错误:存在歧义!
```

也就是说,在使用泛型类的构造类型时,如果因类型参数的替换而导致出现两个标识相同的方法成员,那么程序就不能通过编译。

不过,在两种方法出现重名时,如果其中一种方法的原型使用了类型参数,另一种方法的原型没有使用类型参数,此时不会引起歧义,而程序会总是调用没有使用类型参数的方法。例如下面定义的泛型类 Pair<T>也有两种 SetValue 方法,而且对于构造类型 Pair<int>二者的标识相同,但对 p1 对象只能执行第一个 SetValue 方法:

```
//程序 P11_3
using System;
namespace P11_3
{
    class Program
    {
        static void Main()
        {
            Pair<int> p1 = new Pair<int>(0);
            p1.SetValue(5); //调用 SetValue(int i)
```

```
            Console.WriteLine(p1.Left);  //输出 5
            Console.WriteLine(p1.Right); //输出 0

            Pair<string> p2 = new Pair<string>("");
            p2.SetValue(5); //调用 SetValue(int i)
            p2.SetValue("王小红"); //调用 SetValue(T t)
            Console.WriteLine(p2.Left);
            Console.WriteLine(p2.Right);
        }
    }
    public class Pair<T>
    {
        public int Left;
        public T Right;

        public Pair(T right)
        {
            Right = right;
        }

        public void SetValue(int i)
        {
            Left = i;
        }

        public void SetValue(T t)
        {
            Right = t;
        }
    }
```

11.2.5　泛型的静态成员

在本书 4.1.3 小节中介绍过，类的静态成员不属于类的某个实例，而是属于类的本身所有。泛型类的静态成员则更为特殊，它既不属于泛型类的某个实例，也不属于泛型类，而是属于泛型类的构造类型。

例如下面的代码为泛型类 Pair<L, R>定义了一个静态方法 Swap：

```
public class Pair<L, R>
{
    public L Left;
    public R Right;

    public Pair(L left, R right)
    {
        Left = left;
        Right = right;
    }

    public static void Swap(Pair<L, R> p1, Pair<L, R> p2)
    {
        L t1 = p1.Left;
        p1.Left = p2.Left;
        p2.Left = t1;
        R t2 = p1.Right;
        p1.Right = p2.Right;
        p2.Right = t2;
    }
}
```

那么对于两个相同构造类型的对象 p1 和 p2，Swap(p1, p2)方法前的类型引用既不能是 Pair，也不能是 Pair<L, R>，而应该是 Pair<L, R>的构造类型（即 p1 和 p2 的实际类型）：

```
Pair<char, string> p1 = new Pair<char, string>('T', "张大为");
Pair<char, string> p2 = new Pair<char, string>('S', "王小红");
Pair<char, string>.Swap(p1, p2);
```

对于普通类，一个静态字段在内存中最多只有一份复制；对于泛型类，使用了它的多少种构造类型，其静态字段就在内存中拥有多少份复制。同样，泛型类的静态构造函数会针对每个构造类型执行一次。下面的程序说明了这一点（可将其与程序 P4_5 进行比较）：

```
//程序 P11_4
using System;
namespace P11_4
{
    class Program
    {
        static void Main()
        {
            GA<int> g1 = new GA<int>();
            GA<string> g2 = new GA<string>();
            GA<int> g3 = new GA<int>();
            g1 = new GA<int>();
            g2 = new GA<string>();
        }
    }

    public class GA<T>
    {
        T t = default(T);
        static int objects = 0;
        static int classes = 0;

        public GA()  //实例构造函数
        {
            Console.WriteLine("GA<{0}>对象计数: {1}", typeof(T), ++objects);
        }

        static GA()  //静态构造函数
        {
            Console.WriteLine("GA<{0}>类计数: {1}", typeof(T), ++classes);
        }
    }
}
```

从程序 P11_4 的输出中可以看到，构造类型 GA<int>和 GA<string>的将各自进行对象计数和类计数，二者互不影响：

```
GA<System.Int32>类计数: 1
GA<System.Int32>对象计数: 1
GA<System.String>类计数: 1
GA<System.String>对象计数: 1
GA<System.Int32>对象计数: 2
GA<System.Int32>对象计数: 3
GA<System.String>对象计数: 2
```

11.3 类型限制

在默认情况下，泛型中的类型参数可以被替换为任意类型，这在很多情况下并不是人们所希望的。C#支持在泛型定义中通过 where 关键字来对类型参数进行限制，限制方式包括主要限制、次要限制和构造函数限制。

11.3.1 主要限制

主要限制是指类型参数只能被替换为值类型或引用类型，例如下面定义的 Pair<L, R>类就限制类型参数 L 只能被替换成值类型：

```
public class Pair<L, R> where L : struct
{}
```

在泛型类的定义中，类型限制是在 where 关键字后依次写上类型参数、冒号以及限制方式，其中 struct 关键字表示值类型限制，而 class 关键字表示引用类型限制。那么对于上面定义的泛型类，其类型参数 L 只能被替换成简单值类型、枚举类型或结构类型（在.NET 类库中，前两者在本质上仍属于 struct），而使用下面这样的语句都是不合法的，因为 string 和 object 都是引用类型：

```
Pair<string, string> p1 = new Pair<string, string>("", "王小红"); //错误
Pair<object, int> p2 = new Pair<object, int>(null, 100); //错误
```

下面的定义则表示 Pair<L, R>的限制类型参数 R 只能被替换成引用类型：

```
public class Pair<L, R> where R : class
{
    public L Left;
    public R Right = null;
}
```

注意其中的 R 型字段 Right 的值可被初始化为 null，因为 R 已经被限制为引用类型。

如果要为多个类型参数同时添加限制，那么就应使用多个 where 限制子句，之间通过空格或换行符进行分割，例如：

```
public class Pair<L, R> where L : struct where R : class
{}
```

11.3.2 次要限制

次要限制是将类型参数的目标类型限制为从指定的基类或接口继承，声明方式是在 where 子句的冒号后跟基类或接口的名称。例如下面的泛型类 Pair<L, R>要求类型参数 R 的目标类型必须支持 IComparable 接口：

```
public class Pair<L, R> where R : IComparable
{
    public L Left;
    public R Right;

    public Pair(L left, R right)
    {
        Left = left;
        Right = right;
    }

    public static bool operator >(Pair<L, R> p1, Pair<L, R> p2)
    {
        return (p1.Right.CompareTo(p2.Right) > 0);
```

```
    }
    public static bool operator <(Pair<L, R> p1, Pair<L, R> p2)
    {
        return (p1.Right.CompareTo(p2.Right) < 0);
    }
}
```

由于 IComparable 接口中定义了一个 CompareTo 方法，而任何支持该接口的类型必将实现该方法，所以对 R 型变量就可以调用其 CompareTo 方法，上面的代码就是利用这一点来重载大于和小于操作符的。次要限制能够给程序带来很大的方便，因为只要对类型参数进行了基类和接口限制，就能够对相应的"抽象型"变量调用这些基类和接口的成员。显然，用作类型限制的基类不能是密封类。

因为 C#支持类的单继承和接口的多继承，所以次要限制中最多只能有一个基类，但可以有多个接口，它们之间通过逗号分隔。如果存在基类限制，那么基类应该写在所有接口之前，例如：

```
public class Student { }
public class Pair<L, R> where R : Student, IComparable { }
```

主要限制和次要限制还可以同时存在，例如：

```
public class Pair<L, R> where R : class, IComparable { }
```

但如果已经指定了基类限制，这就隐含了类型参数的目标类型必须是类的要求（结构类型不能从类中派生），那么就不能再进行基于 class 的主要限制。

11.3.3 构造函数限制

构造函数限制也叫 new 限制，它要求类型参数的目标类型必须提供一个无参的构造函数，书写格式是 new 关键字加一对括号。如果还同时存在主要限制或次要限制，那么 new 限制应写在其他所有限制之后。例如下面的 Pair<L, R>类就为类型参数 R 指定了构造函数限制，那么在类的方法代码中就可以调用相应的构造函数来创建 R 型对象：

```
public class Pair<L, R> where R : IComparable, new()
{
    public L Left;
    public R Right;

    public Pair()
    {
        Left = default(L);
        Right = new R();
    }
}
```

11.4 泛型继承

在介绍与泛型有关的继承之前，首先引入开放类型和封闭类型的概念。以二元泛型类 Pair<L, R>为例，在其代码中允许定义的变量类型可分为两类：

- 含有类型参数的类型，如 L、R、Pair<L, R>、Pair<int, R>、Pair<L, string>等，以及以这些类型为元素的数组类型，它们统称为开放类型。
- 不含类型参数的类型，如 int、string、Pair<int, string>等，以及以这些类型为元素的数组类型，它们统称为封闭类型。

从计算机的角度来理解，开放类型含有"待定"的类型参数，还不属于能够真正实现的类型；只有封闭类型才可以创建实例，并拥有内存存储。而泛型类之间的继承有一条基本原则：开放类型不能作为封闭类型的基类。

在引入了泛型的概念后，类之间的继承可以分为以下4种情况。

（1）普通类之间的继承：这属于封闭类型之间的继承。

（2）泛型类继承普通类：这种继承与前一种其实没有什么区别，派生类一样从基类继承各种成员，而派生类自己的泛型功能与基类无关。这属于开放类型从封闭类型继承。

（3）普通类继承泛型类：非泛型的普通类不能直接继承自泛型类，而是只能继承泛型类的封闭构造类型。这也属于封闭类型之间的继承。

（4）泛型类继承泛型类：它要求基类中的类型参数必须都在派生类的定义中出现，形象地说就是基类不能比派生类更"开放"。这属于开放类型之间的继承。

继承的关系就是从一般到特殊的关系。如果基类中出现了类型参数，那么派生类要么继续使用这些类型参数，要么对这些类型参数进行具体化，而不能对其"置之不理"。例如在下面的继承示例中，Triple<A, B, C>使用了泛型基类的两个类型参数，IndexedValue<T>使用了一个类型参数，并将另一个具体化为 int，StringTable 则对基类的两个类型参数都进行了具体化：

```
public class Pair<L, R>
{
    public L Left;
    public R Right;

    public Pair(L left, R right)
    {
        Left = left;
        Right = right;
    }
}

public class Triple<A, B, C> : Pair<A, C>
{
    public B Middle;

    public Triple(A a, B b, C c) : base(a, c)
    {
        this.Middle = b;
    }
}

public class IndexedValue<T> : Pair<int, T>
{
    public IndexedValue(int index, T value) : base(index, value) { }
}

public class StringTable : Pair<string, string>
{
    public StringTable(string s1, string s2) : base(s1, s2) { }
}
```

由于类型参数名仅仅是起指代作用，那么上面的代码可以看成是 Pair<L, R>的派生类 Triple<L, B, R>，Pair<int, R>的派生类 IndexedValue<R>，以及 Pair<string, string>的派生类 StringTable。

派生类可以任意增加自己的类型参数，但必须包含基类的类型参数。例如不能使 StringTable 派生自 Pair<string, R>，也不能使 IndexedValue<T>派生自 Pair<int, R>。

程序 P11_5 定义了链表节点类 LinkNode<T>的派生类——双向链表节点类 BiLinkNode<T>，它继承了基类的抽象字段_data 和属性 Data，并增加了_previous 字段和 Previous 属性，从而能够获得指向前一个节点的链接指针；BiLinkNode<T>还要求其后续节点也是双向链表节点，因此隐藏了基类的_next 字段和 Next 属性（编译时注意添加对程序 P11_2 的引用）。

```csharp
//程序 P11_5
using System;
using P11_2
namespace P11_5
{
    public class BiLinkNode<T> : LinkNode<T>
    {
        protected BiLinkNode<T> previous;
        protected new BiLinkNode<T> next;

        public BiLinkNode<T> Previous   // 获取或设置节点的前一个节点
        {
            get { return previous; }
            set {
                previous = value;
                if (previous != null && previous.Next != this)
                    previous.Next = this;
            }
        }

        public new BiLinkNode<T> Next  // 获取或设置节点的下一个节点
        {
            get { return next; }
            set {
                next = value;
                if (next != null && _next.Next != this)
                    next.previous = this;
            }
        }

        public BiLinkNode(T t) : base(t) { }

        //移至 node 的第 n 个前驱节点
        public static BiLinkNode<T> operator <<(BiLinkNode<T> node, int n)
        {
            BiLinkNode<T> node1 = node;
            for (int i = 0; i < n && node1 != null; i++)
                node1 = node1.previous;
            return node1;
        }
    }
}
```

下面的示例代码创建了依次链接的一系列 BiLinkNode<int>的节点，并且能够自由地访问它们的前驱节点和后续节点：

```csharp
int x = 0;
BiLinkNode<int> node1 = new BiLinkNode<int>(x);
while (x++ < 1000){
    if (x % 3 == 0 && x % 4 == 1 && x % 5 == 2){
        node1.Next = new BiLinkNode<int>(x);
        node1 = node1.Next;
    }
}
while (node1 != null){
    Console.WriteLine(node1.Data);
    node1 <<= 1;
}
```

11.5 泛型接口

11.5.1 泛型接口的定义

在接口中也能使用类型参数来指代抽象数据类型。比如.NET 类库中同时定义了接口 IComparable 和泛型接口 IComparable<T>，前者用于当前对象与一个 object 对象的比较，后者则用于当前对象与另一个同类对象之间的比较：

```
public interface IComparable
{
    int CompareTo(object obj);
}

public interface IComparable<T>
{
    int CompareTo(T t);
}
```

在大部分场合下，显然后者的比较更有意义。例如下面的泛型类定义就要求类型参数 R 的目标类型必须支持比较操作：

```
public class Pair<L, R> where R : IComparable<R>
{
    public L Left;
    public R Right;

    public Pair(L left, R right)
    {
        Left = left;
        Right = right;
    }

    public static bool operator >(Pair<L, R> p1, Pair<L, R> p2)
    {
        return (p1.Right.CompareTo(p2.Right) > 0);
    }

    public static bool operator <(Pair<L, R> p1, Pair<L, R> p2)
    {
        return (p1.Right.CompareTo(p2.Right) < 0);
    }
}
```

泛型接口的类型参数也可以加以限制，此时在 11.3 节中介绍的内容同样适用。例如下面定义的泛型接口 IIndexable<T>就要求其类型参数支持 IPrintable 接口：

```
public interface IPrintable
{
    void Print();
}

public interface IIndexable<T> where T : IPrintable
{
    int GetIndex();
}
```

泛型接口本身也可用于类型限制，但此时接口所使用的类型参数必须在被限制的类型中出现。例如下面泛型类 NumberCollection<V>的类型参数指定了 IIndexable<V>接口限制：

```
public class NumberCollection<V> where V : IPrintable, IIndexable<V>
{ }
```

 在上面的泛型定义中，类型参数 V 要求支持 IIndexable<V>接口，而 IIndexable<V>又要求 V 支持 IPrintable 接口，但在 NumberCollection<V>的类型限制中仍要求写出 IPrintable 接口限制。C#要求在泛型定义中完整地写出每个类型参数的所有限制，而不是由编译器去推断并自动添加限制。

11.5.2 泛型接口的实现

和泛型类的继承类似，泛型接口中的所有类型参数必须出现在其派生类型中，形象地说就是接口不能比派生类更"开放"。例如不能使非泛型接口 IIndexable 从泛型接口 IComparable<T>中派生，但 IIndexable 从 IComparable<int>中派生、IIndexable<T>从泛型接口 IComparable<T>中派生都是允许的。

如果一个类派生自（声明支持）某个泛型接口，它同样必须履行该接口的契约，即支持接口中定义的所有方法。这可以分为以下两种情况：

- 如果类声明支持的是泛型接口本身，而接口方法定义中又使用了类型参数，那么类所实现的接口方法也包含这些类型参数。
- 如果类声明支持的是泛型接口的构造类型，而接口方法定义中又使用了类型参数，那么类所实现的接口方法也要对这些类型参数进行相应的具体化。

例如要使泛型类 Pair<T> 支持 IComparable<T>接口，那么在 Pair<T>中就要提供对方法 CompareTo(T t)的实现；而如果是使非泛型类 Student 支持 IComparable<Student>接口，那么在 Student 中就要提供对方法 CompareTo(Student t)的实现。当然，这些实现既可以是公有成员方法的隐式实现，又可以是接口方法的显式实现。不过在显式实现时接口的类型参数不能省略，例如：

```
public class Pair<T> : IComparable, IComparable<T>
{
    int IComparable.CompareTo(object obj) { … }    //显式实现非泛型接口方法
    int IComparable<T>.CompareTo(T t) { … }        //显式实现泛型接口方法
    public int CompareTo(object obj) { … }         //隐式实现非泛型接口方法
    public int CompareTo(T t) { … }                //隐式实现泛型接口方法
}
```

下面的示例代码为 BiLinkNode<T>类增加了对 IComparable<BiLinkNode<T>>接口的支持，并通过节点之间的前驱或后继关系来比较两个节点对象：

```
public class BiLinkNode<T> : LinkNode<T>, IComparable<BiLinkNode<T>>
{
    protected BiLinkNode<T> _previous;
    protected new BiLinkNode<T> _next;
    //…
    public int CompareTo(BiLinkNode<T> node)   //实现接口方法
    {
        BiLinkNode<T> tNode = this._next;
        int i = 1;
        while (tNode != null){
            if (tNode == this)
                return i;
            tNode = tNode._next;
            i++;
        }
```

```
            tNode = this._previous;
            i = -1;
            while (tNode != null){
                if (tNode == this)
                    return i;
                tNode = tNode._previous;
                i--;
            }
            return 0;
    }
```

在本书第 4 章中介绍过，一个类型中不能有两个标识相同的方法（包括属性和索引函数），即使它们的返回类型不同。例如一个联系人可能有多个电话和 E-mail，但不能对 Contact 对象既索引电话又索引 E-mail 地址：

```
public class Contact
{
    private string[] _phones;
    private Email[] _emails;

    public string this[int index]
    {
        get { return _phones[index]; }
        set { _phones[index] = value; }
    }

    public Email this[int index]   //错误：一个类不能有两个索引函数！
    {
        get { return _emails[index]; }
        set { _emails[index] = value; }
    }
}
```

不过接口的显式实现能够区分两个同名方法；而对于泛型接口，一个类能够同时支持其多个构造类型，并通过显式实现来区分不同的方法成员。下面的程序 P11_6 就定义了一个泛型接口 IItems<T>，并在 Contact 类中支持该接口的 3 个构造类型，进而实现了 3 个索引函数：

```
//程序 P11_6
using System;
namespace P11_6
{
    class Program
    {
        static void Main()
        {
            Contact c = new Contact("张大强");
            c[0] = "010-88664321";
            c[1] = "13011110234";
            ((IItems<Email>)c)[0] = new Email("zhangdaq", "hstu.edu.cn");
            ((IItems<Email>)c)[1] = new Email("dqzhang", "soft.cn");
            ((IItems<Address>)c)[0] = new Address("湖北", "武汉", "科技路", 9);
        }
    }

    public interface IItems<T>
    {
        T this[int index] { get; set; }
    }

    public class Contact : IItems<Email>, IItems<string>, IItems<Address>
```

```csharp
    {
        private string _name;
        private string[] _phones;
        private Email[] _emails;
        private Address[] _addresses;

        public string Name
        {
            get { return _name; }
        }

        public string this[int index]
        {
            get { return _phones[index]; }
            set { _phones[index] = value; }
        }

        Email IItems<Email>.this[int index]
        {
            get { return _emails[index]; }
            set { _emails[index] = value; }
        }

        Address IItems<Address>.this[int index]
        {
            get { return _addresses[index]; }
            set { _addresses[index] = value; }
        }

        public Contact(string name)
        {
            _name = name;
            _phones = new string[5];
            _emails = new Email[3];
            _addresses = new Address[3];
        }
    }

    public struct Email
    {
        public string Username, Domain;

        public Email(string username, string domain)
        {
            Username = username;
            Domain = domain;
        }

        public override string ToString()
        {
            return string.Format("{0}@{1}", Username, Domain);
        }
    }

    public struct Address
    {
        public string Province, City, Street;
        public int No;

        public Address(string province, string city, string street, int no)
        {
            Province = province;
```

```
            City = city;
            Street = street;
            No = no;
        }

        public override string ToString()
        {
            return string.Format("{0}{1}{2}{3}号", Province, City, Street, No);
        }
    }
}
```

11.5.3 避免二义性

和普通接口一样，一个类如果支持多个泛型接口，而这些接口又具有相同的方法，那么可以通过显式实现来区分方法来自于哪个接口，例如：

```
public interface IA<T> where T : new()
{
    T F();
}

public interface IB<T> where T : new()
{
    T F();
}

public class Pair<L, R> : IA<L>, IB<R>
    where L : new()
    where R : new()
{
    L IA<L>.F() { return new L(); }
    R IB<R>.F() { return new R(); }
}
```

一个类不能既支持某个泛型接口又支持其构造类型，因为这会导致重复继承。假设 Pair<T>类同时支持 IComparable<T>接口和 IComparable<string>接口，那么构造类型 Pair<string>实际上就两次继承了 IComparable<string>接口，这在 C#中是不允许的。

如果要使一个类同时支持某个泛型接口的多个构造类型，那么就要保证在所有可能的类型参数替换的情况中都不会出现二义性问题。显然，使一个类同时支持 IComparable<int>接口和 IComparable<string>接口是没有问题的，但如果使 Pair<L, R>同时支持 IComparable<L>和 IComparable<R>就是不允许的，因为只要将 L 和 R 替换为同一个具体类型就会出现二义性。

再如下面定义的 PairA<L, R>类是合法的，而 PairB<L, R>则是不合法的，因为其构造类型 PairB<int, string>会出现二义性的问题：

```
public interface IPair<A, B>
{
    void F(A a, B b);
}

public class PairA<L, R> : IPair<L, string>, IPair<L, int>  //正确
{
    public void F(L l, string s) { }
    public void F(L l, int i) { }
}

public class PairB<L, R> : IPair<L, string>, IPair<int, R>  //错误
{
```

```
    public void F(L l, string s) { }
    public void F(int i, R r) { }
}
```

11.5.4 泛型接口与泛型集合

泛型技术能够在集合类型中发挥重要的作用，.NET 类库的 System.Collections.Generic 命名空间中定义了一组与泛型集合有关的接口，并通过在预定义集合类型中支持不同的接口来提供相关服务。表 11-1 简要地描述了.NET 类库中与集合有关的泛型接口，表 11-2 则介绍了.NET 类库中支持这些接口的泛型集合。

表 11-1　　　　　　　　　　　　.NET 类库中与集合有关的泛型接口

接口	支持的概念描述	继承的接口	主要方法
IEnumerable<T>	支持 foreach 遍历的可枚举泛型	IEnumerable	IEnumerator GetEnumerator() IEnumerator<T> GetEnumerator()
IEnumerator<T>	枚举器，提供对可枚举泛型的遍历操作	IEnumerator	bool MoveNext() void Reset() T Current { get; }
ICollection<T>	泛型对象集合	IEnumerable IEnumerable<T>	bool Add (T) void Clear() bool Contains (T) void CopyTo(T[], int) bool Remove (T) int Count { get; } boolIsReadOnly { get; }
IList<T>	泛型对象列表，支持对象的插入、删除和索引访问	IEnumerable IEnumerable<T> ICollection<T>	void Insert(int, T) int IndexOf(T) void RemoveAt(int) object this[int index] { get; set; }
IDictionary<K, V>	有字典序（键/值对）的泛型集合	IEnumerable IEnumerable<V> ICollection<V>	int Add(K, V) bool ContainsKey (K) boolTryGetValue(K, out V) ICollection<K> Keys { get; } ICollection<V>Values { get; } V this[Kkey] { get; set; }

表 11-2　　　　　　　　　　　　.NET 类库中的主要泛型集合

接口	类型描述	继承的接口
List<T>	泛型列表类，集合大小可动态改变	IEnumerable<T>, ICollection, ICollection, IList<T>
LinkedList<T>	泛型双向链表类	IEnumerable<T>, ICollection
Queue<T>	先进先出的泛型队列类	IEnumerable<T>, ICollection
Stack<T>	先进后出的泛型堆栈类	IEnumerable<T>, ICollection
Dictionary<K, V>	泛型字典（键/值对）集合	ICollection<KeyValuePair<K,V>>,IDictionary<K, V>,IDictionary, ICollection
SortedDictionary<K, V>	有序的泛型字典（键/值对）集合	ICollection<KeyValuePair<K,V>>, IDictionary<K, V>, IDictionary, ICollection
HashSet<T>	泛型哈希集合	ICollection<T>, IEnumerable<T>
SynchronizedCollection<T>	支持异步操作的泛型集合	IEnumerable<T>, ICollection<T>, IList<T>, ICollection

其中一些泛型集合在 System.Collections 命名空间下有对应的非泛型类型，不过泛型支持类型替换，而且要求集合元素为同一目标类型，因此比非泛型集合更加实用、高效和安全。出于兼容性考虑，大部分泛型集合还同时支持泛型和非泛型接口。

下面的代码示范了 Stack<T>和 Queue<T>的简单用法：

```
Stack<int> stack1 = new Stack<int>();
Queue<int> queue1 = new Queue<int>(new int[] { 1, 2, 3, 4, 5 });
foreach (int i in queue1)
    Console.WriteLine(i); //依次输出1、2、3、4、5
while (queue1.Count != 0)
    stack1.Push(queue1.Dequeue());
while (stack1.Count != 0)
    queue1.Enqueue(stack1.Pop());
foreach (int i in queue1)
    Console.WriteLine(i); //依次输出5、4、3、2、1
```

Dictionary<K, V>则用于维护一组键/值对的集合，其元素类型为泛型结构 KeyPairValue<K, V>，每一个元素对象都由一个 K 类型的键（Key）和一个 V 类型的值（Value）共同组成。例如下面的代码创建了一个简单的中英词典，用户可以快速查询中文单词(键)所对应的英文单词(值)：

```
Dictionary<string, string> dic1 = new Dictionary<string, string>();
dic1.Add("苹果", "Apple");
dic1.Add("桔子", "Orange");
dic1.Add("香蕉", "Banana");
dic1.Add("梨", "Pear");
Console.Write("请输如要查询的中文单词: ");
string key = Console.ReadLine();
string value = dic1[key];
if (value != null)
    Console.WriteLine("{0}的英文单词是: {1}", key, value);
else
    Console.WriteLine("字典中没有该单词");
```

程序 P11_7 创建了一个自定义泛型集合——单链表 LinkList<T>，它维护一组依次链接的 LinkNode<T>对象。链表具有头节点 first、尾节点 last 以及当前节点 current；通过支持实现 ICollection<T>和 IList<T>的接口方法，它支持在链表中查找、追加、插入和删除元素。和数组不同，链表不能直接获取指定下标的元素，而是每次都要从头节点开始依次向后遍历，但它的优点是大小可以动态调整。程序的主方法演示了如何创建一个约会（Appointment）链表，并在其中加入和删除约会安排：

```
//程序 P11_7
using System;
using System.Collections;
using System.Collections.Generic;
using P11_2;
namespace P11_7
{
    class Program
    {
        static void Main()
        {
            LinkList<Appointment> l1 = new LinkList<Appointment>();
            l1.Add(new Appointment(new DateTime(2009, 1, 1), "徐海洋"));
            l1.Add(new Appointment(new DateTime(2009, 1, 3), "王亮"));
            l1.Add(new Appointment(new DateTime(2009, 1, 4), "刘静"));
```

```csharp
        l1.Add(new Appointment(new DateTime(2009, 1, 7), "王小燕"));
        l1.Insert(2, new Appointment(new DateTime(2009, 1, 5), "成亮"));
        l1.RemoveAt(3);
        while (l1.MoveNext())
            Console.WriteLine(l1.Current);
    }

    struct Appointment
    {
        public DateTime Time;
        public string Customer;

        public Appointment(DateTime time, string customer)
        {
            Time = time;
            Customer = customer;
        }

        public override string ToString()
        {
            return Time.ToLongDateString() + Customer;
        }
    }
}

public class LinkList<T> : ICollection<T>, IList<T>
{
    protected LinkNode<T> first, last, current;
    public T Current
    {
        get { return current.Data; }
        set { current.Data = value; }
    }

    private int count = 0;
    public int Count
    {
        get { return count; }
    }

    public bool IsReadOnly
    {
        get { return false; }
    }

    public T this[int index]   // 获取指定位置的节点值
    {
        get { return this.GetNode(index).Data; }
        set { this.GetNode(index).Data = value; }
    }

    public LinkList()
    {
        current = last = first = new LinkNode<T>(default(T));
    }

    public void Add(T t)   //向链表尾部加入元素
    {
        last.Next = new LinkNode<T>(t);
```

```csharp
        last = last.Next;
        count++;
    }

    public void Clear() //清空链表
    {
        current = last = first;
        first.Next = null;
    }

    public bool Contains(T t) //检查是否包含某个元素
    {
        return (this.IndexOf(t) >= 0);
    }

    public bool MoveNext() //当前位置移动到下一个元素
    {
        if (current.Next != null){
            current = current.Next;
            return true;
        }
        else
            return false;
    }

    public void Reset() //将当前节点重设为头节点
    {
        current = first;
    }

    public int IndexOf(T t) //查找包含指定值的节点位置
    {
        LinkNode<T> node = first.Next;
        int i = 0;
        while (node != null){
            if (node.Data.Equals(t))
                return i;
            node = node.Next;
        }
        return -1;
    }

    public void Insert(int index, T t) // 在指定位置处插入节点
    {
        LinkNode<T> node = this.GetNode(index - 1);
        LinkNode<T> newNode = new LinkNode<T>(t);
        if (node.Next != null)
            newNode.Next = node.Next;
        node.Next = newNode;
        count++;
    }

    public void RemoveAt(int index) // 删除指定位置的节点
    {
        LinkNode<T> node = this.GetNode(index - 1);
        node.Next = node.Next.Next;
        count--;
    }
```

```csharp
public bool Remove(T t)  // 删除指定的节点
{
    int i = this.IndexOf(t);
    if (i < 0)
        return false;
    this.RemoveAt(i);
    return true;
}

public void CopyTo(T[] array, int index)   //将所有节点数据复制到数组
{
    LinkNode<T> node = first.Next;
    while (node != null&& index < array.Length){
        array[index++] = node.Data;
        node = node.Next;
    }
}

protected LinkNode<T> GetNode(int index)
{
    LinkNode<T> node = first.Next;
    for (int i = 0; i < index; i++){
        if (node == null)
            throw new IndexOutOfRangeException("超出链表末尾");
        node = node.Next;
    }
    return node;
}

IEnumerator<T> IEnumerable<T>.GetEnumerator(){return null;}
IEnumerator IEnumerable.GetEnumerator(){return null;}
}
```

11.6 泛型方法

11.6.1 泛型方法的定义和使用

泛型方法是指使用了类型参数的方法成员。我们已经掌握了如何为泛型类添加泛型的方法成员，这些泛型方法的类型参数会随着泛型的类型参数一同进行实例化。例如下面 CMath<T>类的静态方法 Swap 可用于交换两个 T 类型的变量值：

```csharp
public class CMath<T>
{
    public static voidSwap(ref T x, ref T y)
    {
        T tmp = x;
        x = y;
        y = tmp;
    }

    public static int Round(double d)
    {
        int i = (int)d;
        if (d - i >= 0.5)
```

```
            return i + 1;
        else
            return i;
    }
}
```

这样构造类型 CMath<int>的 Swap 方法就可用于交换两个整数，CMath<string>的 Swap 方法就可用于交换两个字符串。不过该类还定义了一个用于对小数进行四舍五入的 Round 方法，该方法并没有用到类型参数，但每次调用该方法仍需通过构造类型进行，这就会带来不便和混淆。

为此，C#允许为普通类型定义泛型的方法成员，此时类型参数写在方法名之后的一对尖括号中，例如可以将上面的类定义改为如下内容：

```
public class CMath
{
    public static voidSwap<T>(ref T x, ref T y)
    {
        T tmp = x;
        x = y;
        y = tmp;
    }

    public static int Round(double d)
    {
        int i = (int)d;
        if (d - i >= 0.5)
            return i + 1;
        else
            return i;
    }
}
```

那么类型参数 T 就只对 Swap 方法有效，而在调用其他非泛型方法时不需要处理构造类型的问题，例如：

```
double x = 2.5, y = 3.2;
CMath.Swap<double>(ref x, ref y);  //调用泛型方法
int z = CMath.Round(x);  //调用非泛型方法
```

提示　　C#不允许在类的构造函数、属性、事件、索引函数和操作符这些特殊方法成员中使用类型参数。

泛型方法也可以对其类型参数进行限制，包括主要限制、次要限制和构造函数限制。例如可以为 CMath 类添加如下的泛型方法，它要求类型参数的目标类型支持 IComparable<T>接口：

```
public static T Max<T>(T x, T y) where T : IComparable<T>
{
    return (x.CompareTo(y) >= 0) ? x : y;
}

public static T Min<T>(T x, T y) where T : IComparable<T>
{
    return (x.CompareTo(y) <= 0) ? x : y;
}
```

在不引起混淆的情况下，调用泛型方法时可以省略类型参数的实际类型，而由 C#编译器根据传递给方法的参数来进行推断，例如：

```
int a = CMath.Max(25, 30);  //调用 CMath.Max<int>方法
double b = CMath.Max(0.5, 0.8);  //调用 CMath.Max<double>方法
```

11.6.2　泛型方法的重载

和普通方法一样，泛型方法也可以被定义成为虚拟方法、重载方法、抽象方法或密封方法，普通方法的继承和多态性规则同样适用于泛型方法。例如下面的 Searcher 类定义了一个虚拟的泛型方法 Find<T>，它用于在一个 T 型数组中顺序查找一个指定元素；派生类 BiSearcher 中则重载了 Find<T>方法，它是从数组的中间向两头查找：

```
public class Searcher
{
    public virtual int Find<T>(T[] array, T target)
    {
        for (int i = 0; i < array.Length; i++){
            if (array[i].Equals(target))
                return i;
        }
        return -1;
    }
}
public class BiSearcher : Searcher
{
    public override int Find<T>(T[] array, T target)
    {
        for (int i = array.Length / 2, j = array.Length / 2; i < array.Length && j >= 0; i++, j--){
            if (array[i].Equals(target))
                return i;
            if (array[j].Equals(target))
                return j;
        }
        return -1;
    }
}
```

11.6.3　泛型方法与委托

泛型方法也可以通过委托来进行调用。由于在最终调用时类型参数总是要被替换为构造类型，因此可以使用普通委托来调用泛型方法的构造型。例如可以定义下面的委托类型：

```
delegate int DualFunction(int x, int y);
```

那么通过该委托对象就能调用对应的方法：

```
DualFunction fun1 = CMath.Max<int>;
int x = fun1(25, 30);
fun1 = CMath.Min<int>;
int y = fun1(-5, 5);
```

另一种方式是使用泛型委托，其定义和泛型方法类似，也是在委托名称之后的一对尖括号中指定类型参数，例如：

```
delegate T DualFunction<T>(T x, T y);
```

这样经过泛型委托和泛型方法中类型参数的同步替换，就能够使用泛型委托的构造型来调用泛型方法的构造型，例如：

```
DualFunction<int> fun1 = CMath.Max<int>;
int x = fun1(25, 30);
DualFunction<double> fun2 = CMath.Max<double>;
double y = fun2(-2.5, 0.5);
```

.NET 类库中定义了一组泛型委托，大多数情况下我们可以直接应用这些委托，而不需要每次自定义委托类型。其中名称为 Action 的委托用于封装无返回值的方法，它有多种重载形式，如

Action<T>要求方法具有一个类型为 T 的参数，如 Action<T1,T2>要求有一个两个类型分别为 T1 和 T2 的参数，如 Action<T1,T2,T3>要求三个参数，等等。如果没有参数，则可以使用非泛型的委托 Action。

名为 Func 的委托用于封装有返回值的方法，如 Func<TR>要求方法无参数、返回类型为 TR，Func<T,TR>要求有一个参数，Func<T1,T2,TR>要求有两个参数，等。那么上面的代码可使用 Func<T1,T2,TR>改写为：

```
Func<int,int,int> fun1 = CMath.Max<int>;
int x = fun1(25, 30);
Func<double,double,double> fun2 = CMath.Max<double>;
double y = fun2(-2.5, 0.5);
```

此外，名为 Predicate 的委托专门用于封装返回值为 bool 类型的方法，它同样通过多种重载形式来支持不同个数的参数，比如 Predicate<T>，Predicate<T1,T2>，等等。

下面的程序给出了一个将泛型委托作为方法参数进行传递的例子：通过将不同的 Func<T, T, T>委托对象传递给泛型方法 Aggr<T>，就可以对不同类型的数组实现求和、积、最大值、最小值等聚合运算，程序的输出如图 11-1 所示。

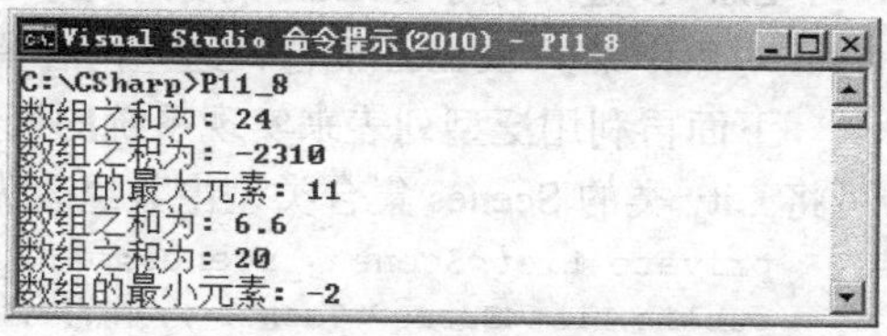

```
//程序 P11_8
using System;
namespace P11_8
{
    class Program
    {
        static void Main()
        {
            int[] a1 = { 3, 5, -2, 11, 7 };
            Console.WriteLine("数组之和为: " + CMath.Aggr(a1, (x, y) => x + y));
            Console.WriteLine("数组之积为: " + CMath.Aggr(a1, (x, y) => x * y));
            Console.WriteLine("数组的最大元素: " + CMath.Aggr(a1, Math.Max));
            double[] a2 = { -2, -0.5, 1.6, 2.5, 5 };
            Console.WriteLine("数组之和为: " + CMath.Aggr(a2, (x, y) => x + y));
            Console.WriteLine("数组之积为: " + CMath.Aggr(a2, (x, y) => x * y));
            Console.WriteLine("数组的最小元素: " + CMath.Aggr(a2, Math.Min));
        }
    }

    public class CMath
    {
        public static T Max<T>(T x, T y) where T : IComparable<T>
        {
            return (x.CompareTo(y) >= 0) ? x : y;
        }
        public static T Min<T>(T x, T y) where T : IComparable<T>
        {
            return (x.CompareTo(y) <= 0) ? x : y;
        }
        public static T Aggr<T>(T[] array, Func<T, T, T> fun)
        {
            T t = array[0];
            for (int i = 1; i < array.Length; i++)
                t = fun(t, array[i]);
            return t;
        }
    }
}
```

图 11-1　程序 P11_8 的输出结果

11.7 案例研究——旅行社管理系统中的泛型集合

11.7.1 使用泛型列表 List<T>

对于经常需要添加和删除元素的集合来说,用 ArrayList 或 List<T>来实现都很方便;如果集合中的元素都是同一类型(或是同一类型的派生类),使用 List<T>具有更好的性能。旅行社管理系统中就存在多个这样的集合。

例如第 10 章中为 StaffForm 和 Form1 定义的_staffs 字段,其类型都可以从 ArrayList 换为 List<Staff>:
```
private List<Staff> _staffs;
```
在改变了字段类型及其构造语句之后,即使不改动其他代码,程序仍可以编译运行,这是因为 ArrayList 和 List<T>实现的接口方法契约基本一致。

List<T>还可用于在 Line 对象中维护 Packages 集合,在 Package 对象中维护 Tours 集合,在 Tour 对象中维护 Scenes 集合,这些内容留给读者自行实现。

下面再利用泛型列表来实现系统的景点资料维护功能。如果景点资料可能会发生变化,那么可将 City 类的 Scenes 集合类型由数组改为列表:
```
private List<Scene> _scenes;
public List<Scene> Scenes //景点集合
{
    get{return _scenes;}
}
```

接下来为 TravelWin 项目添加一个新的 Windows 窗体 SceneForm,按图 11-2 所示内容设计窗体,并在"类别"、"状态"和"学历"组合框的 Items 集合中分别加入对应的枚举值。用户可在该窗体中选择省份和城市,并维护城市的景点集合。SceneForm 的下面两个字段分别表示当前所选的省份和城市:

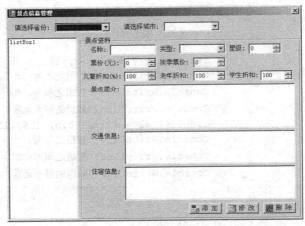

图 11-2 景点信息管理窗体 SceneForm

```
private Province _province;
private City _city;
```

窗体在启动时载入所有省份信息;用户选择一个省份后,城市组合框中将载入该省的所有城市信息;用户选择一个城市后,左侧列表框将列出该城市的所有景点信息:
```
private void SceneForm_Load(object sender, EventArgs e)
{
    foreach (Province p in Province.GetAll())
        cmbProvince.Items.Add(p);
}

private void cmbProvince_SelectedIndexChanged(object sender, EventArgs e)
{
    _province = (Province)cmbProvince.SelectedItem;
    listBox1.Items.Clear();
    cmbCity.Items.Clear();
    foreach (City c in _province.Cities)
        cmbCity.Items.Add(c);
}
```

```csharp
private void cmbCity_SelectedIndexChanged(object sender, EventArgs e)
{
    _city = (City)cmbCity.SelectedItem;
    listBox1.Items.Clear();
    foreach (Scene s in _city.Scenes)
        listBox1.Items.Add(s);
}
```

当用户选择一个景点后,该景点的详细信息就显示在窗体右侧的各个控件中:

```csharp
private void listBox1_SelectedIndexChanged(object sender, EventArgs e)
{
    if (listBox1.SelectedItem == null)
        return;
    Scene s = (Scene)listBox1.SelectedItem;
    tbName.Text = s.Name;
    cmbType.SelectedIndex = (int)s.Type;
    nudStar.Value = s.Star;
    nudPrice.Value = s.Price;
    nudOffSeasonPrice.Value = s.OffSeasonPrice;
    nudChlDiscount.Value = s.ChlDiscount * 100;
    nudOldDiscount.Value = s.OldDiscount * 100;
    nudStuDiscount.Value = s.StuDiscount * 100;
    tbIntroduction.Text = s.Introduction;
    tbTrafficInfo.Text = s.TrafficInfo;
    tbLodgeInfo.Text = s.LodgeInfo;
}
```

"添加"、"修改"和"删除"按钮分别用于景点列表的添加、修改和删除,其实现思路和 **StaffForm** 窗体基本类似。下面仅给出添加功能的实现代码:

```csharp
private void btnAdd_Click(object sender, EventArgs e)
{
  if (cmbCity.SelectedItem == null){
    MessageBox.Show("尚未选择城市", "提示", MessageBoxButtons.OK, MessageBoxIcon.Warning);
    cmbCity.Focus();
    return;
  }
  if (tbName.Text == ""){
    MessageBox.Show("名称不能为空", "提示", MessageBoxButtons.OK, MessageBoxIcon.Warning);
    tbName.Focus();
    return;
  }
  if (cmbType.SelectedIndex == -1){
    MessageBox.Show("请选择景点类型", "提示", MessageBoxButtons.OK, MessageBoxIcon.Warning);
    cmbType.Focus();
    return;
  }
  for (int i = 0; i < listBox1.Items.Count; i++)
  if (tbName.Text == listBox1.Items[i].ToString()){
      MessageBox.Show("存在同名景点", "提示", MessageBoxButtons.OK, MessageBoxIcon.Warning);
      tbName.Focus();
      return;
  }
  Scene s = new Scene(tbName.Text);
  s.Type = (SceneType)cmbType.SelectedIndex;
  s.Star = (byte)nudStar.Value;
  s.Price = nudPrice.Value;
  s.OffSeasonPrice = nudOffSeasonPrice.Value;
  s.ChlDiscount = nudChlDiscount.Value / 100;
  s.OldDiscount = nudOldDiscount.Value / 100;
  s.StuDiscount = nudStuDiscount.Value / 100;
  s.Introduction = tbIntroduction.Text;
```

```csharp
    s.TrafficInfo = tbTrafficInfo.Text;
    s.LodgeInfo = tbLodgeInfo.Text;
    s.City = _city;
    _city.Scenes.Add(s);
    listBox1.Items.Add(s);
}
```

11.7.2 泛型优先级队列

在第 10 章中还实现了游客报名优先级的处理功能。按照泛型的概念，报名对象只有在相互之间才能进行优先级比较，因此 Enrollment 所支持的 IComparable 接口应该是泛型版本：

```csharp
public class Enrollment : IComparable<Enrollment>// 报名类
{
    //…
    public int CompareTo(Enrollment enr)
    {
        return _priority - enr._priority;
    }
}
```

在此基础上同样可以修改 EnrollmentQueue 的定义，在其内部使用 List<Enrollment>替代 ArrayList 来维护报名队列。不过，更为有效的做法是定义一个泛型的优先级队列 PriorityQueue<T>，它要求队列元素 T 必须可比较。这样将类型参数替换为 Enrollment 就实现了报名队列的功能，而在其他的程序中还能够泛型队列的不同构造类型。

下面给出了这样一个泛型优先级队列的定义代码：

```csharp
// 优先级队列类
public class PriorityQueue<T> : IEnumerable<T>, ICollection<T>
    where T : IComparable<T>
{
    private List<T> _queue;

    public int Count
    {
        get { return _queue.Count; }
    }

    public PriorityQueue()
    {
        _queue = new List<T>();
    }

    public bool Contains(T t)
    {
        return _queue.Contains(t);
    }

    public T Peek()
    {
        return _queue[_queue.Count - 1];
    }

    public void Enqueue(T t)
    {
        int i = 0;
        while (i < _queue.Count && t.CompareTo(_queue[i]) > 0)
            i++;
        _queue.Insert(i, t);
    }

    public T Dequeue()
```

```
    {
        T t = _queue[_queue.Count - 1];
        _queue.RemoveAt(_queue.Count - 1);
        return t;
    }

    public bool Remove(T t)
    {
        return _queue.Remove(t);
    }

    public void Clear()
    {
        _queue.Clear();
    }

    public bool IsReadOnly
    {
        get { return false; }
    }

    public void CopyTo(T[] array, int index)
    {
        _queue.CopyTo(array, index);
    }

    public IEnumerator<T> GetEnumerator()
    {
        return _queue.GetEnumerator();
    }

    IEnumerator IEnumerable.GetEnumerator()
    {
        return _queue.GetEnumerator();
    }

    void ICollection<T>.Add(T t)
    {
        this.Enqueue(t);
    }
}
```

.NET 类库中一些接口的泛型和非泛型版本在声明上存在一定差异,如 ICollection<T> 要求实现 Add、Remove 和 Clear 方法,ICollection 则不作要求。再如 IList 的 Add 方法返回一个整数值,表示新元素加入的位置,而 IList<T> 的 Add 方法则没有返回类型。读者在具体使用时(特别是在一个类型中同时支持泛型和非泛型接口时)要注意区别。

这样在程序中使用到报名队列的地方,都可以使用 PriorityQueue<Enrollment> 来取代 EnrollmentQueue。

11.8 小　　结

C#中的泛型使用类型参数来表示抽象数据类型;在创建泛型的构造类型的对象时,类型参数可被替换为不同的具体类型,这就能够在泛型定义中实现操作与类型的分离,从而达到更高层次上的代码重用。类型限制是对替换类型参数的具体类型范围做出限定,这对提高泛型程序的安全性十分重要。引入泛型之后,C#的继承机制也变得更为丰富和完善,但一条基本原则是封闭类型不能从开放类型中继承。

11.9 习题

1. 简述类型参数在泛型中的用途。
2. 看看下面的泛型类定义中是否包含错误：
   ```
   public class A<T>
   {
       protected T m_value;
       protected B<T> m_B;
       public B<T> B
       {
           get{return m_B;}
       }
       public class B<S>
       {
           protected S m_value;
       }
   }
   ```
3. 写出下面程序的运行结果。
   ```
   class Program
   {
       static void Main()
       {
           B<int> b1 = new B<int>();
           B<double> b2 = new B<double>();
           b1.Output<int>(2);
           b2.Output<int>(2);
           b1.Output<double>(2.5);
           b2.Output<double>(2.5);
       }
   }

   public class A
   {
       public void Output<T>(T t1)
       {
           Console.WriteLine("{0} : A's {1}", t1, t1.GetType());
       }
   }

   public class B<T> : A
   {
       public new void Output<S>(S s1)
       {
           if (s1.GetType() == default(T).GetType())
               base.Output(s1);
           else
               Console.WriteLine("{0} : B's {1}", s1, s1.GetType());
       }
   }
   ```
4. 利用程序 P11_7 定义的链表类，实现多项式相加的功能。基本思路为：用结构 Term 表示多项式的项，其字段 coe 和 exp 分别表示该项的系数和指数；一个多项式就是一个 LinkList<Term> 类型的链表对象。
5. 使用 System.Collections.Generic 命名空间中的泛型列表 List<T> 来改写程序 P10_5。
6. 将第 10 章习题 7 中的二叉树类型 BinaryTree 扩展为泛型版本。
7. 使用 List<User> 来改写旅行社的用户管理功能；再定义一个 Hotel 类并创建一个 HotelForm 窗体，在其中使用 List<Hotel> 来实现酒店管理。最后为这些列表对象增加持久化特性。

第 12 章
C#中的泛型模式：可空类型和迭代器

本章介绍了泛型技术在.NET 类库中的两个典型应用：可空类型和迭代器，其中可空类型是指在值类型的基础上增加对空值 null 的支持，而迭代器则是对自定义类型实现 foreach 遍历的基础。在面向对象的分析和设计过程中，要考虑哪些对象属性可定义为可空类型，还有哪些集合型对象需要实现迭代器。

12.1 可空类型

12.1.1 可空类型：值类型+null

考虑下面的学生类 Student，其中要维护的信息包括学生的姓名、生日、性别和年级：

```
public class Student
{
    public string Name;
    public DateTime Birthday;
    public bool Gender;
    public int Grade;
}
```

如果不指定带参数的构造函数，那么在创建每一个 Student 对象时，其字段会被赋予类型的默认值，如 Birthday 的值为公元元年 1 月 1 日、Grade 的值为 0，这在现实生活中都是不合理的。一种解决办法是在定义时或在构造函数中给字段赋一个初始值，例如：

```
Birthday = new DateTime(1990, 1, 1);
Gender = false;
Grade = 1;
```

不过这种方式仍存在局限性，因为通常没有足够的理由认为学生的生日应当默认为 1990 年 1 月 1 日、性别应当默认为女；如果用户在创建对象后忘了给特定字段赋值，这些默认值就可能导致错误。

更为合理的解决方式是区分"未赋值"和"已赋值"这两种情况。对于引用类型，null 值就可以表示一个变量未赋值，或者说值为空。但 C#的值类型不能取 null 值，那么就需要定义新的类型，将值类型和空值封装在一起。下面的代码就定义了一个 NullableInt 结构，它通过将整数值放在字段 value 中，并通过一个布尔型字段 hasValue 来指示变量是否被赋予了一个整数值：

```
public struct NullableInt
{
    private int value;
    public int Value
    {
```

```csharp
    get {
        if (hasValue)
            return value;
        else
            throw new InvalidOperationException("可空整数尚未赋值");
    }
}

private bool hasValue;
public bool HasValue
{
    get { return hasValue; }
}

publicNullableInt(int i)
{
    value = i;
    hasValue = true;
}

public override string ToString()
{
    if (!hasValue)return value.ToString();
    elsereturn"";
}
}
```

按此思路,我们可以为其他值类型也定义相应的封装结构,如 DoubleOrNull、DateTimeOrNull 等,它们的实现方式基本相同,只不过是内部的 value 字段的类型不同。这就让我们自然而然地想到了泛型技术,即定义一个"可空泛型"Nullable<T>,它使用类型参数 T 来指代 value 字段的类型:

```csharp
public struct Nullable<T> where T : struct
{
    private T value;
    public T Value
    {
        get{
            if (hasValue)
                return value;
            else
                throw new InvalidOperationException("可空类型尚未赋值");
        }
    }

    private bool hasValue;
    public bool HasValue
    {
        get { return hasValue; }
    }

    public Nullable(T t)
    {
        value = t;
        hasValue = true;
    }

    public T GetValueOrDefault()
    {
        return value;
    }
```

```
    public override string ToString()
    {
        if (hasValue) return value.ToString();
        else return "";
    }

    public static implicit operator Nullable<T>(T t)
    {
        return new Nullable<T>(t);
    }

    public static explicit operator T(Nullable<T> nt)
    {
        return nt.Value;
    }
}
```

.NET 类库中就提供了这样一个 Nullable<T>泛型结构，其功能及实现方式和上述代码基本类似，不过它还重载了 object 类的 Equals、GetHashCode 和 GetType 方法，具体内容此处不再描述。对于 Nullable<T>的构造类型，T 所指代的类型也称之为可空类型的基础类型，如 int 就是 Nullable<int>的基础类型。由于对类型参数 T 指定了主要限制，因此只能将其替换为值类型，而不能使用 Nullable<object>、Nullable<string>这样的构造类型（引用类型本身就可以取 null 值，所以也没有必要这么做）。

同样，Nullable<T>的 HasValue 属性指示当前的结构变量是否已被赋值，Value 属性表示变量被赋予的值。再来看其 GetValueOrDefault 方法，它在变量具有基础值时返回 value 字段值，否则返回 T 的默认值。那么方法的执行代码可以写成：

```
if (hasValue)
    return value;
else
    return default(T);
```

不过，在使用无参构造函数创建变量时，字段值 value 已经被赋予了 T 的默认值，因此上述代码等价于一行 "return value;"。此外，上面的程序还重载了从值类型 T 到 Nullable<T>的隐式类型转换操作符，以及从 Nullable<T>到值类型 T 的显式类型转换操作符，那么就可以使用下面这样的代码对可空类型变量直接进行赋值：

```
Nullable<int> i1 = 3;
Nullable<DateTime> dt1 = new DateTime(1989, 10, 1);
```

还可以使用如下代码从 Nullable<T>变量中提取所需的值（但在值为空时会引发异常）：

```
int i2 = (int)i1;
DateTime dt2 = (DateTime)dt1;
```

C#为可空类型提供了专门的语法支持，允许将一个 null 值赋值给一个 Nullable<T>变量（这对于一般结构是不允许的），以表示变量取空值，那么下面两行代码是等价的：

```
Nullable<int> i1 = null;
Nullable<int> i1 = new Nullable<int>();
```

为方便可空类型的使用，C#还提供了声明和初始化的简写形式，即允许将 Nullable<T>的构造类型简写为 "T?" 的形式，比如 "int?" 表示 Nullable<int>，"DateTime?" 表示 Nullable<DateTime>，等等。看下面的代码示例：

```
int? i1 = null;
double? d1 = 2.5;
DateTime? dt1 = DateTime.Now;
DateTime? dt2 = new DateTime?(dt1);
```

将 null 赋值给可空类型变量仅仅是 C#的一种替代语法，这和将 null 赋值给引用类型的变量是完全不同的：前者创建了一个可空类型变量，不过其 HasValue 属性值为 false；后者则表示整个对象为空，在内存中没有实际的存储空间。

在下面程序中，Student 类的生日、性别、年级、班级等属性定义为可空类型，其 GetInformation 方法则只返回已被赋值的属性信息：

```
//程序 P12_1
using System;
using System.Text;
namespace P12_1
{
    class Program
    {
        static void Main()
        {
            Student[] students = { "王小红", "周军", "方小白", "吴杨" };
            students[1].Birthday = new DateTime(1992, 12, 10);
            students[1].Gender = true;
            students[2].Grade = 2;
            students[2].Class = 3;
            students[3].Gender = false;
            students[3].Grade = 1;
            foreach (Student s in students)
                Console.WriteLine(s.GetInformation());
        }
    }

    public class Student
    {
        private string name;
        public string Name
        {
            get { return name; }
        }

        public DateTime? Birthday;
        public bool? Gender;
        public int? Grade;
        public int? Class;

        public Student(string name)
        {
            this.name = name;
        }

        public string GetInformation()
        {
            StringBuilder sb1 = new StringBuilder("姓名:");
            sb1.Append(name);
            if (Birthday.HasValue){
                sb1.Append(" 年龄");
                sb1.Append(DateTime.Now.Year - Birthday.Value.Year);
            }
            if (Gender.HasValue)
                sb1.Append(Gender.Value ? " 男 " : " 女 ");
            if (Grade.HasValue){
                sb1.Append(Grade.Value);
```

```
                sb1.Append("年级");
            }
            if (Class.HasValue){
                sb1.Append(Class.Value);
                sb1.Append("班");
            }
            return sb1.ToString();
        }

        public override string ToString()
        {
            return name.ToString();
        }

        public static implicit operator Student(string name)
        {
            return new Student(name);
        }
    }
}
```

程序 P12_1 的输出结果如下：

```
姓名:王小红
姓名:周军年龄 17 男
姓名:方小白 2 年级 3 班
姓名:吴杨女 1 年级
```

.NET 类库将可空类型定义为一种结构，因此可空类型的变量是直接包含数据，而不是指向目标数据的引用。不过由于良好的设计，Nullable<T>看上去更像是值类型和空值 null 的组合。为了突出这一点，C#还规定：如果泛型中的某个类型参数指定了值类型限制，那么它不能被替换为可空类型，因为可空类型的变量取值为 null 时会破坏值类型限制。例如下面代码中创建 p1 变量的语句是错误的，因为泛型 Pair<L, R>的类型参数 L 限制为值类型，此时将其替换为 Nullable<int>就是不合法的：

```
class Program
{
    static void Main()
    {
        Pair<int?, string> p1 = new Pair<int?, string>(3, "王小红"); //错误!
    }
}

public class Pair<L, R> where L : struct
{
    public L Left;
    public R Right;

    public Pair(L left, R right)
    {
        Left = left;
        Right = right;
    }
}
```

同理，Nullable<T>本身的类型参数也不能被替换成可空类型，即不能使用 int??或 Nullable<Nullable<double>>这样的嵌套可空类型。

12.1.2 可空类型转换

Nullable<T>的类型定义中说明了可空类型到其基础类型的转换原则：
- 基础类型 T 可隐式地转换到对应的可空类型 Nullable<T>。
- 可空类型 Nullable<T>可显式地转换到对应的基础类型 T，不过在可空类型取空值时会引发异常。

在此基础上，与可空类型有关的其他类型转换操作可以分为以下 3 类：

（1）从非可空类型 S 到可空类型 Nullable<T>的（隐式或显式）转换，可看作是先从 S 到 T 进行（隐式或显式）转换，再从 T 到 Nullable<T>进行隐式转换。例如：

```
int i1 = 3;
double? d1 = i1; //隐式转换
double d2 = 3.14;
int? i2 = (int?)d2; //显式转换
```

特别地，从 object 类型到可空类型 Nullable<T>可进行拆箱转换：当 object 类型值不为 null 时，这相当于先从 object 类型拆箱转换到 T 类型，再从 T 转换到 Nullable<T>；而当 object 类型值为 null 时，这相当于直接给 Nullable<T>赋空值。例如：

```
object obj1 = 3;
int? i1 = (int?)obj1; //i1 = 3
object obj2 = null;
int? i2 = (int?)obj2; //i2 = null (i2.HasValue 为 false)
```

（2）从可空类型 Nullable<T>到非可空类型 S 的（显式）转换，可看作是先从 Nullable<T>到 T 进行显式转换，再从 T 到 S 进行（隐式或显式）转换。例如：

```
float? f1 = 3.14f;
int i1 = (int)f1; //显式转换
double d1 = (double)f1; //显式转换
```

特别地，从可空类型 Nullable<T>到 object 类型可进行装箱转换：当可空类型值不为 null 时，这相当于先从 Nullable<T>转换到 T 类型，再从 T 装箱转换到 object 类型；而当可空类型值为 null 时，这相当于直接给 object 类型赋空值。例如：

```
double? d1 = 0.5;
object obj1 = d1; //obj1 = 0.5
double? d2 = null; ;
object obj2 = d2; //obj2 = null
```

（3）C#专门规定：从可空类型 Nullable<T>到可空类型 Nullable<S>可进行（隐式或显式）转换，前提是从基础类型 T 到 S 可进行（隐式或显式）转换。例如：

```
float? f2 = 3.14f;
int? i2 = (int)f2; //显式转换
double? d2 = f2; //隐式转换
```

此外还应注意：除了将常量 null 赋值给一个可空类型的变量外，不允许在可空类型和 object 以外的任何其他引用类型之间进行转换。例如下面的代码都是不合法的，尽管其中各个变量的取值都是 null：

```
int? i1 = null;
string s1 = i1; //错误！
string s2 = null;
int? i2 = s2; //错误！
```

12.1.3 操作符提升

C#为可空类型专门提供了一种操作符提升机制：如果某个操作符能够作用于某个值类型，那么它同样能够作用于对应的可空类型，即操作符的作用范围从基础类型"提升"到了可空类型。下面的代码就示范了加法、乘法、相等和不等操作符提升到 Nullable<int>上的效果：

```
int? i1 = 2, i2 = 3;
int? i3 = i1 + i2;  //i3 = 5
int? i4 = i1 * i2;  //i4 = 6
bool b1 = (i1!= i2); // b1 = true;
bool b2 = (i3 == i4); // b2 = false;
bool b3 = (i3 + 1 == i4); // b3 = true;
```

在上面的代码中，可空类型变量包含的都是基础值，那么操作符提升后的结果与操作符作用在基础类型上的结果完全相同。而当可空类型变量包含空值时，操作符提示的效果可以分为以下几种情况：

- 两个可空类型变量均为 null 时二者相等，只有一个为 null 时二者不等。
- 对于其他关系操作符，表达式的返回值总是为 false。
- 对于其他操作符，表达式的返回值总是为 null。

下面的代码示例说明了关系操作符作用于可空类型上的操作结果：

```
int? i1 = null;
int? i2 = null;
Console.WriteLine(i1 == i2);  //输出 True
Console.WriteLine(i1 <= i2 || i1 >= i2);  //输出 False
i2 = 3;
Console.WriteLine(i1 == i2);  //输出 False
Console.WriteLine(i1 < i2 || i1 > i2);  //输出 False
```

不过上述操作符提升规则存在一种例外，那就是可空布尔类型 Nullable<bool>上的与操作和或操作：

- 对于与操作"&"，只要有一个操作数的值为 false，表达式的值就为 false；而只要有一个操作数的值为 true，表达式的值就是另一个操作数的值；两个操作数都为 null 时，表达式的值为 null。
- 对于或操作符"|"，只要有一个操作数的值为 true，表达式的值就为 true；而只要有一个操作数的值为 false，表达式的值就是另一个操作数的值；两个操作数都为 null 时，表达式的值为 null。

也就是说，无论是 Nullable<bool>还是 bool 类型，它们都满足布尔运算的"短路效应"。

12.2 遍历和迭代

12.2.1 可遍历类型和接口

C#的 foreach 语句用于循环遍历某个集合中的元素，用户不需要了解集合的内部结构，也不用事先判断元素的个数。前面介绍过，如果某个类型支持 IEnumerable 接口或 IEnumerable<T>泛型接口，那么就认为该类型是"可遍历的"，就可以对其进行 foreach 循环遍历。C#中数组的基类型 Array，以及 ArrayList、Queue、Stack 等集合类型都支持 IEnumerable 接口，而 List<T>、Queue<T>、Stack<T>、Dictionary<K, V>等泛型集合都支持 IEnumerable<T>接口，因此对它们都可以进行 foreach 遍历，例如：

```
int[] x = { 1, 3, 5, 7, 9 };
foreach (int i in x)
    Console.WriteLine(i);  //依次输出 1、3、5、7、9
ArrayList al = new ArrayList(x);
foreach (int i in al)
    Console.WriteLine(i);  //依次输出 1、3、5、7、9
```

```csharp
Stack<int> s = new Stack<int>(x);
foreach (int i in s)
    Console.WriteLine(i);  //依次输出 9、7、5、3、1
```

如果希望自定义的类型也能支持 foreach 循环遍历，那么可声明其支持 IEnumerable 或 IEnumerable<T>接口，并实现接口的 GetEnumerator 方法。例如程序 P10_6 中的 Phones 就支持 ICollection 接口，并间接支持其父接口 IEnumerable，同时提供了对 GetEnumerator 方法的实现：

```csharp
public class Phones : ICollection
{
    private string[] phones;
    private int count = 0;

    public int Count
    {
        get { return count; }
    }

    public string this[PhoneType type]
    {
        get{
            return phones[(int)type];
        }
        set
        {
            int index = (int)type;
            if (phones[index] == null && value != null)
                count++;
            else if (phones[index] != null && value == null)
                count--;
            phones[index] = value;
        }
    }

    public Phones()
    {
        phones = new string[5];
    }

    public void CopyTo(Array array, int index)
    {
        foreach (string s in phones)
            if (s != null)
                array.SetValue(s, index++);
    }

    IEnumerator IEnumerable.GetEnumerator()
    {
        for (int i = 0; i < 5; i++)
            if (phones[i] != null)
                yield return phones[i];
    }
    //...
}
```

其 GetEnumerator()方法中的 yield return 语句用于依次生成一组元素，那么使用 foreach 语句遍历得到的元素顺序与该语句生成元素的顺序一致。下面的代码就对一个 Phones 进行了 foreach 遍历，之后将依次输出集合中的家庭电话、办公电话、手机和传真号码（与 PhoneType 的枚举排

列顺序一致):
```
Phones phones = new Phones();
phones[PhoneType.手机] = "1331121122";
phones[PhoneType.办公电话] = "66116622";
phones[PhoneType.家庭电话] = "88338844";
phones[PhoneType.传真] = "66116623";
foreach (string s in phones)
    Console.WriteLine(s);
```

尽管可以通过 ICollection 这样的接口来间接支持 IEnumerable 或 IEnumerable<T>接口,在定义可遍历类型时还是应该显式声明支持这两个接口,这也有助于提高程序的可读性,特别是方便开发人员之间的交互。

这里 Phones 对象遍历输出的是一组 string 元素,那么可以使其支持接口 IEnumerable<string>,并实现对应的 GetEnumerator 方法;此外,出于兼容性的考虑,泛型接口 IEnumerable<T>也继承了非泛型接口 IEnumerable,因此可遍历类型还要再提供一个非泛型版本的 GetEnumerator 方法,不过在一般情况下二者可以采用相同的代码,例如:

```
public class Phones : ICollection, IEnumerable<string>
{
    private string[] phones;

    //...

    IEnumerator<string> IEnumerable<string>.GetEnumerator()
    {
        for (int i = 0; i < 5; i++)
            if (phones[i] != null)
                yield return phones[i];
    }

    IEnumerator IEnumerable.GetEnumerator()
    {
        return ((IEnumerable<string>)this).GetEnumerator();
    }
}
```

可遍历类型也可以隐式实现接口方法,但由于接口 IEnumerable 和 IEnumerable<T>的 GetEnumerator 方法仅仅是返回类型不同,因此不能对两个接口方法同时进行隐式实现。更常见的做法是通过公有成员隐式实现 IEnumerable<T>.GetEnumerator 方法,并在 IEnumerable.GetEnumerator 方法中再调用该方法:

```
public class Phones : ICollection, IEnumerable<string>
{
    private string[] phones;

    //...

    public IEnumerator<string> GetEnumerator()
    {
        for (int i = 0; i < 5; i++)
            if (phones[i] != null)
```

```
            yield return phones[i];
    }

    IEnumerator IEnumerable.GetEnumerator()
    {
        return this.GetEnumerator();
    }
}
```

12.2.2 迭代器

迭代器（Iterator）是一种面向对象的设计模式，其基本思想是：集合对象只负责维护其中的各个元素，而对元素的访问则通过定义一个新的迭代器对象来进行；迭代器对象负责获取集合中的元素，并允许按照特定的顺序来遍历这些元素。

在 .NET 中，IEnumerator 和 IEnumerator<T> 就是对迭代器的抽象，IEnumerable 和 IEnumerable<T>则表示可通过迭代器来遍历的集合类型。事实上，.NET 的中间语言 IL 并不支持 foreach 循环，而是由 C#编译器将程序代码中的 foreach 语句转换为 while 循环或 for 循环结构。以下面的循环语句为例：

```
foreach (string s in phones)
    Console.WriteLine(s);
```

那么程序会首先调用可遍历对象的 GetEnumerator 方法，由该方法返回一个迭代器对象，并通过它来执行遍历过程，那么可以用如下代码来表示实际执行的程序内容：

```
IEnumerator<string> enumerator = phones.GetEnumerator();
while (enumerator.MoveNext())
    Console.WriteLine(enumerator.Current);
```

其中，MoveNext 方法用于推动迭代器向前遍历，成功得到下一个元素就返回 true，否则返回 false；Current 属性则用于返回当前正在遍历的元素。IEnumerator/IEnumerator<T>接口还有一个 Reset 方法，它用于重置迭代器以便重新开始遍历。

如果要自定义一个迭代器类型，那就需要使其支持 IEnumerator 或 IEnumerator<T>接口，并实现相应的接口方法（请参见表 10-1 和表 11-1）。例如下面的程序就定义了一个迭代器 MyEnumerator<T>，用于对一个 T 型数组进行遍历访问；而在可遍历类型 Phones 所实现的 GetEnumerator 方法并没有直接输出元素，而是返回一个 MyEnumerator<string>对象以支持遍历：

```
//程序 P12_2
using System;
using System.Collections;
using System.Collections.Generic;
namespace P12_2
{
    class Program
    {
        static void Main()
        {
            Phones phones = new Phones();
            phones[PhoneType.手机] = "1331121122";
            phones[PhoneType.办公电话] = "66116622";
            phones[PhoneType.家庭电话] = "88338844";
            phones[PhoneType.传真] = "66116623";
            foreach (string s in phones)
                Console.WriteLine(s);
        }
```

```csharp
}

public class MyEnumerator<T> : IEnumerator<T>
{
    private int current;
    internal T[] array;

    public T Current
    {
        get { return array[current]; }
    }

    object IEnumerator.Current
    {
        get { return array[current]; }
    }

    public MyEnumerator(T[] array)
    {
        this.array = array;
        this.current = -1;
    }

    public bool MoveNext()
    {
        if (++current == array.Length)
            return false;
        Console.WriteLine("MoveNext");
        return true;
    }

    public void Reset()
    {
        current = -1;
    }

    void IDisposable.Dispose() { }
}

public enum PhoneType
{
    家庭电话, 办公电话, 手机, 小灵通, 传真
}

public class Phones : IEnumerable<string>
{
    private string[] phones;

    public string this[PhoneType type]
    {
        get { return phones[(int)type]; }
        set { phones[(int)type] = value; }
    }

    public Phones()
    {
        phones = new string[5];
```

```
        public IEnumerator<string> GetEnumerator()
        {
            return new MyEnumerator<string>(phones);
        }
        IEnumerator IEnumerable.GetEnumerator()
        {
            return this.GetEnumerator();
        }
    }
```

那么在使用 foreach 语句遍历 Phones 对象时,就会自动调用其 GetEnumerator 方法来得到迭代器对象,并在 foreach 循环的每一步调用迭代器的 MoveNext 方法并读取其 Current 属性值。程序 P12_2 的输出如图 12-1(a)所示。

 同样,出于兼容性的考虑,泛型接口 IEnumerator<T>也继承了非泛型接口 IEnumerator,因此 MyEnumerator<T>需要实现两个版本的 Current 属性(两个接口的 MoveNext 和 Reset 方法标识一致,可以都通过公有方法隐式实现)。此外 IEnumerator<T>还继承了 System 命名空间下的 IDisposable 接口,因此需要为其方法 Dispose 提供实现。

如果要使 MyEnumerator<T>在遍历过程中略过空元素,那么只需将其 MoveNext 方法改为如下内容,而修改后程序 P12_2 的输出将如图 12-1(b)所示:

```
public bool MoveNext()
{
    if (++current == array.Length)
        return false;
    if (array[current] == null)
        MoveNext();
    Console.WriteLine("MoveNext");
    return true;
}
```

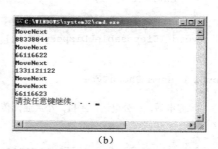

(a)　　　　　　　　　　　　　　　(b)

图 12-1　程序 P12_2 的输出结果

迭代器对象可以有 4 种状态:准备(Before)、运行(Running)、挂起(Suspending)和完成(After),其基本执行过程分为以下几个步骤(如图 12-2 所示):

(1)开始 foreach 循环,通过可遍历对象的 GetEnumerator 方法创建迭代器对象,此时其处于准备状态;

(2)在循环的每一步,调用迭代器的 MoveNext 方法;

(3)如果方法返回 true,那么迭代器进入运行状态,否则结束循环;

(4)如果循环中止或结束,那么迭代器转为完成状态,否则迭代器转为挂起状态;

（5）在挂起状态下，迭代器将从 Current 属性返回当前遍历元素，而后转第（2）步。

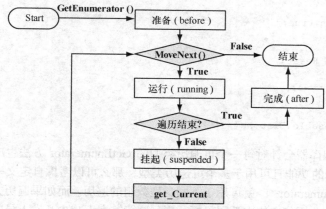

图 12-2　迭代器的状态转换流程

12.2.3　迭代器代码

对比程序 P10_6 和程序 P12_2 中可遍历类型 Phones 的 GetEnumerator 方法，后者的确返回了一个 IEnumerator<T>对象，这符合 IEnumerable<T>接口中的方法定义；而前者并没有明显看到方法的返回值，而仅仅是使用 yield return 语句来描述遍历生成的元素：

```
IEnumerator IEnumerable.GetEnumerator()
{
    for (int i = 0; i < 5; i++)
        if (phones[i] != null)
            yield return phones[i];
}
```

这是 C#为迭代器模式提供的一种简化实现方式，即不需要开发人员显式定义一个迭代器类型，而是直接在可遍历类型的 GetEnumerator 方法中描述遍历过程，C#编译器会自动创建一个默认的迭代器来完成遍历工作。例如为上面代码生成的隐含迭代器的定义可用如下内容表示：

```
public class CEnumerator : IEnumerator<string>
{
    private int current;
    internal string[] phones;

    public string Current
    {
        get { return phones[current]; }
    }

    object IEnumerator.Current
    {
        get { return phones[current]; }
    }

    public CEnumerator(string[] phones)
    {
        this.phones = phones;
        current = -1;
    }

    public bool MoveNext()
    {
```

```
            if (++current == 5)
                return false;
            else
                return true;
        }
        public void Reset()
        {
            current = -1;
        }
        void IDisposable.Dispose() { }
    }
```

也就是说，C#编译器会针对每一个可遍历类型的 GetEnumerator 方法生成一个迭代器。如果希望迭代器具有特殊的功能且可用于多个可遍历类型，那么可以考虑自定义一个迭代器，例如程序 P12_2 中的 MyEnumerator<T>就可用于不同类型数组的遍历。而如果遍历方式较为简单，那么就可以通过 yield return 语句来依次返回目标元素，这能够大大减少开发人员的工作量。不过这两种方式实现的迭代器在运行时的效果没有明显的区别。

在 GetEnumerator 方法中还可以使用 yield break 语句来随时中止遍历。例如下面的代码会在遍历到空元素时立刻停止遍历（而不是跳过空元素）：

```
IEnumerator IEnumerable.GetEnumerator()
{
    for (int i = 0; i < 5; i++)
    {
        if (phones[i] != null)
            yield return phones[i];
        else
            yieldbreak;
    }
}
```

如果代码中出现了 yield return 或 yield break 语句，C#编译器就将为包含该代码的方法返回一个迭代器（IEnumerator 或 IEnumerator<T>的构造类型），该方法中的代码也叫做迭代器代码。和普通方法不同，迭代器代码不是从开始到结束一次性执行完毕，而是随着遍历的过程而分步执行。C#专门规定：

- 迭代器代码所在的方法不能使用任何引用参数或输出参数。
- 迭代器代码中不能出现单独的 return 语句。

yield break 语句会停止遍历，其后的代码不会被执行，这一点和 break 语句类似；而 yield return 语句只是生成遍历元素，其后的代码还会被执行，这一点和 return 语句不同。

12.2.4 使用多个迭代器

不同的遍历方式还可以出现在同一个可遍历类型中。看下面的程序示例：

```
//程序 P12_3
using System;
using System.Collections;
using System.Collections.Generic;
namespace P12_3
{
    class Program
    {
        static void Main()
```

```csharp
    {
        Phones phones = new Phones();
        phones[PhoneType.手机] = "1331121122";
        phones[PhoneType.办公电话] = "66116622";
        phones[PhoneType.家庭电话] = "88338844";
        phones[PhoneType.传真] = "66116623";
        Console.WriteLine("默认遍历:");
        foreach (string s in phones)
            Console.WriteLine(s);
        Console.WriteLine("\n非默认遍历:");
        foreach (string s in phones.BreakingEnumerator())
            Console.WriteLine(s);
    }
}

public enum PhoneType
{
    家庭电话, 办公电话, 手机, 小灵通, 传真
}

public class Phones : IEnumerable<string>
{
    private string[] phones;

    public string this[PhoneType type]
    {
        get { return phones[(int)type]; }
        set { phones[(int)type] = value; }
    }

    public Phones()
    {
        phones = new string[5];
    }

    public IEnumerator<string> GetEnumerator()
    {
        for (int i = 0; i < 5; i++){
            if (phones[i] != null){
                Console.Write((PhoneType)i);
                yield return phones[i];
            }
        }
    }

    public IEnumerable<string> BreakingEnumerator()
    {
        for (int i = 0; i < 5; i++){
            if (phones[i] != null){
                Console.Write((PhoneType)i);
                yield return phones[i];
            }
            else
                yield break;
```

```
        }
    }
    IEnumerator IEnumerable.GetEnumerator()
    {
        return this.GetEnumerator();
    }
}
```

那么 C#编译器将为 Phones 类型创建两个迭代器，其中基于 GetEnumerator 方法创建的叫做默认迭代器，而基于 BreakingEnumerator 方法方法创建的则是一个非默认迭代器。注意非默认迭代器的返回类型不是 IEnumerator 或 IEnumerator<T>，而是 IEnumerable 或 IEnumerable<T>。而在 foreach 语句中使用非默认迭代器时，不能只写出可遍历对象，还需要调用其实现迭代器的方法（因为 foreach 语法要求在 in 关键字之后是一个可遍历类型）：

```
foreach (string s in phones.BreakingEnumerator())
    Console.WriteLine(s);
```

从程序 P12_3 的输出中可以看到，默认迭代器将略过空的电话号码，非默认迭代器则在遇到空值时停止遍历：

```
默认遍历：
家庭电话 88338844
办公电话 66116622
手机 1331121122
传真 66116623

非默认遍历：
家庭电话 88338844
办公电话 66116622
手机 1331121122
```

一个可遍历类型只能有一个默认迭代器，但可以实现多个非默认迭代器。例如可以为 Phones 类型再增加如下的成员方法，它所实现的非默认迭代器将反向遍历 phones 数组中的电话号码：

```
public IEnumerable<string> ReverseEnumerator()
{
    for (int i = 4; i >= 0; i--){
        if (phones[i] != null){
            Console.Write((PhoneType)i);
            yield return phones[i];
        }
    }
}
```

12.2.5 自我迭代

考虑程序 P11_7 中的 LinkList<T>类，它所支持的 IList<T>接口继承了 IEnumerator<T>接口，那么在其中可通过如下的 GetEnumerator 方法代码来支持 foreach 遍历：

```
public IEnumerator<T> GetEnumerator()
{
    while (current.Next != null){
        yield return current.Data;
        current = current.Next;
```

```
        }
    }
    IEnumerator IEnumerable.GetEnumerator()
    {
        return this.GetEnumerator();
    }
```

不过，分析 LinkList<T>的定义代码可以发现，该类实际上已经提供了 Current 属性、MoveNext 方法和 Reset 方法，那么只要再增加如下两个非泛型的接口方法，它就满足了迭代器 IEnumerator 和 IEnumerator<T>的要求：

```
object IEnumerator.Current
{
    get { return current.Data; }
}

void IDisposable.Dispose() { }
```

这样就可以在 LinkList<T>类的定义中声明对 IEnumerator<T>接口的支持，而相应的 GetEnumerator 方法只需简单地返回对象本身即可：

```
IEnumerator<T> IEnumerable<T>.GetEnumerator()
{
    return this;
}

IEnumerator IEnumerable.GetEnumerator()
{
    return this;
}
```

也就是说，此时 LinkList<T>既是可遍历类型又是迭代器。该技术对于既需要维护当前元素又要维护相邻元素关系的集合类型非常有用。下面的程序就定义了一个数列类 NumberSequence，它使用委托类型 LongFunction 来定义数列相邻项之间的递推关系，并在 MoveNext 方法中进行递推关系，这样就能方便地生成和遍历数列项。程序主方法就基于 NumberSequence 生成和遍历了一个等差数列和一个等比数列：

```
using System;
using System.Collections;
using System.Collections.Generic;
namespace P12_4
{
    class Program
    {
        static void Main()
        {
            NumberSequence ns1 = new NumberSequence(1, 10, delegate(long x) { return x + 3; });
            Console.WriteLine("等差数列");
            foreach (long x in ns1)
                Console.WriteLine(x);
            NumberSequence ns2 = new NumberSequence(1, 10, delegate(long x) { return x * 3; });
            Console.WriteLine("等比数列");
            foreach (long x in ns2)
                Console.WriteLine(x);
        }
    }

    public delegate long LongFunction(long x);

    public class NumberSequence : IEnumerable<long>, IEnumerator<long>
    {
```

```csharp
    protected int length, position = 0;
    public int Position
    {
        get { return position; }
    }
    public int Length
    {
        get { return length; }
    }

    protected long first, current = 0;
    public long First
    {
        get { return first; }
        set { first = value; }
    }
    public long Current
    {
        get { return current; }
    }
    object IEnumerator.Current
    {
        get { return current; }
    }

    private LongFunction recur;

    public NumberSequence(long first, int length, LongFunction recur)
    {
        this.current = this.first = first;
        this.length = length;
        this.recur = recur;
    }

    public virtual bool MoveNext()
    {
        if (position < length){
            current = recur(current);
            position++;
            return true;
        }
        return false;
    }

    public virtual void Reset()
    {
        position = 0;
        current = first;
    }

    public IEnumerator<long> GetEnumerator()
    {
        return this;
    }

    IEnumerator IEnumerable.GetEnumerator()
    {
        return this;
    }

    public void Dispose() { }
}
}
```

12.3 案例研究——旅行社管理系统中的可空值与迭代器

12.3.1 旅行社业务对象中的可空值

在分析和设计对象时，要考虑对象的哪些值类型的属性可能需要取空值。首先考虑景点类 Scene，系统中一些景点的信息可能会不完整，如不知道景点票价和折扣；但这种情况下取默认值而不是应用可空类型更为合适，如票价默认为 0（免费），折扣默认为 100%（无折扣）：

```
public Scene(string name)  //构造函数
{
    _name = name;
    _chlDiscount = _oldDiscount = _stuDiscount = 1;
}
```

再考虑职员对象，系统要求每名职员的信息齐全，那么其各个属性也不适合采用可空类型。不过游客对象则可能缺少生日信息，而且儿童游客也可能没有身份证件，那么可将对应的属性类型改为可空类型：

```
private DateTime? _birthday;
public DateTime? Birthday  //生日
{
    get { return _birthday; }
    set { _birthday = value; }
}

private IDCard? _idCard;
public IDCard? IdCard  //身份证件
{
    get { return _idCard; }
    set { _idCard = value; }
}
```

还有就是要考虑对象状态演变过程。例如对于 Package 对象，经理或主管在开发一个组团方案时，可能还没有确定旅行团的价格或人数限制，但其值不适合取默认值 0，因此不允许创建和保存对象也会带来很多不便，那么可以将其也定义为可空类型：

```
private decimal? _price, _chlPrice;  //成人和儿童价格
public decimal? Price
{
    get { return _price; }
    set { _price = value; }
}
public decimal? ChlPrice
{
    get { return _chlPrice; }
    set { _chlPrice = value; }
}
```

对于枚举型的属性，既可以将其定义为枚举所对应的可空类型，也可以为枚举增加一个特殊取值来表示空值或不确定的情况，实际使用时可根据需要来选择，不过在程序中采用后一种方式的效率更高。

定义了可空类型的对象属性后，在将属性值显示在 Windows 窗体控件上时也要注意区分。属性值为空时，在文本框上可以显示空字符串，而在另外一些控件上则可能需要显示默认值，例如将价格信息显示在一个 NumericUpDown 控件中：

```
nudPrice.Value = _package.Price.GetValueOrDefault();
```

特别的，DateTime 结构的默认值为 0 年 0 月 0 日，而 DateTimePicker 控件的时间显示范围在 1753~9998 年之间，此时要么修改控件的时间范围，要么取控件的极值，例如：

```
if (_customer.Birthday != null)
    dtpBirthday.Value = _customer.Birthday;
else
    dtpBirthday.Value = dtpBirthday.MaxDate;
```

不过当用户保存信息时，通常应对这些情况给出提示，以防止用户忘了输入本来准备输入的内容，例如：

```
if (dtpBirthday.Value && dtpBirthday.MaxDate)
    if (MessageBox.Show("还未设置生日数据，您确认要继续吗?", "警告", MessageBoxButtons.YesNo,
MessageBoxIcon.Warning) == DialogResult.No)
        _customer.Birthday = null;
```

12.3.2 遍历游客集合

在 10.5.2 小节中定义的 CustomerCollection 类型支持 IEnumerable 接口，并通过调用 ArrayList 对象的 GetEnumerator 方法来得到迭代器对象。引入泛型后，可将其 _collection 字段的类型改为 List<Customer>，并使该类型同时支持 IEnumerable 和 IEnumerable<Customer>接口，而实现这两个接口的迭代器方法代码则仍然是一致的：

```
public sealed class CustomerCollection : IEnumerable, IList, IEnumerable<Customer>,
ICollection<Customer>, IList<Customer>
{
    private List<Customer> _collection;

    //…

    public IEnumerator<Customer>GetEnumerator()
    {
        return _collection.GetEnumerator();
    }

    IEnumerator IEnumerable.GetEnumerator()
    {
        return _collection.GetEnumerator();
    }
}
```

系统有时候也需要单独遍历游客集合中的某类游客对象。那么可以为 CustomerCollection 添加如下成员，它们分别用于遍历集合中的普通游客、儿童游客、老年游客和学生游客：

```
public IEnumerator<Customer> GetCommonEnumerator()
{
    for (int i = 0; i < _collection.Count; i++)
        if (_collection[i].Type == CustomerType.Common)
            yield return _collection[i];
}

public IEnumerator<Customer> GetChildEnumerator()
{
    for (int i = 0; i < _collection.Count; i++)
        if (_collection[i].Type == CustomerType.Child)
            yield return _collection[i];
```

```csharp
public IEnumerator<Customer> GetOldEnumerator()
{
    for (int i = 0; i < _collection.Count; i++)
        if (_collection[i].Type == CustomerType.Old)
            yield return _collection[i];
}

public IEnumerator<Customer> GetStudentEnumerator()
{
    for (int i = 0; i < _collection.Count; i++)
        if (_collection[i].Type == CustomerType.Student)
            yield return _collection[i];
}
```

12.4 小　　结

借助于泛型技术，C#语言中增加了对可空类型的支持，即任何一个值类型都可以扩展为一个可空类型，这样其变量既能够取基础类型的值，也能够取空值 null。在引入了可空类型之后，在对象设计时就要考虑哪些值类型的属性可能需要取空值，而在编程过程中要注意与可空类型有关的类型转换。

在开发自定义的集合类型时，为了方便对集合组元素的遍历访问，可以为其定义迭代器代码，如果希望支持多种的遍历则还可以定义多个迭代器。迭代器技术是在 C#语言级别上实现的一种面向对象的设计模式。

12.5 习　　题

1. 对于可空类型的变量，访问其哪些成员时会引发异常？
2. 写出下面代码段的输出结果：

```
int? x = null, y = 0, z = 1;
x /= y; y /= z; z /= x;
Console.WriteLine("x = {0}", x.HasValue ? x : "null");
Console.WriteLine("y = {0}", y.HasValue ? y : "null");
Console.WriteLine("z = {0}", z.HasValue ? z : "null");
```

3. 进一步改进第 11 章习题 6 中的程序，通过不同的迭代器来实现二叉树的前序、中序和后序遍历。
4. 针对 12.2.4 小节为 Phones 类型实现的 BreakingEnumerator 和 ReverseEnumerator 方法，写出它们所对应的隐含迭代器的定义代码。
5. 编写程序，测试 12.3.2 小节为 CustomerCollection 实现的不同迭代器。
6. 阅读面向对象设计模式的有关文献，了解除了迭代器外还有哪些常用的设计模式。

第 13 章 WPF 应用程序设计

第 7 章中介绍了基于 Windows Form 界面的程序开发技术，本章将介绍另一种风格的 Windows 界面技术——WPF（Windows Presentation Fundation）。它基于高性能的 DicrectX 绘图引擎，提供了大量精美的窗体元素和控件，并可实现更加赏心悦目的图形和动画等多媒体功能。基于 .NET Framework，WPF 界面和 Windows Form 界面还能够方便地进行集成。

13.1 WPF 窗体和控件

13.1.1 创建一个 WPF 程序

在 Visual Studio 开发环境中新建项目，选择项目模板为"WPF 应用程序"，输入项目名称并单击"确定"按钮，Visual Studio 就会自动创建程序项目，其中包含一个默认主窗体 MainWindow。在解决方案资源管理器中可以看到，窗体的设计文件是 MainWindow.xaml，C#源代码文件则是 MainWindow.xaml.cs。从源代码可知，每一个 WPF 窗体都继承自基类 Window，该类在 System.Windows 命名空间下定义。

打开窗体的设计视图，就可以从工具箱中向窗体拖放各种 WPF 控件。Visual Studio 创建的 WPF 窗体中包含一个 Grid 元素，它是一个控件容器，新加入的控件都会成为该元素的子元素。Grid 元素默认没有命名；如果要在程序代码中访问此元素对象，可在属性窗口顶部的文本框中输入一个名称（如 grid1），如图 13-1（a）所示；在窗口下方的属性列表中还可以设置其他的常用属性，如选中 ShowGridLines 属性（值为 true）会在 Grid 中显示网格线；切换到事件选项卡，在事件列表可以方便地创建事件处理方法，如图 13-1（b）所示。这些与 Windows Form 程序开发都是类似的。

(a)　　　　　　　　　　(b)

图 13-1　在属性窗口中设置控件名称、属性和事件

这里我们希望该 WPF 程序实现类似于程序 P7_3 的功能。Grid 元素的 RowDefinitions 和 ColumnDefinitions 分别表示网格中的行集合和列集合；将新的 RowDefinition 对象加入到 RowDefinitions 集合，或是将新的 ColumnDefinition 对象加入到 ColumnDefinitions 集合，就在网格中加入了相应的行或列。下面的代码就将网格 grid1 设为 3 行 4 列，在每一个单元格中放置一个 Label 控件，并为每个标签设置了不同的光标样式：

```
//程序 P13_1(MainWindow.xaml.cs)
using System.Windows;
using System.Windows.Controls;
using System.Windows.Input;
namespace P13_1
{
    public partial class MainWindow : Window
    {
        public MainWindow()
        {
            InitializeComponent();
            for (int i = 0; i < 3; i++)
                grid1.RowDefinitions.Add(new RowDefinition());
            for (int i = 0; i < 4; i++)
                grid1.ColumnDefinitions.Add(new ColumnDefinition());
            Cursor[] cursors = { Cursors.Arrow, Cursors.ArrowCD, Cursors.Cross,
                Cursors.Hand, Cursors.Help, Cursors.IBeam, Cursors.No, Cursors.Pen, Cursors.ScrollAll,
                Cursors.SizeAll, Cursors.UpArrow, Cursors.Wait };
            Label[] labels = new Label[12];
            for (int i = 0; i < 12; i++)
            {
                labels[i] = new Label();
                labels[i].Content = cursors[i].ToString();
                labels[i].Cursor = cursors[i];
                grid1.Children.Add(labels[i]);
                Grid.SetRow(labels[i], i / 4);
                Grid.SetColumn(labels[i], i % 4);
            }
        }
    }
}
```

从上述代码中可以看到，实现类似的功能时，WPF 窗体和控件的用法和 Windows Forms 差别不大，只不过类型的成员内容可能存在一些差别。如在 WPF 中，元素的子元素集合是通过属性 Children（而非 Controls）来访问的，Label 控件的文本内容则是通过属性 Content（而非 Text）来设置的。WPF 的 Cursor 类型属于 System.Windows.Input 命名空间，它支持的光标种类比 Windows Forms 中更多。另外值得注意的一点是，Gird 元素是通过 SetRow 和 SetColumn 方法来分别设置其子控件位于哪一行和哪一列（默认为首行首列）。程序 P13_1 的输出结果示例如图 13-2 所示。

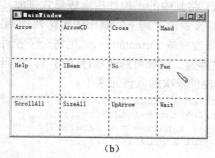

图 13-2　程序 P13_1 的示例输出结果

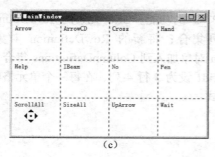 

图 13-2　程序 P13_1 的示例输出结果（续）

13.1.2　窗体和布局

一个 WPF 窗体只能包含一个元素，这称为窗体的根元素，例如程序 P13_1 的主窗体的根元素就是 Grid 容器。如果要在窗体中使用多个控件，那么这些控件应当放在其他的容器当中。WPF 中也定义了一个面板类 Panel，不过它是一个抽象类，包括 Grid 在内的几个最常用的控件容器都是 Panel 的派生类，其继承结构如图 13-3 所示。

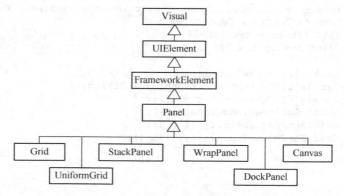

图 13-3　WPF 中的容器元素继承示意图

 在 Windows Form 中，所有 System.Windows.Forms.Control 的派生类型都称为"控件"。而在 WPF 中，所有用户界面元素有一个共同的基类 UIElement，因此我们将其都称为"元素"；只有那些 System.Windows.Controls.Control 的派生类型才称为"控件"。

UniformGird 是和 Gird 类似的网格型容器，不过它强制要求所有的单元格都具有相同的尺寸。StackPanel 将其中的子控件按堆栈形式排放：在默认情况下，每个子控件占据与面板客户区相同的宽度，并按照由上到下的顺序依次排列；如果将 StackPanel 的 Orientation 属性值设置为 Horizontal（属性类型为枚举 Orientation），则每个子控件占据与面板客户区相同的高度，并按照由左到右的顺序依次排列。例如可修改程序 P13_1，将当前窗体的根元素设为一个 StackPanel 元素，再将各个 Label 控件依次加入该元素中：

```
StackPanel stackPanel1 = new StackPanel();
this.Content = stackPanel1;
Label[] labels = new Label[12];
for (int i = 0; i < 12; i++)
{
    labels[i] = new Label() { Cursor = cursors[i], Content = cursors[i].ToString() };
```

```
        stackPanel1.Children.Add(labels[i]);
}
```

图 13-4（a）和（b）分别演示了当面板的 Orientation 属性值为 Vertical（默认）和 Horizontal 时的效果。

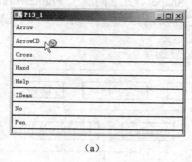 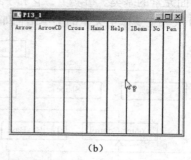

图 13-4　StackPanel 容器效果示例

WrapPanel 默认将其中的子控件按从左到右的顺序依次排列，如果到达面板的右边沿则自动换行。该元素也有一个 Orientation 属性（默认值为 Horizontal），将其值设置为 Vertical 时，子控件则按从上到下的顺序依次排列，如果到达面板的底边沿则自动换列。将上面程序的窗体根元素换成 WrapPanel 后，这两种方式排列的控件效果分别如图 13-5（a）和（b）所示。

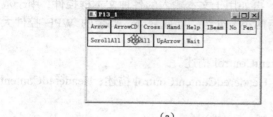 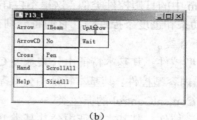

图 13-5　WrapPanel 容器效果示例

DockPanel 允许子控件停靠在面板某一侧面，停靠方式通过一个 Dock 枚举值指定，取值包括 Top、Bottom、Left 和 Right。下面的代码将各个 Label 控件按照向左和向上两种停靠方式交替排列，输出结果如图 13-6 所示。

```
DockPanel dockPanel1 = new DockPanel();
this.Content = dockPanel1;
Label[] labels = new Label[12];
for (int i = 0; i < 12; i++)
{
    labels[i] = new Label() { Cursor = cursors[i],
Content = cursors[i].ToString() };
    labels[i].BorderThickness = new Thickness(1);
    labels[i].BorderBrush = Brushes.Black;
    dockPanel1.Children.Add(labels[i]);
    DockPanel.SetDock(labels[i], (i % 2 == 0) ? Dock.Left : Dock.Top);
}
```

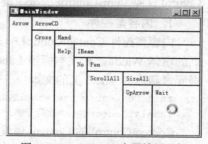

图 13-6　DockPanel 容器效果示例

Canvas 元素又叫画布，它使用二维坐标来对子控件进行定位，这比较适用于图形图像的绘制（详见后面的 13.3 节和 13.4 节中的介绍）。对于由标签、文本框、按钮等常用控件组成的窗体，通常 Grid 元素更适合作为窗体的根元素，也可以组合运用 Grid、StackPanel、WrapPanel 和 DockPanel 来对窗体布局进行更加精巧的安排。

13.1.3 控件内容模型

System.Windows.Controls 命名空间下的 Control 类定义了 WPF 控件的公共基类，不过它并不是所有用户界面元素的基类。事实上，它和 Panel、Shape 等元素同属于 FrameworkElement 的派生类。Control 及其主要派生类型的继承结构如图 13-7 所示。

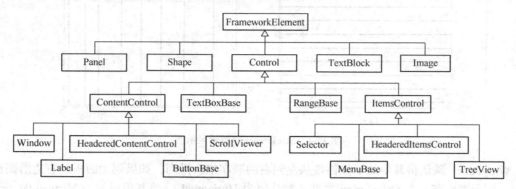

图 13-7　Control 及其主要派生类型的继承示意图

Control 的派生控件很多，如 DatePicker 和 Calendar 都是它的直接派生类，它们的用法与 Windows Form 中的日历控件类似，这里不再详述。再如用于文本输入的各种文本框控件，用于范围选择的滚动条和进度条控件等，它们将分别在后两小节介绍。除此之外，剩余的 WPF 控件大致可分为以下 4 类。

- 内容型控件：其基本特征由公共基类 ContentControl 描述。
- 标题-内容型控件：其基本特征由公共基类 HeaderedContentControl 描述；HeaderedContentControl 又是 ContentControl 的派生类。
- 项集型控件：其基本特征由公共基类 ItemsControl 描述。
- 标题—项集型控件：其基本特征由公共基类 HeaderedItemsControl 描述；HeaderedItemsControl 又是 ItemsControl 的派生类。

1. 内容型控件

内容型控件 ContentControl 中只能包含一项内容，其 Content 属性就表示控件的内容。前面用到的窗体 Window 和文本标签 Label 都是内容型控件，例如下面两行代码分别设置了当前窗体的根元素和标签 label1 上的文本：

```
this.Content = new StackPanel();
label1.Content = "欢迎光临!";
```

ContentControl 的 Content 属性的类型为 object，这意味着可以把任意对象设置为控件的内容。WPF 在处理控件内容时，如果内容对象属于用户界面元素（UIElement 的派生类），那么将在控件中显示内容界面；否则就显示内容对象的 ToString 方法所得到的文本。

在 WPF 中，按钮 Button、复选框 ChechBox、单选框 RadioButton 也都是内容型控件，而且它们有着一个共同的基类 ButtonBase。例如下面的代码基于指定图像文件创建了一个 Image 对象，并将其设置为按钮 button1 的 Content 属性，这样按钮上就会显示指定的图像内容：

```
Image img = new Image();
img.Source = new BitmapImage(new System.Uri("C:\\Pictures\\1.png"));
button1.Content = img;
```

另一个常用的内容型控件是 ScrollViewer，如果把其他某个元素作为其 Content 属性，那么 ScrollViewer 就会支持该元素的滚动显示。

2. 标题—内容型控件

除了 Content 属性外，标题—内容型控件 HeaderedContentControl 还提供了一个 Header 属性，表示控件的标题项，其类型也是 object。分组框 GroupBox 就是一个典型的标题—内容型控件。下面的代码创建了一个 GroupBox 对象，将一个 Button 控件设置为其标题，将一个 StackPanel 设置为其内容，这样得到的分组框外观如图 13-8 所示。

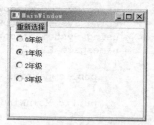

图 13-8　分组框控件示例

```
StackPanel stp = new StackPanel();
for (int i = 0; i < 4; i++)
    stp.Children.Add(new RadioButton() { Content = i.string() + "年级" });
GroupBox groupBox1 = new GroupBox();
groupBox1.Header = new Button() { Content = "重新选择" };
groupBox1.Content = stp;
this.Content = groupBox1;
```

扩展器 Expander 是 WPF 特有的一个标题—内容型控件，其标题部分左侧有一个小箭头按钮，单击按钮可以展开或隐藏控件的内容。下面的代码就创建了两个 Expander 控件，其界面外观如图 13-9 所示。

```
StackPanel stp1 = new StackPanel();
StackPanel stp2 = new StackPanel();
for (int i = 1; i <= 4; i++)
{
    stp1.Children.Add(new RadioButton() { Content = "大" + i.ToString() });
    stp2.Children.Add(new RadioButton() { Content = "研" + i.ToString() });
}
Expander exp1 = new Expander() { Header = "本科生", Content = stp1 };
Expander exp2 = new Expander() { Header = "研究生", Content = stp2 };
```

　　　　　　(a)　　　　　　　　　　　　(b)

图 13-9　扩展器控件示例

3. 项集型控件

顾名思义，项集型控件 ItemsControl 中可以包含一组项的集合，通过其 Items 属性可访问此集合。ItemsControl 有一个抽象的派生类 Selector，它代表了用户可从中选择项的控件，其 SelectedIndex 属性对应选择项的索引，SelectedItem 则对应所选择的项。在 WPF 中，组合框 ComboBox、列表框 ListBox，甚至标签页 TabControl 都是 Selector 的派生类。

以 TabControl 控件为例，其每一个标签页对应一个 TabItem 项，而 TabItem 又是一个标题—内容型控件：Header 属性对应标签页的标题，Content 属性对应标签页上的内容。下面的程序在窗体上放置了一个 TabControl 控件，并将 13.1.1 小节中用到的各种 Panel 控件分别加入到各个标签页中。程序的输出结果如图 13-10 所示。

```
//程序 P13_2(MainWindow.xaml.cs)
using System.Windows;
```

```csharp
using System.Windows.Controls;
using System.Windows.Input;
using System.Windows.Media;
namespace P13_2
{
    public partial class MainWindow : Window
    {
        public MainWindow()
        {
            InitializeComponent();
            Grid grid1 = new Grid() { Name = "Grid" };
            StackPanel stackPanel1 = new StackPanel() { Name = "StackPanel" };
            WrapPanel wrapPanel1 = new WrapPanel() { Name = "WrapPanel" };
            DockPanel dockPanel1 = new DockPanel() { Name = "DockPanel" };
            Panel[] panels = { grid1, stackPanel1, wrapPanel1, dockPanel1 };
            TabControl tabControl1 = new TabControl();
            this.Content = tabControl1;
            for (int i = 0; i < panels.Length; i++)
                tabControl1.Items.Add(new TabItem() {Header = panels[i].Name, Content = panels[i]} );
            Cursor[] cursors = { Cursors.Arrow, Cursors.ArrowCD, Cursors.Cross, Cursors.Hand,
Cursors.Help, Cursors.IBeam, Cursors.No, Cursors.Pen, Cursors.ScrollAll,
Cursors.SizeAll, Cursors.UpArrow, Cursors.Wait };
            // 设置各个 Panel 元素内容的代码见 13.1.1 小节，此处省略……
        }
    }
}
```

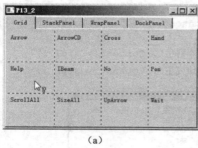
(a)

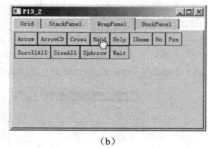
(b)

图 13-10　TabControl 控件示例

ItemsControl 的 Items 集合中的元素类型也是 object，因此可以将不同类型的界面元素加入控件的项集中。下面的代码就依此将 1 个字符串、1 个 TextBox 控件、3 个 Image 控件和 1 个 DatePicker 控件加入一个 ListBox 当中，其界面效果如图 13-11 所示。

图 13-11　列表框控件示例

```csharp
listBox1.Items.Add("列表框控件示例");
listBox1.Items.Add(new TextBox() { Width = 200 });
Image img = new Image() {Source = new BitmapImage(new Uri("Phone.png"));
listBox1.Items.Add(img);
img = new Image() { Source = new BitmapImage(new Uri("Phone1.png")) };
listBox1.Items.Add(img);
img = new Image() { Source = new BitmapImage(new Uri("Phone2.png")) };
listBox1.Items.Add(img);
listBox1.Items.Add(new DatePicker());
```

ItemsControl 控件还提供了一个 ItemsSource 属性，可以将任何一个支持 IEnumerable 接口的对象赋值给该属性，这样对象集合中的所有项就会自动绑定到 ItemsControl 控件的 Items 集合上。

WPF 中也支持树型视图 TreeView 和列表视图 ListView,不过前者是 ItemsControl 的直接派生类,后者则是 ListBox 的派生类。

4. 标题—项集型控件

标题—项集型控件 HeaderedItemstControl 在 ItemsControl 的基础上增加了一个 Header 属性,即控件包含一个标题项和一组项集。典型的标题—项集型控件有工具栏 ToolBar、菜单项 MenuItem,以及树型视图的节点项 TreeViewItem。比如 MenuItem 的标题就是菜单项,项集就是其子菜单项的集合。再如 TreeViewItem 的标题就是当前树节点,项集就是其子节点的集合。

树型视图 TreeView 的 Items 属性中可以加入任意元素,但只有类型为 TreeViewItem 的元素才可以作为树节点使用(比如展开和收缩),其他的文本和图像等元素只能静态显示。类似的,在 ListView 的 Items 属性中,只有类型为 ListViewItem 的元素才可作为标准列表项。

下面的程序就使用 WPF 中的 TreeView 和 ListView 控件来改写了第 7 章中的程序 P7_9,可以看到其中 TreeViewItem 和 ListViewItem 的用法比 Windows Form 中更加灵活:

```
//程序 P13_3(MainWindow.xaml.cs)
using System;
using System.Collections.Generic;
using System.Windows;
using System.Windows.Controls;
namespace P13_3
{
    public partial class MainWindow : Window
    {
        private School school;

        public MainWindow()
        {
            InitializeComponent();
            Office o1 = new Office("计算机基础", "王军", "杨小勇", "何平", "姜涛");
            Office o2 = new Office("软件工程", "马建国", "陈君", "刘小燕");
            Office o3 = new Office("信息安全", "冯尧", "李建军", "张涛");
            Department d1 = new Department("计算机", o1, o2, o3);
            Office o4 = new Office("自动控制", "吴自力", "陈峰", "薛小龙");
            Office o5 = new Office("工业设计", "吴淑华", "方坤", "何力", "蔡聪");
            Department d2 = new Department("机电工程", o4, o5);
            Office o6 = new Office("信息管理", "赵民", "盛小楠", "徐小平");
            Office o7 = new Office("工商管理", "张敏", "李玲", "吕倩", "高剑");
            Department d3 = new Department("经济管理", o6, o7);
            school = new School("交通大学", d1, d2, d3);
        }

        private void Window_Loaded(object sender, RoutedEventArgs e)
        {
            TreeViewItem root = new TreeViewItem() { Header = school };
            treeView1.Items.Add(root);
            foreach (Department d in school.Departments)
            {
                TreeViewItem item = new TreeViewItem() { Header = d };
                root.Items.Add(item);
                foreach (Office o in d.Offices)
```

```
            item.Items.Add(new TreeViewItem() { Header = o });
        }
    }

    private void treeView1_SelectedItemChanged(object sender, RoutedPropertyChangedEventArgs<object> e)
    {
        listView1.Items.Clear();
        if (treeView1.SelectedItem == null)
            return;
        TreeViewItem item = (TreeViewItem)treeView1.SelectedItem;
        if (item.Header is School)
            foreach (Department d in school.Departments)
                listView1.Items.Add(d);
        else if (item.Header is Department)
            foreach (Office o in ((Department)(item.Header)).Offices)
                listView1.Items.Add(o);
        else if (item.Header is Office)
            foreach (string s in ((Office)item.Header).Teachers)
                listView1.Items.Add(s);
    }

    // School, Department, Office 的类型定义与程序 P7_9 中相同，此处省略......
}
```

13.1.4 文本框控件

Control 有一个抽象的派生类 TextBoxBase，它是普通文本框 TextBox 和富文本框 RichTextBox 的公共基类。TextBoxBase 的 Text 属性表示文本框中的文本；如果文本内容被修改则会引发其 TextChanged 事件。此外，TextBoxBase 还提供了一系列方法来模拟文本框上的编辑操作，比如复制 Copy、剪切 Cut、粘贴 Paste、撤销 Undo 和重复 Redo 等。

除了支持一般的文本编辑外，WPF 中的 RichTextBox 的主要作用是显示流文档的内容。FlowDocument 类对流文档进行了封装，一个文档由一组块元素 Block（包括段落元素 Paragraph、列表元素 List、表格元素 Table 等）组成，Block 中又可以包含一系列嵌入元素 Inline（包括文本元素 Run、可浮动元素 Floater 等）。将一个 FlowDocument 元素赋值给 RichTextBox 控件的 Document 属性，就可以在文本框中显示文档。下面的程序就创建并显示了一个流文档，其运行效果如图 13-12 所示。

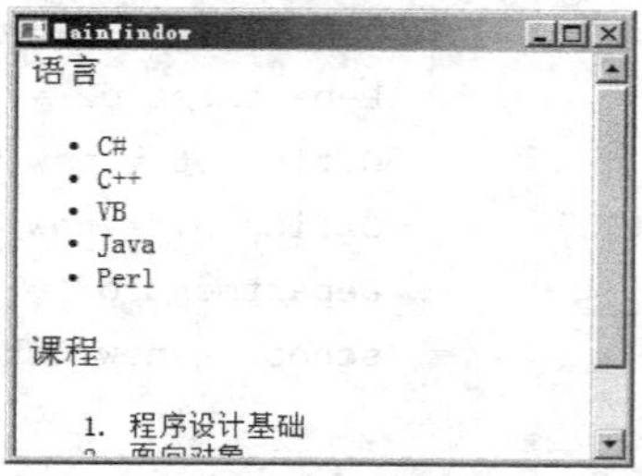

```
//程序 P13_4(MainWindow.xaml.cs)
using System.Windows;
using System.Windows.Controls;
using System.Windows.Documents;
namespace P13_4
{
    public partial class MainWindow : Window
    {
        public MainWindow()
        {
            InitializeComponent();
            RichTextBox rtb = new RichTextBox();
            rtb.Document = new FlowDocument();
            Paragraph p1 = new Paragraph(new Run("语言")) { FontSize = 15 };
            rtb.Document.Blocks.Add(p1);
            string[] lans = { "C#", "C++", "VB", "Java", "Perl" };
            List l1 = new List();
```

图 13-12 在 RichTextBox 控件中显示流文档

```
        foreach (string s in lans)
            l1.ListItems.Add(new ListItem(new Paragraph(new Run(s))));
        rtb.Document.Blocks.Add(l1);
        Paragraph p2 = new Paragraph(new Run("课程")) { FontSize = 15 };
        rtb.Document.Blocks.Add(p2);
        string[] cs = { "程序设计基础", "面向对象", "数据库应用", "Web 应用" };
        List l2 = new List() { MarkerStyle = TextMarkerStyle.Decimal };
        foreach (string s in cs)
            l2.ListItems.Add(new ListItem(new Paragraph(new Run(s))));
        rtb.Document.Blocks.Add(l2);
        ScrollViewer sView = new ScrollViewer();
        sView.Content = rtb;
        this.Content = sView;
    }
}
```

13.1.5 范围控件

滑块 Slider、滚动条 ScrollBar 和进度条 ProgressBar 都允许用户在一个范围内进行调节；在 WPF 中，这三个控件有一个共同的抽象基类 RangeBase，其 Minimum 和 Maximum 属性分别表示范围的下限和上限，Value 属性表示当前值，这 3 个属性的都是 double 类型。

在下面的示例程序，我们结合一个时钟组件 DispatcherTimer 来动态地控制一个进度条控件。程序主窗体中放置了一个进度条控件 progressBar1，其 Minimum 和 Maximum 属性值分别取默认值 0 和 100。

组件 DispatcherTimer 又称为定时器，它通过 Interval 属性指定一个时间间隔；当定时器开启时（IsEnabled 属性值设为 true），它就会每隔指定时间引发一次 Tick 事件，通过该事件的处理方法就能够使程序定时执行相应的操作。下面的程序将每隔 200 毫秒使进度条向前填充 5 个单位，填满之后又清空进度条并重新开始，程序的输出如图 13-13 所示。

```
//程序 P13_5(MainWindow.xaml.cs)
using System;
using System.Windows;
using System.Windows.Threading;
namespace P13_5
{
    public partial class MainWindow : Window
    {
        public MainWindow()
        {
            InitializeComponent();
            DispatcherTimer timer1 = new DispatcherTimer();
            timer1.Tick += new EventHandler(timer1_Tick);
            timer1.Interval = TimeSpan.FromMilliseconds(200);
            timer1.IsEnabled = true;
        }
        void timer1_Tick(object sender, EventArgs e)
        {
            if (progressBar1.Value == 100)
                progressBar1.Value = 0;
            else
                progressBar1.Value += 5;
            this.Title = string.Format("当前进度 {0}%", progressBar1.Value);
        }
    }
}
```

图 13-13　程序 P13_5 的输出结果示例

13.2 使用 XAML 设计界面

后台程序代码和前台设计界面相分离，这是提高 GUI 应用程序开发效率的一种有效方式。在 Visual Studio 中，Windows Form 窗体的 C#源代码和设计内容分别保存在.cs 和 Designer.cs 文件中。而对于 WPF 窗体，其源代码和设计内容分别保存在.xaml.cs 和.xaml 文件中；特别是其设计界面的描述语言不再是 C#，而是 XAML（extensible application markup language，可扩展应用程序标记语言）。

XAML 的最大优点在于它是一种自描述的语言，简单易读且逻辑性强，便于程序设计人员和图形界面设计人员之间的交流。例如程序 P13_4 的窗体界面也可以改用如图 13-14 所示的 XAML 文档来实现。

图 13-14　程序 P13_5 的输出结果示例

Visual Studio 的 XAML 文档视图和设计视图之间可方便地进行切换，还可以以拆分方式同时显示两个视图。修改 XAML 文档内容，设计视图会自动更新；在设计视图上进行控件拖放等操作，XAML 文档内容也会自动更新。

13.2.1　XAML 文档和元素

XAML 语言遵循 XML 规范，每一个 XAML 文档只有一个根元素，一个元素中可以嵌套另一个元素。对于每一个 WPF 窗体，其设计文件的根元素为 Window。例如从图 13-14 中可以看到，MainWindow.xaml 文档以标记<Window>开始，以标记</Window>结束；该根元素通过 xmlns 特性指定了文档中要引用的命名空间，并通过 class 特性指定了元素对应的类名是 MainWindow（class 的前缀"x:"表示该类属于当前程序的命名空间），这样编译器就会将 MainWindow.xaml 文档与 C#源代码中定义的 MainWindow 类一同处理。

Visual Studio 中创建的 WPF 程序项目还包含一个 App.xaml 文档，以及与之配套的 App.xaml.cs 源代码文件。App.xaml 文档描述了整个 WPF 程序的设计框架，其大致内容如下：

```
<Application x:Class="P13_4.App"
```

```
        xmlns="http://schemas.microsoft.com/winfx/2006/xaml/presentation"
        xmlns:x="http://schemas.microsoft.com/winfx/2006/xaml"
        StartupUri="MainWindow.xaml">
    <Application.Resources>
    </Application.Resources>
</Application>
```

可知该文件以 Application 元素为根元素，其中也指定了引用的命名空间以及对应的程序类。此外，Application 元素还具有一个 StartupUri 特性，它用于指定程序启动时显示的主窗体。

作为"可扩展应用程序标记语言"，XAML 文档中的每一个 XML 元素都会被映射到一个 .NET 对象，如 Application 元素对应 WPF 程序对象，Windows 元素对应 WPF 窗体对象。在图 13-13 所示的 XAML 文档中，Window 元素依次嵌套了 ScrollViewer、RichTextBox、FlowDocument 等元素，它们分别被映射到相应的 WPF 控件对象。

13.2.2 元素属性和事件

XAML 元素可以直接设置其简单属性值，比如可通过如下方式设置一个文本框控件的尺寸和字体大小：

```
<TextBox Name="textBox1" Height="80" Width="200" FontSize="15"></TextBox>
```

这等价于使用如下 C#源代码来设置控件的属性值（其中 Height 和 Width 属性继承自 FrameworkElement，FontSize 属性继承自 Control）：

```
textBox1.Height = 80;
textBox1.Width = 200;
textBox1.FontSize = 15;
```

不同之处在于，在 XAML 中设置的简单属性值总是文本格式的（值内容要放在一对双引号之间），编译器会自动根据文本内容来生成相应类型的值。在很多情况下，XAML 中的属性值描述方式会更为简洁，例如可通过如下代码设置一个 WrapPanel 控件采用水平排列的方式，在属性值中不用写出完整的枚举值"Orientation.Vertical"：

```
<WrapPanel Orientation="Vertical"></WrapPanel>
```

有些 WPF 对象的属性较为复杂，如属性值还具有自己的属性，通过 XAML 简单属性难以进行描述。此时可将复杂属性作为元素的嵌套元素，例如 FrameworkElement 的 Margin 属性表示界面元素的外边距，其类型是 Thickness 结构，该结构又分别通过 Top、Bottom、Left、Right 这 4 个属性指定了 4 个方向的宽度。下面的代码演示了如何通过嵌套属性元素的方式来设置一个 Grid 的外边距：

```
<Grid>
  <Grid.Margin>
    <Thickness Left="3" Right="3" Top="2" Bottom="2" />
  </Grid.Margin>
</Grid>
```

此时属性元素名称前要加上其所属父元素的前缀，如"Grid.Margin"。

XAML 还有一种属性叫做附加属性，这种属性不是元素自身固有的属性，而是取决于元素所在的外部环境。例如将一个控件放在 Grid 容器中，要指定控件位于网格的哪一行和哪一列，就可以使用附加属性。下面的示例代码就设置了一个 2 行 3 列的 Grid 容器，并在第 2 行的第 1 列和第 3 列分别放置了一个按钮控件：

```
<Grid>
  <Grid.RowDefinitions>
    <RowDefinition Height="80" />
    <RowDefinition />
  </Grid.RowDefinitions>
  <Grid.ColumnDefinitions>
```

```xml
    <ColumnDefinition />
    <ColumnDefinition />
    <ColumnDefinition Width="100" />
  </Grid.ColumnDefinitions>
  <Button Content="确定" Grid.Row="1" />
  <Button Content="取消" Grid.Row="1" Grid.Column="2" />
</Grid>
```

在上述代码中,只有当 Button 元素嵌套在 Grid 元素中时,附加属性 Grid.Row 和 Grid.Column 才会发挥作用。不过即使没有 Grid 父元素,上述代码也不会出错,此时附加属性会被自动忽略。

在 XAML 中,元素的属性不仅可以直接设置固定值,还可以通过表达式与另一个元素的某个属性值相关联,这称为绑定表达式。

假设窗体中有一个名为 silder1 的滑块控件,那么下面的代码能够使 Label 控件的文本字体随着滑块控件的值的变化而变化:

```xml
<Label FontSize="{Binding ElementName=slider1, Path=Value}"/>
```

其中{Binding ElementName=slider1, Path=Value}就是一个绑定表达式:表达式内容放在一对大括号中,以区别于普通的属性值;表达式以关键字 Binding 开头,之后通过 ElementName 项指定要绑定的元素,并通过 Path 项指定要绑定的属性路径。这等价于在滑块控件的 ValueChanged 事件处理方法中执行如下的 C#代码:

```
label1.FontSize=slider1.Value;
```

在绑定表达式的 Path 项中还可以指定多层嵌套的属性。在下面的程序示例中,ListBox 控件中的每个列表项都通过 FontFamily 属性指定了不同的字体;除了通过滑块来调节字体大小外,Label 控件还将 FontFamily 属性和列表框当前选项的字体族绑定在一起:

```xml
//程序 P13_6(MainWindow.xaml)
<Window x:Class="P13_6.MainWindow"
    xmlns="http://schemas.microsoft.com/winfx/2006/xaml/presentation"
    xmlns:x="http://schemas.microsoft.com/winfx/2006/xaml"
    Title="MainWindow" Height="200" Width="300">
  <DockPanel>
    <Slider Name="slider1" Minimum="10" Maximum="60" Width="200" DockPanel.Dock="Top" />
    <ListBox Name="listBox1">
      <ListBoxItem FontFamily="宋体">宋体</ListBoxItem>
      <ListBoxItem FontFamily="楷体">楷体</ListBoxItem>
      <ListBoxItem FontFamily="隶书">隶书</ListBoxItem>
      <ListBoxItem FontFamily="黑体">黑体</ListBoxItem>
    </ListBox>
    <Label Content="程序字体" FontSize="{Binding ElementName=slider1, Path=Value}"
FontFamily="{Binding ElementName=listBox1, Path=SelectedItem.FontFamily}"/>
  </DockPanel>
</Window>
```

运行该程序,通过左右拖动滑块可改变文本标签的字体大小,通过列表框中则可以为文本标签选择字体样式,如图 13-15 所示。在该程序中我们没有写一行 C#代码,而只是通过 XAML 代码就实现了所需的功能。

XAML 元素声明中还可以将元素的事件关联到事件处理程序,其格式是:事件名="事件处理方法名"。例如下面的 XAML 代码表示将按钮的 Click 事件绑定到 button_Click 方法:

```xml
<Button Content="提交" Click="button_Click" />
```

当然,这要求在窗体类中定义了这样的方法,且方法的签名符合事件处理方法的要求:

```
private void button_Click(object sender, RoutedEventArgs e)
{
    //…
}
```

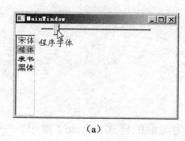
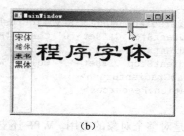

（a）　　　　　　　　　　　　　　（b）

图 13-15　程序 P13_6 的输出结果示例

13.2.3　资源和样式

1. 资源

C#代码中可以重用各种类型和对象。在 XAML 文档中，可以将需要重用的元素定义为资源（Resource），从而方便地在文档的其他地方进行重用。

举一个常见的例子，一个窗体中可能有多个控件要使用相同的背景色，那么可以把背景色所对应的画刷定义为资源。具体做法是在 Windows 根元素下增加一个所属的 Resources 子元素，而后在其中定义待重用的画刷元素（画刷的详细用法将在 13.3 节中介绍）：

```
<Window.Resources>
    <LinearGradientBrush x:Key="lgBrush">
        <GradientStop Offset='0' Color='Red' />
        <GradientStop Offset='.5' Color='Yellow' />
        <GradientStop Offset='1' Color='White' />
    </LinearGradientBrush>
</Window.Resources>
```

定义为资源的元素需要通过特性 Key 指定键值。在某个控件中要使用该背景色，可使用 StaticResource 表达式来设置控件的 Background 属性值，并在表达式的 ResourceKey 项中指定资源的键值。例如下面的 XAML 代码为两个 Label 控件和两个 Button 控件都应用了预设的背景色，应用效果如图 13-16 所示。

```
<Label Content="C#" Background="{StaticResource ResourceKey=lgBrush}"/>
<Label Content="C++" Background="{StaticResource ResourceKey=lgBrush}"/>
<Button Content="演示" Background="{StaticResource ResourceKey=lgBrush}"/>
<Button Content="返回" Background="{StaticResource ResourceKey=lgBrush}"/>
```

图 13-16　资源应用示例

设置属性的 StaticResource 表达式也可以换成 DynamicResource 表达式，例如：

```
<Label Content="C#" Background="{DynamicResourceResourceKey=lgBrush}"/>
```

二者的区别在于：StaticResource 表达式所引用的资源保持不变，而 DynamicResource 表达式所引用的资源可以动态修改，但后者的运行效率不如前者。

如果一个程序包含多个窗体，且希望这些窗体中的控件都能够重用某个资源，那么可以将资源的定义提升到应用程序的 XAML 文档中，例如：

```
<Application.Resources>
  <LinearGradientBrush x:Key="lgBrush">
    <GradientStop Offset='0' Color='Red' />
    <GradientStop Offset='.5' Color='Yellow' />
    <GradientStop Offset='1' Color='White' />
  </LinearGradientBrush>
</Application.Resources>
```

2. 样式

一般资源是对整个对象的重用。WPF 还支持更为灵活的样式（Style）重用，一个样式可以看成是任意一组属性值的组合。例如下面的代码定义了两个样式，第一个样式设置控件字体加粗、前景色为白色、背景色为蓝色，第一个样式设置控件字体倾斜、边框色为红色：

```
<Window.Resources>
  <Style x:Key="style1">
    <Setter Property="Control.FontWeight" Value="Bold" />
    <Setter Property="Control.Foreground" Value="White" />
    <Setter Property="Control.Background" Value="Blue" />
  </Style>
  <Style x:Key="style2">
    <Setter Property="Control.FontStyle" Value="Italic" />
    <Setter Property="Control.BorderBrush" Value="Red" />
  </Style>
</Window.Resources>
```

可见 Style 元素中可以包含多个 Setter 元素，每个 Setter 元素设置一个属性，其中 Property 特性指定要设置的目标属性，Value 特性指定要设置的属性值。如果属性值是一个复杂对象，那么可采用嵌套属性元素的方式来进行设置，例如：

```
<Setter Property="Control.Margin">
  <Setter.Value>
    <Thickness Left="3" Right="3" Top="2" Bottom="2" />
  </Setter.Value>
</Setter>
```

对其他元素应用样式，只需要在 StaticResource 表达式中指定样式的键值，并将表达式设置给元素的 Style 属性即可。例如下面的代码将上面定义的第一个样式应用到两个 Label 控件，将第二个样式应用到两个 Button 控件，应用效果如图 13-17 所示。

```
<Label Content="C#" Style="{StaticResource style1}" />
<Label Content="C++" Style="{StaticResource style1}" />
<Button Content="演示" Style="{StaticResource style2}" />
<Button Content="返回" Style="{StaticResource style2}" />
```

样式中也可以设置事件处理方法，这样应用该样式的控件将共享同一个事件处理方法。Style 通过子元素 EventSetter 来设置事件，其格式是：Event="事件名", Handler="事件处理方法名"，例如：

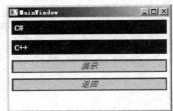

图 13-17 样式应用示例

```
<Style x:Key="style2">
  <Setter Property="Control.FontStyle" Value="Italic" />
  <EventSetter Event="Control.Click" Handler="button_Click" />
</Style>
```

XAML 的 Style 元素还提供了一个 TargetType 特性，通过它可以为特定类型的控件自动应用样式。例如下面定义的样式就指定了 TargetType 特性值为 "Button"：

```
<Style x:Key="style2" TargetType="Button">
```

```
        <Setter Property="Control.FontStyle" Value="Italic" />
        <EventSetter Event="Control.Click" Handler="button_Click" />
    </Style>
```

这样该样式会自动应用到当前窗体的所有按钮上，按钮的定义中并不需要进行样式设置。但是，如果某个按钮使用 StaticResource 表达式指定了另一个样式，它就将使用新样式来替代自动样式。特别的，如果有的按钮不需要应用样式，那么应显式指定其样式属性值为空。例如在下面的代码中，前两个按钮将自动应用上面定义的样式 style2，第三个按钮应用样式 style1，最后一个按钮则不应用任何样式：

```
<Button Content="演示" />
<Button Content="返回" />
<Button Content="重复" Style="{StaticResource style1}" />
<Button Content="取消" Style="{x:Null}" />
```

3. 触发器

除了事件关联外，样式中还可以使用触发器（Trigger）来定义可重用的行为模式。Style 元素的 Triggers 子元素中可以包含一组 Trigger 元素，其中每个 Trigeer 元素通过 Property 和 Value 特性来指定触发条件，并通过一个嵌套的 Setter 元素来指定触发器的设置工作。例如下面的样式中定义了一个触发器，其触发条件为控件的 IsFocused 属性值为 True（即控件获得焦点），条件满足时就设置控件的边框色为红色：

```
<Style x:Key="emStyle">
  <Style.Triggers>
    <Trigger Property="Control.IsFocused" Value="True">
      <Setter Property="Control.BorderBrush" Value="Red" />
    </Trigger>
  </Style.Triggers>
</Style>
```

下面的程序定义了一个样式 eStyle，它包含两个触发器：第一个在鼠标移到控件上方时触发，此时设置控件字体为红色加粗；第二个在按钮控件被单击时触发，此时设置按钮的边框色为红色。窗体中的两个 Label 控件和两个 Button 控件均应用了该样式：

```
//程序 P13_7(MainWindow.xaml)
<Window x:Class="P13_7.MainWindow"
    xmlns="http://schemas.microsoft.com/winfx/2006/xaml/presentation"
    xmlns:x="http://schemas.microsoft.com/winfx/2006/xaml"
    Title="MainWindow" Height="200" Width="300">
  <Window.Resources>
    <Style x:Key="eStyle">
      <Setter Property="Control.FontStyle" Value="Italic" />
      <Style.Triggers>
        <Trigger Property="Control.IsMouseOver" Value="True">
          <Setter Property="Control.Foreground" Value="Red" />
          <Setter Property="Control.FontWeight" Value="Bold"/>
        </Trigger>
        <Trigger Property="Button.IsPressed" Value="True">
          <Setter Property="Button.BorderBrush" Value="Blue" />
        </Trigger>
      </Style.Triggers>
    </Style>
  </Window.Resources>
  <StackPanel>
    <Label Content="C#" Style="{StaticResource eStyle}" Margin="5" />
    <Label Content="C++" Style="{StaticResource eStyle}" Margin="5" />
    <Button Content="演示" Style="{StaticResource eStyle}" Margin="5" />
    <Button Content="返回" Style="{StaticResource eStyle}" Margin="5" />
```

```
</StackPanel>
</Window>
```
运行该程序，通过移动鼠标和单击按钮可以看到样式应用的效果，如图 13-18 所示。特别的，由于样式中第二个触发器的触发条件为 Button.IsPressed，因此它对非按钮控件不起作用。

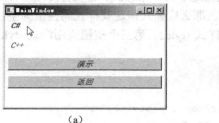
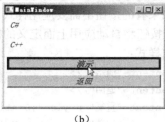

(a)　　　　　　　　　　　　(b)

图 13-18　程序 P13_7 的输出结果示例

13.3　绘制图形

13.3.1　画刷

WPF 窗体控件的基类 Control 提供了很多外观属性，比如表示前景色的 Foreground、表示背景色的 Background、表示边框色的 BorderBrush 等。但和 Windows Form 控件不同，这些与颜色相关的属性类型不是 Color，而是 Brush。

Brush 是 System.Windows.Media 命名空间中的一个抽象类，它有以下几种派生类。
- SolidColorBrush：使用单一固定颜色的画刷。
- GradientBrush：使用渐变色的画刷，该类又有 LinearGradientBrush 和 RadialGradientBrush 两个派生类，分别表示线性渐变色和径向渐变色的画刷。
- TileBrush：使用图块进行绘制的画刷，该类又有 DrawingBrush、ImageBrush 和 VisualBrush 三个派生类，它们分别使用 Drawing 图形、Image 图像和可视化元素进行绘制。
- BitmapCacheBrush：使用位图缓存进行绘制的画刷。

如前所述，在 XAML 中使用固定色画刷来设置前景色、背景色、边框色等属性，只需要为属性指定相应的颜色就可以了，例如，下面的代码设置按钮的前景色和背景色分别为蓝色和黄色：

```
<Button Name="btn" Content="确定" Foreground="Blue" Background="Yellow" />
```
在 C#程序代码中则需要创建相应的 SolidColorBrush 对象，例如：
```
btn.Foreground = new SolidColorBrush(Colors.Blue);
btn.Background = new SolidColorBrush(Colors.Yellow);
```
或是从 Brushes 类的静态属性中获得相应的 SolidColorBrush 对象，例如：
```
btn1.Foreground = Brushes.Blue;
btn1.Background = Brushes.Yellow;
```
如果不想填充控件的背景，可设置其背景色为画刷 Brushes.Transparent，那么控件除内容以外的部分就都是透明的（这和白色画刷 Brushes.White 是不同的）。另一种替代办法是设置控件的不透明度属性 Opacity（该属性继承自 UIElement），其值为 0 时表示完全透明，值为 1 时表示完全不透明。

渐变色画刷是指在不同颜色之间连续变化的画刷，它在使用时需要指定两种或两种以上的颜色。例如下面的 XAML 代码为按钮 btn1 的背景色指定了一个渐变色画刷，其填充色是由红到黄再到蓝的线性变化：

```
<Button Name="btn1" Margin="5">
  <Button.Background>
    <LinearGradientBrush>
      <GradientStop Offset='0' Color='Red' />
      <GradientStop Offset='.5' Color='Yellow' />
      <GradientStop Offset='1' Color='Black' />
    </LinearGradientBrush>
  </Button.Background>
</Button>
```

下面的 C#代码则创建了一个径向渐变色画刷，并将其应用到另一个按钮 btn2：

```
RadialGradientBrush brush = new RadialGradientBrush();
brush.GradientStops.Add(new GradientStop(Colors.Red, 0.0));
brush.GradientStops.Add(new GradientStop(Colors.Yellow, 0.5));
brush.GradientStops.Add(new GradientStop(Colors.Black, 1.0));
btn2.Background = brush;
```

采用上述方式设置的两个按钮的背景色如图 13-19 所示。

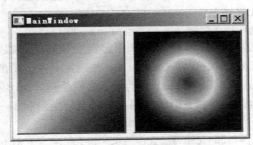

图 13-19　渐变色画刷示例

使用 TileBrush 时，可通过 Stretch 属性指定图块的填充方式，默认值 Fill 表示填满目标区域，值为 Uniform 时可保持纵横比不变；还可通过 TileMode 属性指定平铺方式，默认值 None 表示不平铺，值为 Tile 时表示平铺，图块平铺的尺寸和度量方式可分别通过 Viewport 和 ViewportUnits 属性指定。下面的代码使用图像 smiley.jpg 作为画刷内容，并采用不同的填充和平铺方式应用于两个按钮，其效果如图 13-20 所示。

```
<Button Margin="5">
  <Button.Background>
    <ImageBrush ImageSource="smiley.jpg" Stretch="Uniform" />
  </Button.Background>
</Button>
<Button Grid.Column="1" Margin="5">
  <Button.Background>
    <ImageBrush ImageSource="smiley.jpg" TileMode="Tile" Viewport="0,0 48,48" ViewportUnits="Absolute" />
  </Button.Background>
</Button>
```

图 13-20　图块画刷示例

13.3.2 形状

从图 13-7 中可以看到，FrameworkElement 有一个抽象的派生类 Shape，它的派生类包括线条 Line、矩形 Rectangle、椭圆 Ellipse、多线条 Polyline、多边形 Polygon、路径 Path 等 WPF 预定义形状的公共基类。这些形状都属于 WPF 界面的基本元素，其使用方式在很大程度上与窗体控件是类似的，例如形状也有自己的位置和尺寸，可放置在其他布局容器中，能够响应鼠标和键盘事件等。这是因为它们同样继承了 Visual、UIElement 和 FrameworkElement 这些基类。

Shape 本身提供了各种形状公共的属性，其中常用的是以下三个属性。

- Fill：形状内部的填充画刷。
- Stroke：形状边框的绘制画刷。
- StrokeThickness：形状边框的宽度。

最简单的形状是线条 Line，线条的起点位置由属性 X1 和 Y1 确定，终点位置由属性 X2 和 Y2 确定。Line 的 Fill 属性不起作用，必须设置其 Stroke 属性才能显示线条。例如下面的 XAML 代码就绘制了一条蓝色的水平线条：

```
<Line X1="0" Y1="270" X2="619" Y2="270" Stroke="Blue"/>
```

矩形 Rectangle 通过属性 Width 和 Height 来指定宽度和高度。如果形状是放在 Canvas 容器中，那么可通过附加属性 Canvas.Left 和 Canvas.Top 来指定其左上角的横坐标和纵坐标。此外，如果设置了 Rectangle 的 RadiusX 和 RadiusY 属性，那么所绘制的矩形就是圆角矩形，圆角的椭圆半径由这两个属性值共同确定。例如下面的代码在画布上分别绘制了一个普通矩形和一个圆角矩形，效果如图 13-21 所示。

```
<Rectangle Canvas.Left="100" Canvas.Top="200" Height="70" Width="100" Stroke="Black"/>
<Rectangle Canvas.Left="300" Canvas.Top="200" Height="70" Width="100" Stroke="Black" RadiusX="10" RadiusY="10"/>
```

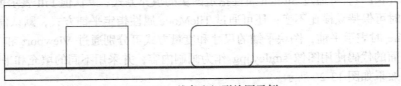

图 13-21　线条和矩形绘图示例

椭圆 Ellipse 也是通过属性 Width 和 Height 来指定宽度和高度，即横轴和纵轴方向上的直径长度。当这两个属性值相等时，绘制出来的就是一个圆形。例如下面的代码在画布上方添加了一个红色的圆，此时，画布上的内容将如图 13-22 所示。

```
<Ellipse Canvas.Left="500" Canvas.Top="50" Name="ellipse1" Fill="Red" Height="50" Width="50" />
```

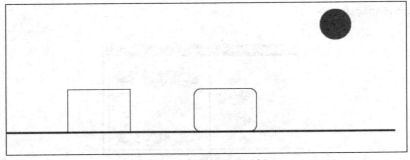

图 13-22　椭圆绘图示例

Polyline 和 Polygon 都是通过一组点来绘制一系列首尾相连的线条，这些点通过 Points 属性来指定。不同之处在于，Polygon 会将最后一个点和第一个点也连接起来，从而形成一个封闭的多边形。在 XAML 代码中设置 Points 属性值时，只需依次列出这些点的坐标，各点之间以空格分隔，每个点的横坐标和纵坐标之间则以逗号分隔。例如下面的代码分别绘制了一个 Polyline 和 Polygon 对象，从如图 13-23 所示的结果中可以看出两者的区别。

```
<Polyline Stroke="Black" Fill="Cyan" Points="100,205 90,205 150,165 210,205 200,205" />
<Polygon Stroke="Black" Fill="Gray" Points="290,205 310,165 390,165 410,205" />
```

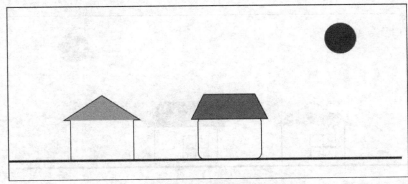

图 13-23　多线条和多边形绘制示例

路径 Path 是最为复杂的一种形状元素，它可用于绘制各种复杂的几何图形。Path 有一个 Data 属性，其类型是几何图形 Geometry。但 Geometry 是一个抽象类，其派生类有以下几种。

- LineGeometry：直线几何对象。
- RectangleGeometry：矩形几何对象。
- EllipseGeometry：椭圆几何对象。
- CombinedGeometry：任意两个几何对象的组合。
- GeometryGroup：任意多个几何对象的集合。
- PathGeometry：任意路径构成的几何对象。

上述类型是"轻量级"的几何图形，它们不支持鼠标和键盘事件，也不处理边框和填充等外观属性，因此对象运行的性能更优于 Line、Rectangle 等形状对象。将一个几何对象赋值给 Path 的 Data 属性，几何对象中的内容会按照 Path 所指定的外观进行统一绘制。例如下面定义的 Path 对象就包含了 4 个矩形，它们都使用蓝色边框和黄色填充：

```
<Path Canvas.Left="115" Canvas.Top="235" Stroke="Blue" Fill="Yellow">
  <Data>
    <GeometryGroup x:Key="Win">
      <RectangleGeometry Rect="0,0 20,20"/>
      <RectangleGeometry Rect="0,20 20,20"/>
      <RectangleGeometry Rect="20,0 20,20"/>
      <RectangleGeometry Rect="20,20 20,20"/>
    </GeometryGroup>
  </Data>
</Path>
```

如果某个图形需要在程序中多次使用，那么可将相应的几何对象定义为资源，例如可将上面代码中的 GeometryGroup 对象放在窗体的资源中进行定义：

```
<Window.Resources>
<GeometryGroup x:Key="Win">
<RectangleGeometry Rect="0,0 20,20"/>
<RectangleGeometry Rect="0,20 20,20"/>
```

```
        <RectangleGeometry Rect="20,0 20,20"/>
        <RectangleGeometry Rect="20,20 20,20"/>
      </GeometryGroup>
    </Window.Resources>
```

下面的代码定义了两个 Path 对象,它们的 Data 属性都设置为资源中的 GeometryGroup 对象,加入它们之后画布上的内容如图 13-24 所示。

```
    <Path Canvas.Left="115" Canvas.Top="220" Stroke="Blue" Fill="Yellow" Data="{StaticResource Win}" />
    <Path Canvas.Left="315" Canvas.Top="220" Stroke="Blue" Fill="Yellow" Data="{StaticResource Win}" />
```

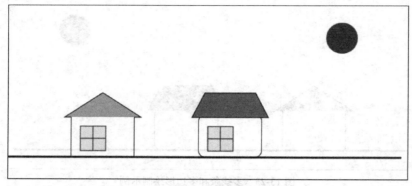

图 13-24　路径绘图示例

使用 XAML 代码来绘制形状,其优点是简单明了,并且在设计过程中就能获得"所见即所得"的效果。当然,如果要通过复杂的选择或循环流程来控制形状的绘制,那还是需要使用 C# 程序代码。比如下面的代码就通过循环语句为画布右上方的"太阳"加上了一系列"光线",程序的输出结果如图 13-25 所示。

```
//程序 P13_8(MainWindow.xaml.cs)
using System;
using System.Windows;
using System.Windows.Controls;
using System.Windows.Media;
using System.Windows.Shapes;
namespace P13_8
{
    public partial class MainWindow : Window
    {
        public MainWindow()
        {
            InitializeComponent();
            Line[] lights = new Line[12];
            double x0 = Canvas.GetLeft(ellipse1) + ellipse1.Width / 2;
            double y0 = Canvas.GetTop(ellipse1) + ellipse1.Height / 2;
            double r = ellipse1.Width / 2;
            for (int i = 0, a = 0; i < 12; i++)
            {
                lights[i] = new Line();
                lights[i].X1 = x0 + 1.2 * r * Math.Sin(a * Math.PI / 180);
                lights[i].Y1 = y0 + 1.2 * r * Math.Cos(a * Math.PI / 180);
                lights[i].X2 = x0 + 1.8 * r * Math.Sin(a * Math.PI / 180);
                lights[i].Y2 = y0 + 1.8 * r * Math.Cos(a * Math.PI / 180);
                lights[i].Stroke = Brushes.Red;
                canvas1.Children.Add(lights[i]);
                a += 30;
```

```
            }
          }
        }
      }
```

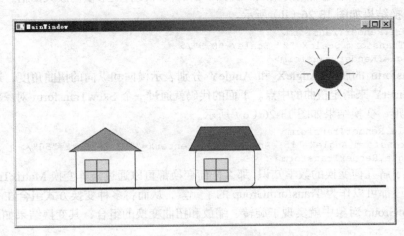

图 13-25　程序 P13_8 的完整输出结果

13.3.3　图形变换

WPF 中的所有用户界面元素都支持平移、旋转、缩放等几何变换。针对变换，UIElement 类型定义了一个属性 RenderTransform，其类型为 System.Windows.Media 命名空间中抽象类 Transform，它具体的派生类有以下几种。

- TranslateTransform：表示平移变换。
- RotateTransform：表示旋转变换。
- ScaleTransform：表示放大或缩小变换。
- SkewTransform：表示扭曲变换。
- MatrixTransform：由变换矩阵指定的变换。
- TransformGroup：多个变换的组合。

使用 TranslateTransform 对象进行平移变换时，其属性 X 和 Y 分别表示横坐标和纵坐标的位移。例如下面的代码创建了一个宽 80、高 40 的矩形，其初始坐标为(20,40)；通过嵌套 RenderTransform 属性元素进行变换后，实际坐标就变成了(50,30)，如图 13-26（a）所示。

```
<Rectangle  Name="rect1"Stroke="Blue"  Width="80"  Height="40"  Canvas.Left="20"
Canvas.Top="40">
    <Rectangle.RenderTransform>
      <TranslateTransform X="30" Y="-10"/>
    </Rectangle.RenderTransform>
</Rectangle>
```

RotateTransform 的 Angle 属性表示旋转的角度（顺时针方向），CenterX 和 CenterY 则共同指定了旋转的中心点。例如将上面矩形的 RenderTransform 属性改为如下内容，它就会以左上角为圆心顺时针旋转 30°，结果如图 13-26（b）所示。

```
<Rectangle.RenderTransform>
  <RotateTransform Angle="30" />
</Rectangle.RenderTransform>
```

如果再设置旋转的圆心位置为（60,60），则结果如图 13-26（c）所示。

```
<Rectangle.RenderTransform>
  <RotateTransform Angle="30" CenterX="60" CenterY="60" />
```

```
</Rectangle.RenderTransform>
```

ScaleTransform 通过属性 ScaleX 和 ScaleY 来分别指定横向和纵向的缩放比例，还可通过属性 CenterX 和 CenterY 来指定缩放的中点。例如下面的代码设置矩形横向放大两倍，纵向缩小为原来的一半，变换结果如图 13-26（d）所示。

```
<Rectangle.RenderTransform>
  <ScaleTransform ScaleX="2" ScaleY="0.5"/>
</Rectangle.RenderTransform>
```

SkewTransform 的属性 AngleX 和 AngleY 分别表示横向和纵向的扭曲角度，还可通过属性 CenterX 和 CenterY 来指定扭曲的中点。下面的代码就通过一个 SkewTransform 对象将矩形扭曲为一个平行四边形，变换结果如图 13-26（e）所示。

```
<Rectangle.RenderTransform>
  <SkewTransform AngleX="15" AngleY="30" CenterX="50" CenterY="50"/>
</Rectangle.RenderTransform>
```

如果深入了解几何变换的数学知识，那么任何变换都可以通过矩阵变换 MatrixTransform 来实现。上述变换还都可以作为 TransformGroup 的子元素，从而将多种变换方式组合在一起。下面定义的 TransformGroup 对象中就实现了旋转、缩放和扭曲变换的组合，其变换结果如图 13-26（f）所示。

```
<Rectangle.RenderTransform>
  <TransformGroup>
    <RotateTransform Angle="30" />
    <ScaleTransform ScaleY="0.5" />
    <SkewTransform AngleX="15" />
  </TransformGroup>
</Rectangle.RenderTransform>
```

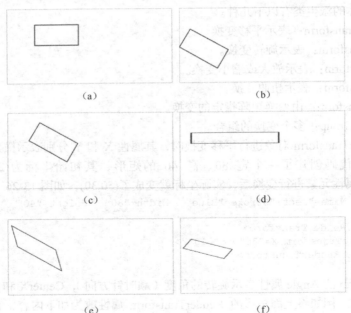

图 13-26　图形变换效果示例

13.3.4　打印输出

WPF 窗体界面上的任何内容都可以输出到打印机。System.Windows.Controls 命名空间中提供了一个 PrintDialog 类，表示打印对话框。创建了 PrintDialog 对象后，调用其 ShowDialog 方法可显示打印对话框，用户可在其中选择打印机、设置打印选项，以及执行打印命令。

PrintDialog 提供了以下两种方法来实现打印功能：

- PrintVisual 方法，它能够打印任意 Visual 对象中的内容。
- PrintDocument 方法，它能够打印任何的文档对象。

例如对于程序 P13_8，可为窗体添加键盘处理事件，当用户按下组合键 Ctrl+P 时就显示打印对话框，然后将画布对象作为参数传递给 PrintVisual 方法（方法的第二个参数为打印任务的文字说明），就可以打印整个画布上的绘图内容。

```
private void Window_PreviewKeyDown(object sender, KeyEventArgs e)
{
    if (e.Key == Key.P && Keyboard.Modifiers == ModifierKeys.Control)
    {
        PrintDialog dlg = new PrintDialog();
        if (dlg.ShowDialog() == true)
            dlg.PrintVisual(canvas1, "WPF 打印示例");
    }
}
```

 对于 WPF 中的对话框和消息框，其 ShowDialog 方法的返回值为可空布尔类型 bool?，这一点要注意和 Windows Form 的对话框和消息框相区别。

假设打印输出到 XPS 文档，那么得到的结果如图 13-27 所示。

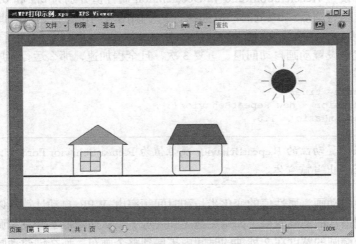

图 13-27　打印效果示例

13.4　动画和多媒体

13.4.1　基于属性的动画

使用动画能够大大增强用户界面的视觉效果。普通的动画技术是基于计时器的，即每隔指定的时间让界面元素自动发生变化。例如程序 P13_5 实际上就演示了这样的一个动画，它通过 DispatcherTimer 组件来周期性地触发事件，并在事件处理方法中修改控件内容。基于计时器的动画技术需要开发人员进行具体细节的控制，因而在创建复杂动画时往往会导致很大的代码量。

在 Windows Form 中也有一个名为 Timer 的计时器组件，它的使用方法与 DispatcherTimer 基本相同。

WPF 提供了另一种基于属性的动画技术。应用这种动画，就是在指定的时间间隔内、使元素的某个属性值在指定范围内发生变化。假设画布上有一个名为 ball 的椭圆对象，下面的代码能够使其纵坐标的值在 1 秒时间内由 0 变化到 120，这样运行程序就能看到椭圆在窗体中匀速下落：

```
DoubleAnimation ani = new DoubleAnimation() { From = 0, To = 120 };
ani.Duration = TimeSpan.FromSeconds(1);
ball.BeginAnimation(Canvas.TopProperty, ani);
```

上述代码中使用的动画对象 ani 的类型是 DoubleAnimation。在 WPF 中，所有动画的基类都是 System.Windows.Media.Animation 命名空间下的抽象类 Animatable，它的不同派生类用于不同形式的动画。DoubleAnimation 就是基于 double 类型的属性的动画。画布中的元素都具有属性 Canvas.TopProperty，属性类型为 double，因此这种改变元素纵坐标的动画应使用 DoubleAnimation 对象。创建了动画对象之后，应用动画的方法是调用元素的 BeginAnimation 方法（从 UIElement 继承），该方法的第一个参数是要变化的属性，第二个参数就是 Animatable 动画对象。

在上述代码中可以看到，DoubleAnimation 通过属性 From 和 To 来设置属性的起始值和终止值，通过属性 Duration 设置变化的时间间隔。默认情况下，属性值是在 From 和 To 之间均匀变化，通过动画对象的 AccelerationRatio 和 Deceleration 属性可设置加速和减速，它们的值是以百分比的形式表示的。例如设置 AccelerationRatio 和 Deceleration 属性值分别为 0.3 和 0.4，那么动画在前 30% 的时间会加速执行，后 40% 的时间减速执行，中间的时间段保持匀速。此外，动画对象的 AutoReverse 属性表示动画播放完之后是否回退，RepeatBehavior 属性表示动画重复的次数。

例如下面的代码设置动画自动回退、重复 3 次，且全程加速，那么运行程序可看到椭圆 3 次加速垂直下落和弹回：

```
ani.AutoReverse= true;
ani.RepeatBehavior = new RepeatBehavior(3);
ani.AccelerationRatio = 1.0;
```

如果设置动画的 RepeatBehavior 属性值为 RepeatBehavior.Forever，动画则会持续播放，直至退出当前窗体。

使用基于属性的动画，属性值的变化步长和时间频率由 WPF 自动计算，以便根据当前系统的显示性能来达到最佳的动画效果（如避免拖尾和闪烁），从而降低了开发和调试的难度。

类似的，要使椭圆的宽度在 2 秒钟的时间内扩展到整个画布，那么创建动画的程序代码如下（From 属性值未设置则默认为当前值），其动画效果如图 13-28 所示。

```
DoubleAnimation ani = new DoubleAnimation() { To = canvas1.ActualWidth };
ani.Duration = TimeSpan.FromSeconds(2);
ball.BeginAnimation(Ellipse.WidthProperty, ani);
```

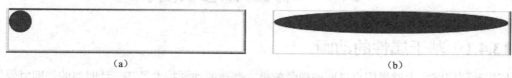

图 13-28　基于属性的图形动画示例

上述动画代码中，BeginAnimation 方法所关联的属性都不是类型的一般成员属性，而是 WPF 中的一种特殊属性——依赖项属性（Dependency Property）。这种属性定义为类型的静态成员，其命名规范是在一般属性名后面加上"Property"，如 Canvas.TopProperty 和 Ellipse.WidthProperty。不过在 XAML 代码中，"Property"部分可以省略。前面提到的附加属性也是依赖项属性的一种。通过依赖项属性能够使程序对属性值的变化自动作出响应，这是实现基于属性的动画的关键。

其他常用的基于属性的动画类型还有 Int32Animation、ByteAnimation、ColorAnimation、StringAnimation、MatrixAnimation 等，实际使用时应根据属性的不同类型来做出选择。

13.4.2 故事板和事件触发器

许多动画不仅仅是限于单个属性值的变化。例如要使一个椭圆斜向移动，就需要同时改变其横坐标和纵坐标的值。这时就要用到故事板 Storyboard，其 Children 属性能够包含一组动画对象的集合，而 Storyboard 对象能够协调控制这些动画的播放。

下面的示例代码就创建了两个 DoubleAnimation 对象，它们分别用于修改椭圆对象 ball 的横坐标和纵坐标；Storyboard 对象在 Children 集合中加入了这两个动画对象，并通过调用 Begin 方法来播放整个动画：

```
DoubleAnimation ani1 = new DoubleAnimation() { From = 0, To = 270 };
DoubleAnimation ani2 = new DoubleAnimation() { From = 0, To = 120 };
ani1.Duration = ani2.Duration = TimeSpan.FromSeconds(2);
ani1.AutoReverse = ani2.AutoReverse = true;
ani1.RepeatBehavior = ani2.RepeatBehavior = RepeatBehavior.Forever;
Storyboard sb = new Storyboard();
sb.Children.Add(ani1);
sb.Children.Add(ani2);
Storyboard.SetTarget(ani1, ball);
Storyboard.SetTargetProperty(ani1, new PropertyPath(Canvas.LeftProperty));
Storyboard.SetTarget(ani2, ball);
Storyboard.SetTargetProperty(ani2,new PropertyPath(Canvas.TopProperty));
sb.Begin();
```

使用故事板时，需要将为 Children 集合中的每个动画设置目标元素和属性，这分别是通过 Storyboard 的静态方法 SetTarget 和 SetTargetProperty 来实现的。

如果不涉及复杂的程序控制逻辑，那么动画功能也可以使用 XAML 代码来实现。在 13.2.3 小节中介绍了触发器的概念。一般情况下，XAML 中的动画总是通过一种特殊的事件触发器来播放的。事件触发器的类型名为 EventTrigger，它通过 RoutedEvent 特性指定触发事件，并通过 Actions 特性指定与事件相关联的动作。对于动画，最常用的动作是 BeginStoryBoard，在其中放上一个 StoryBoard 对象，就表示"事件发生时执行故事板中的动画"。

事件触发器可以在界面元素的 Triggers 集合中定义。例如下面的 XAML 代码定义了一个 Rectangle 对象，在其 Triggers 集合中定义了一个事件触发器，它在 MouseDown 事件发生时执行 BeginStoryBoard 动作。代码的 StoryBoard 对象中定义了三个 DoubleAnimation 动画对象：前两个负责移动矩形，后一个则旋转矩形。程序代码的运行效果如图 13-29 所示。

```
//程序 P13_9(MainWindow.xaml 片段)
<Rectangle    Name="rect"    Stroke="Blue"    Fill="Red"    Width="30"    Height="30"
RenderTransformOrigin="0.5,0.5">
  <Rectangle.RenderTransform>
    <RotateTransform />
  </Rectangle.RenderTransform>
  <Rectangle.Triggers>
    <EventTrigger RoutedEvent="Shape.MouseDown">
      <EventTrigger.Actions>
        <BeginStoryboard>
          <Storyboard>
            <DoubleAnimation  From="0"  To="270"  Duration="0:0:2"  AutoReverse="True"
```

```xml
RepeatBehavior ="Forever" Storyboard.TargetName="rect" Storyboard.TargetProperty="(Canvas.
Left)" />
            <DoubleAnimation From="0" To="120" Duration="0:0:2" AutoReverse="True"
RepeatBehavior ="Forever" Storyboard.TargetName="rect" Storyboard.TargetProperty="(Canvas.
Top)" />
            <DoubleAnimation From="0" To="360" Duration="0:0:0.2" AutoReverse="True"
RepeatBehavior ="Forever" Storyboard.TargetName="rect" Storyboard.TargetProperty="RenderTransform.
Angle" />
          </Storyboard>
        </BeginStoryboard>
      </EventTrigger.Actions>
    </EventTrigger>
  </Rectangle.Triggers>
</Rectangle>
```

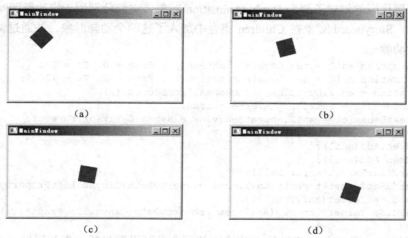

图 13-29 程序 P13_9 的动画效果示例

在 XAML 代码中设置动画对象的播放时间 Duration，格式是"时:分:秒"，比如"0:1:30"表示 1 分 30 秒，"0:0:0.1"表示 0.1 秒；设置 Storyboard.TargetProperty 时，如果目标属性是像 Canvas.Top 这样的附加属性，那么属性必须放在一对括号中。

当然，事件触发器及其中的动画也可以定义在资源中，这样预定义的动画效果就能够方便地应用到不同的界面元素上。例如按照下面的代码定义样式资源，其中的动画就能够自动应用到窗体中的所有矩形元素上：

```xml
<Style x:Key="aniStyle" TargetType="Rectangle">
    <Style.Triggers>
    <EventTrigger RoutedEvent="Shape.MouseDown">
    <EventTrigger.Actions>
    <BeginStoryboard>
    <Storyboard>
    <!--故事板中的动画...-->
      …
    </Storyboard>
    </BeginStoryboard>
    </EventTrigger.Actions>
    </EventTrigger>
    </Style.Triggers>
</Style>
```

13.4.3 基于路径的动画

WPF 还提供了另一种基于路径的动画技术，此类动画对象通过 PathGeometry 属性来设置一条路径，这样目标元素就会沿着指定的路径移动。前面介绍过，PathGeometry 对象可以包含各种直线和曲线构成的路径，由此可以实现其他方法很难实现的动画控制效果。

假设要使一个控件沿着一条曲线路径移动，那么可先创建一个 PathGeometry 对象来绘制曲线，再创建两个 DoubleAnimationUsingPath 动画对象，它们的 PathGeometry 属性值均为指定的 PathGeometry 对象，Source 属性值分别设为"X"和"Y"（分别和路径每一点的横坐标和纵坐标相关联），其他属性设置和 DoubleAnimation 动画对象类似。

在下面的示例程序中，画布上放置了两个 Image 控件，它们分别显示地球和太阳的图标。程序在窗体资源中定义了一个 PathGeometry 对象，并在窗体启动时触发两个故事板对象：一个使用 DoubleAnimation 动画水平移动太阳，另一个使用 DoubleAnimationUsingPath 动画沿路径移动地球。该程序的输出结果如图 13-30 所示。

```xml
//程序 P13_10(MainWindow.xaml)
<Window x:Class="P13_10.MainWindow"
    xmlns="http://schemas.microsoft.com/winfx/2006/xaml/presentation"
    xmlns:x="http://schemas.microsoft.com/winfx/2006/xaml"
    Title="P13_10" Height="270" Width="480">
  <Window.Resources>
    <PathGeometry x:Key="path">
      <PathFigure StartPoint="15,108">
        <ArcSegment Point="215,108" Size="100,80" RotationAngle="180" SweepDirection="Clockwise" />
        <ArcSegment Point="415,108" Size="100,80" RotationAngle="180" SweepDirection="Counterclockwise" />
        <ArcSegment Point="215,108" Size="100,80" RotationAngle="180" SweepDirection="Counterclockwise" />
        <ArcSegment Point="15,108" Size="100,80" RotationAngle="180" SweepDirection="Clockwise" />
      </PathFigure>
    </PathGeometry>
  </Window.Resources>
  <Window.Triggers>
    <EventTrigger RoutedEvent="Window.Loaded">
      <EventTrigger.Actions>
        <BeginStoryboard>
          <Storyboard>
            <DoubleAnimation  Storyboard.TargetName="sun"  Storyboard.TargetProperty="(Canvas.Left)" From="115" To="315" Duration="0:0:3" AutoReverse="True" RepeatBehavior="Forever"/>
          </Storyboard>
        </BeginStoryboard>
        <BeginStoryboard>
          <Storyboard>
            <DoubleAnimationUsingPath  Storyboard.TargetName="earth"  Storyboard.TargetProperty="(Canvas.Left)" Duration="0:0:6" RepeatBehavior="Forever" PathGeometry="{StaticResource path}" Source="X" />
            <DoubleAnimationUsingPath  Storyboard.TargetName="earth"  Storyboard.TargetProperty="(Canvas.Top)" Duration="0:0:6" RepeatBehavior="Forever" PathGeometry="{StaticResource path}" Source="Y" />
          </Storyboard>
        </BeginStoryboard>
      </EventTrigger.Actions>
    </EventTrigger>
  </Window.Triggers>
  <Canvas>
    <Image Canvas.Left="115" Canvas.Top="100" Source="sun.gif" />
    <Image Canvas.Left="15" Canvas.Top="108" Source="earth.gif" />
    <Path Stroke="Black" Data="{StaticResource path}">
      <Path.RenderTransform>
```

```
        <TranslateTransform X="10" Y="10"/>
      </Path.RenderTransform>
    </Path>
  </Canvas>
</Window>
```

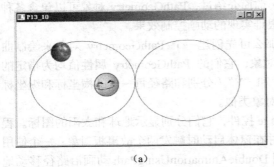
(a)

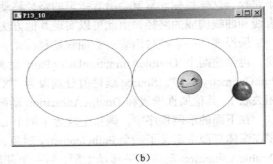
(b)

图 13-30　基于路径的图形动画示例

13.4.4　播放多媒体文件

Windows 应用程序都可以使用 System.Media 命名空间下的 SoundPlayer 类来播放声音。例如下面的代码就能够在程序中播放音频文件 Ringin.wav（如果文件不在当前目录，则需要在构造函数中指定完整路径）：

```
SoundPlayer player = new SoundPlayer("Ringin.wav");
player.Play();
```

WPF 在 System.Windows.Controls 命名空间下提供了一个 SoundPlayerAction 类，它对 SoundPlayer 进行了封装，以便使用 XMAL 代码来播放声音。SoundPlayerAction 是 TriggerAction 的一个派生类，可将该元素作为事件触发器的行为对象，并通过其 Source 属性来设置要播放的音频文件。例如下面的代码将在"播放"按钮被单击时播放音频文件 Ringin.wav：

```
<Button Content="播放">
    <Button.Triggers>
    <EventTrigger RoutedEvent="Button.Click">
    <EventTrigger.Actions>
    <SoundPlayerAction Source="Ringin.wav"/>
    </EventTrigger.Actions>
    </EventTrigger>
    </Button.Triggers>
</Button>
```

上面两个类型都只能播放一般的波形声音。要播放 MP3 等格式的压缩音频，以及 WMV、MPEG、MP4 等各种类型的视频文件，可使用 System.Windows.Media 命名空间下的 MediaPlayer 类。下面的代码就创建了一个 MediaPlayer 对象来播放视频文件 Wolf.wmv：

```
MediaPlayer player = new MediaPlayer();
player.Open(new Uri("Wolf.wmv"));
player.Play();
```

对应的，System.Windows.Controls 命名空间下的 MediaElement 类也对 MediaPlayer 进行了封装。在 XMAL 代码中放置一个 MediaElement 元素，通过其 Source 属性指定要播放的音频或视频文件，那么启动窗体就会自动进行播放。例如：

```
<MediaElement Name="mediaElement1" Source="Wolf.wmv" />
```

如果要通过事件触发器来进行播放，那么可先设置其 LoadedBehavior 属性值为 Mannual（默认值为 Play 表示自动播放）：

```
<MediaElement Name="media1" LoadedBehavior="Manual" />
```

然后在事件触发器的行为中启动一个故事板，在故事板中嵌套一个 MediaTimeline 元素，设置其附加属性 Storyboard.TargetName 的值为 MediaElement 元素名，这样就能够在事件触发时进行播放媒体文件：

```
<EventTrigger RoutedEvent="Button.Click">
  <EventTrigger.Actions>
  <BeginStoryboard Name="PlayMediaElement">
  <Storyboard>
  <MediaTimeline Storyboard.TargetName="media1" Source="Wolf.wmv" />
  </Storyboard>
  </BeginStoryboard>
  </EventTrigger.Actions>
</EventTrigger>
```

各种类型的 Windows 应用程序一般都支持播放声音和视频的功能。WPF 程序所提供的优势在于，它能够将布局、剪裁、渲染、几何变换等技术与媒体播放功能无缝集成起来，从而实现精彩纷呈的界面效果。

下面的程序演示了这样一个例子，窗体上是一个两行两列的网格，其中左上角的单元格使用一个 MediaElement 控件来播放指定视频，其余三个单元格则各放置了一个矩形；由于 MediaElement 也是 Visual 元素，三个矩形均使用基于 MediaElement 的 VisualBrush 来进行填充，这样三个矩形中也能看到同步播放的视频内容；再通过不同的几何变换，三个矩形又将其中显示的内容分别进行了水平翻转、垂直翻转，以及水平垂直翻转。该程序的视频播放效果如图 13-31 所示。

```
//程序 P13_11(MainWindow.xaml)
<Window x:Class="P13_11.MainWindow"
xmlns="http://schemas.microsoft.com/winfx/2006/xaml/presentation"
xmlns:x="http://schemas.microsoft.com/winfx/2006/xaml"
Title="P13_11" Height="480" Width="640" Loaded="Window_Loaded">
<UniformGrid Rows="2" Columns="2">
  <Viewbox Stretch="Fill">
    <MediaElement Name="media1" Source="Wolf.wmv" />
  </Viewbox>
  <Rectangle Name="rightRect" Stroke="DarkGray" StrokeThickness="1">
  <Rectangle.Fill>
  <VisualBrush Visual="{Binding ElementName=media1}">
  <VisualBrush.RelativeTransform>
  <ScaleTransform ScaleX="-1" CenterX="0.5" />
  </VisualBrush.RelativeTransform>
  </VisualBrush>
  </Rectangle.Fill>
  </Rectangle>
  <Rectangle Name="downRect" Stroke="DarkGray" StrokeThickness="1">
  <Rectangle.Fill>
  <VisualBrush Visual="{Binding ElementName=media1}">
  <VisualBrush.RelativeTransform>
  <ScaleTransform ScaleY="-1" CenterY="0.5" />
  </VisualBrush.RelativeTransform>
  </VisualBrush>
  </Rectangle.Fill>
  </Rectangle>
  <Rectangle Name="DragonRect" Stroke="DarkGray" StrokeThickness="1">
  <Rectangle.Fill>
  <VisualBrush Visual="{Binding ElementName=media1}">
  <VisualBrush.RelativeTransform>
  <ScaleTransform ScaleX="-1" CenterX="0.5" ScaleY="-1" CenterY="0.5" />
  </VisualBrush.RelativeTransform>
  </VisualBrush>
  </Rectangle.Fill>
```

```
</Rectangle>
</UniformGrid>
</Window>
```

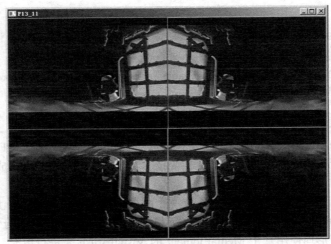

图 13-31　视频播放效果示例

13.5　案例研究——旅行社管理系统的 WPF 界面

13.5.1　构建系统主界面

本节将基于 WPF 构建旅行社内部管理子系统，相应的 WPF 项目名称为 TravelWPF。如图 13-32 所示，除了顶部的菜单栏外，系统主窗口 MainWindow 主要由树型视图和列表视图组成。左侧的 TreeView 控件以各个地区为根节点，以下的子节点依次为地区中的旅游线路和线路的组团方案；选中某个树节点后，ListView 控件中将显示当前对象所包含的下级对象集合；当选中的节点为某个组团方案（叶节点）时，ListView 中将显示该方案的所有旅行团对象。

为了方便数据管理和维护，这里希望把 TreeView 和 ListView 控件和相关的业务对象关联起来。注意 TreeViewItem 是标题—内容型控件，ListViewItem 是内容型控件，在构造 TreeView 和 ListView 视图的过程中，可以把 Line、Package、Tour 等对象赋值给视图项的 Header 或 Content 属性，从而在用户操作过程中方便地访问这些业务对象。

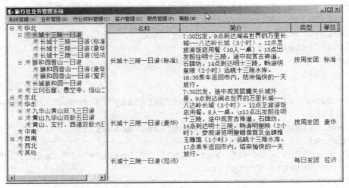

图 13-32　旅行社管理系统主界面

首先为 MainWindow 增加一个 List<Line>类型的字段，其中存放所有线路的集合，并在窗体的 Load 事件处理方法中对其进行初始化：

```csharp
private List<Line> _lines;

private void Form1_Load(object sender, EventArgs e)
{
    _lines = new List<Line>(Line.GetAll());
    InitializeComponent();
     //…
}
```

在窗口启动时，通过如下代码将各个 Line 对象按所属地区加入树型视图中，再在下一级加入每个 Line 所含的各个 Package 对象：

```csharp
private void Window_Loaded(object sender, RoutedEventArgs e)
{
    TreeViewItem item1, item2;
    foreach (Line l in _lines)
    {
        TreeViewItem parent = (TreeViewItem)treeView1.Items[(int)l.Area];
        item1 = new TreeViewItem() { Header = l };
        parent.Items.Add(item1);
        foreach (Package p in l.Packages)
        {
            item2 = new TreeViewItem() { Header = p };
            item1.Items.Add(item2);
        }
    }
}
```

当用户选中某个树节点后，程序对节点中包含的业务对象类型进行判断，并在列表视图中显示当前对象的下级对象集合：

```csharp
private void treeView1_SelectedItemChanged(object sender, RoutedPropertyChangedEventArgs<object> e)
{
    listView1.Items.Clear();
    TreeViewItem tvItem = ((TreeViewItem)treeView1.SelectedItem);
    if (tvItem.Items.Count > 0)
    {
        foreach (TreeViewItem item in tvItem.Items)
            listView1.Items.Add(new ListViewItem() { Content = item.Header });
        if (tvItem.Header is Line)
            listView1.Style = (Style)this.FindResource("PackageViewStyle");
        else
            listView1.Style = (Style)this.FindResource("LineViewStyle");
    }
    else
    {
        Package package = tvItem.Header as Package;
        if (package == null)
            return;
        foreach (Tour t in package.Tours)
            listView1.Items.Add(new ListViewItem() { Content = t });
        listView1.Style = (Style)this.FindResource("TourViewStyle");
    }
}
```

注意列表视图的样式是随着业务对象类型的变化而变化的。例如在树型视图中选中一个组团方案，列表视图中将显示该方案的所有旅行团对象，此时采用的视图样式定义如下：

```xml
<Style TargetType="ListView" x:Key="TourViewStyle">
```

```xml
<Setter Property="View">
  <Setter.Value>
    <GridView>
      <GridView.Columns>
        <GridViewColumn Header="编号" DisplayMemberBinding ="{Binding Path=Id}" />
        <GridViewColumn Header="开始时间" DisplayMemberBinding ="{Binding Path=StartTime}" />
        <GridViewColumn Header="结束时间" DisplayMemberBinding ="{Binding Path=EndTime}" />
        <GridViewColumn Header="状态" DisplayMemberBinding ="{Binding Path=State}" />
      </GridView.Columns>
    </GridView>
  </Setter.Value>
</Setter>
</Style>
```

13.5.2 新建、修改和删除业务对象

接下来提供对 Line、Package、Tour 等业务对象的操作功能，这主要通过列表视图上的快捷菜单来实现。WPF 中的快捷菜单可以作为任意其他控件的嵌套属性元素，例如将下面的 XAML 代码嵌入到 ListView 元素定义中，那么在列表视图上单击鼠标右键，就会弹出一个快捷菜单，其中包含设置"新建"、"修改"和"删除"3 个菜单项：

```xml
<ListView.ContextMenu>
  <ContextMenu Name="cMenu" Visibility="Hidden" Opened="cMenu_Opened">
    <MenuItem Name="cMenuNew" Header="新建(_N)" Click="cMenuNew_Click" />
    <MenuItem Name="cMenuModify" Header="修改(_M)" Click="cMenuModify_Click" />
    <MenuItem Name="cMenuDelete" Header="删除(_D)" Click="cMenuDelete_Click" />
  </ContextMenu>
</ListView.ContextMenu>
```

注意只有在列表视图选择了某个对象后才能进行修改和删除操作。在执行删除操作时，需要先判断当前的业务对象类型，而后删除该对象以及相应的视图项：

```csharp
private void cmenuDelete_Click(object sender, EventArgs e)
{
    if (MessageBox.Show("此操作不可恢复！确认要执行删除吗?","警告", MessageBoxButton.YesNo,
MessageBoxImage.Warning) == MessageBoxResult.No)
        return;
    object o = ((ListViewItem)listView1.SelectedItem).Content;
    if (o is Line)
        _lines.Remove((Line)o);
    else if (o is Package)
        ((Package)o).Line.Packages.Remove((Package)o);
    else if (o is Tour)
        ((Tour)o).Package.Tours.Remove((Tour)o);
    TreeViewItem tvItem = (TreeViewItem)treeView1.SelectedItem;
    foreach (TreeViewItem item in tvItem.Items)
        if (item.Header == o){
            tvItem.Items.Remove(item);
            break;
        }
    listView1.Items.Remove(listView1.SelectedItem);
}
```

下面考虑新建和修改业务对象的操作。前面章节中实现的 UserForm、StaffForm 等都是在一个窗体中维护一组对象，在其中支持对象的新建、修改和删除操作。对于 Line、Package、Tour 这些关键业务对象，它们更适合在一个窗体中操作一个对象；而新建对象和修改现有对象的操作界面基本相同，这两项操作可共用一个窗体类。

以旅行线路为例，对应窗体 LineWindow 的界面如图 13-33 所示，其左侧显示了线路的基本信息，右侧给出了线路上的景点列表。窗体中定义了通过一个 _line 字段：当使用无参构造函数来创建窗体时，该字段值为 null，窗体用于新建线路；使用带参构造函数时，字段值表示当前线路对象。

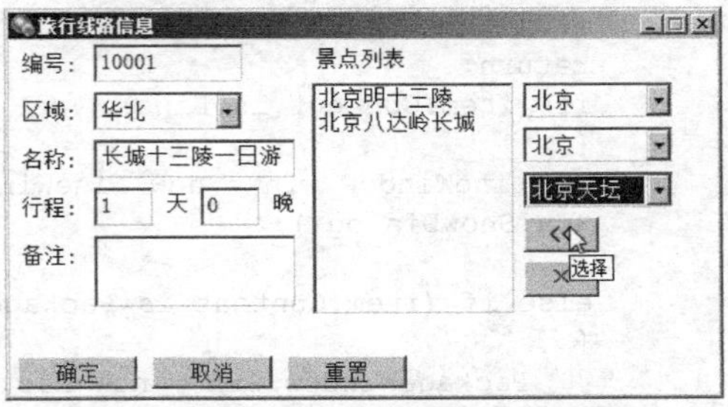

```
Line _line = null;  //要处理的线路对象

public LineWindow() //构造窗体以新建 Line 对象
{
    InitializeComponent();
    tbId.Text = "自动编号";
    tbId.IsReadOnly = true;
}
public LineWindow(Line line)
//构造窗体以编辑现有 Line 对象
{
    _line = line;
    InitializeComponent();
    this.LoadInfo();
}
```

图 13-33 "旅行线路信息"窗体界面

上面带参构造函数调用的 LoadInfo 方法用于将现有 Line 对象的信息显示在窗体中：

```
public void LoadInfo()
{
    tbId.Text = _line.Id.ToString();
    tbName.Text = _line.Name;
    tbDays.Text = _line.Days.ToString();
    tbNights.Text = _line.Nights.ToString();
    cmbArea.SelectedIndex = (int)_line.Area;
    if (_line.Remark != null)
        tbRemark.Text = _line.Remark;
    foreach (Scene s in _line.Scenes)
        listBox1.Items.Add(s);
}
```

当用户在窗体中单击"确定"按钮时，程序根据 _line 字段是否为 null 来选择新建线路对象或修改现有对象信息；单击"重置"按钮时，程序清空输入或载入当前对象信息。具体的事件处理方法代码不再列出。

组团方案窗体 PackageWindow 和旅行团窗体 TourWindow 的界面分别如图 13-34 和图 13-35 所示。除了内容项不同外，二者的功能实现方式与 LineWindow 大体相同，具体代码请参见配书程序。

图 13-34 "组团方案信息"窗体界面

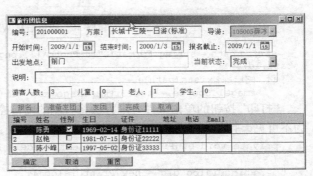

图 13-35 "旅行团信息"窗体界面

在主窗体的快捷菜单中,选择"新建"和"修改"命令,程序将打开相应的窗体以进行信息维护。以"修改"菜单项命令为例,其事件处理方法代码如下:

```csharp
private void cMenuModify_Click(object sender, RoutedEventArgs e)
{
    ListViewItem item = (ListViewItem)listView1.SelectedItem;
    if (item == null)
        return;
    if (item.Content is Line)
    {
        LineWindow win = new LineWindow((Line)item.Content);
        win.ShowDialog();
    }
    else if (item.Content is Package)
    {
        PackageWindow win = new PackageWindow((Package)item.Content);
        win.ShowDialog();
    }
    else if (item.Content is Tour)
    {
        TourWindow win = new TourWindow((Tour)item.Content);
        win.ShowDialog();
    }
}
```

13.5.3 信息打印输出

10.5.1 小节创建了职员信息管理窗体 StaffForm,这里使用如图 13-36 所示的 WPF 窗口来替代其功能,并增加职员照片显示和信息打印的功能。设职员的照片都存放在应用程序所在路径的 Photo 子目录下,格式为 .png 文件,且文件名就是职员的 Id 号。那么将下面的代码加到窗体列表框的选项改变事件中,就可以在 Image 控件中显示职员的照片:

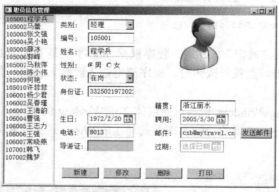

图 13-36 职员信息 WPF 窗体界面

```csharp
string path = string.Format(@"{0}Photo\{1}.png", AppDomain.CurrentDomain.BaseDirectory,
_staff.Id);
    if (File.Exists(path))
        image1.Source = new BitmapImage(new Uri(path));
```

"打印"按钮的单击事件处理方法将右侧的网格控件输出到打印机(不希望打印的控件可先设为不可见,打印完毕后再恢复为可见):

```csharp
private void btnPrint_Click(object sender, RoutedEventArgs e)
{
    PrintDialog dlg = new PrintDialog();
    if (dlg.ShowDialog() == true)
```

```
        {
            sPanel1.Visibility = btnSendMail.Visibility = Visibility.Hidden;
            dlg.PrintVisual(gridA, "职员信息");
            sPanel1.Visibility = btnSendMail.Visibility = Visibility.Visible;
        }
}
```

程序的打印输出效果如图 13-37 所示。

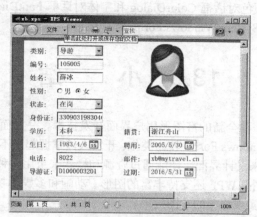

图 13-37 打印职员信息

13.5.4 Windows Form 集成

在 WPF 程序项目中也可以创建和使用 Window Form 窗体，但在实际应用中，这种情况出现得较少。更常见的情况是需要在 WPF 程序中重用已有的 Window Form 窗体。例如我们在第 7 章和第 8 章中创建了旅行社信息管理窗体 BasicForm、用户登录窗体 LoginForm 和用户管理窗体 UserForm，这里不希望将其替换为新的 WPF 窗口，那么直接使用这些现有窗体的基本操作步骤如下：

（1）为 TravelWPF 项目添加 System.Windows.Forms 和 System.Drawing 这两个程序集的引用。

（2）将现有 Window Form 窗体的文件（包括.cs 源程序文件、.designer.cs 设计文件，以及.resx 资源文件）复制到 TravelWPF 项目所在目录下。

（3）将 Window Form 窗体作为现有项添加到 TravelWPF 项目中。

（4）在 WPF 程序中通过相应的命名空间来访问 Window Form 窗体对象。

要在程序启动时显示用户登录窗体，那么可为 MainWindow 添加相关的字段成员，并在构造函数中创建和显示 TravelWin.LoginForm 窗体对象：

```
private User[] _users;
private List<Line> _lines;

public MainWindow()
{
    _users = User.GetAll();
    TravelWin.LoginForm form = new TravelWin.LoginForm(_users);
    if (form.ShowDialog() != System.Windows.Forms.DialogResult.OK)
    {
        this.Close();
        return;
    }
```

```
        _lines = new List<Line>(Line.GetAll());
        InitializeComponent();
}
```

编译运行程序，这时需要通过 LoginForm 窗体完成登录后才能进入程序主界面。显示旅行社信息管理窗体和用户管理窗体的方式也是类似的。

此外，相比 Windows Form，目前的 WPF 版本还缺乏一些控件和功能，比如不支持打开文件对话框 OpenFileDialog、颜色对话框 ColorDialog 和字体对话框 FontDialog 等。要在 WPF 程序中调用这些功能，也需要与 System.Windows.Forms 程序集进行交互。

13.6 小　　结

WPF 是 .NET 平台上一种全新的富图形用户界面开发技术。一方面，WPF 的基本控件类型、属性和事件模型等都和 Windows Forms 较为类似，开发人员掌握起来不会有太大困难；另一方面，WPF 可使用 XAML 语言来设计界面，还能通过资源和样式来重用各种界面元素，从而大大方便了界面的管理和维护。此外，WPF 还支持丰富的图形、动画和多媒体功能，这得益于其底层使用高性能的 DicrectX 引擎来实现精美的绘图效果。

13.7 习　　题

1. 简述 WPF 的控件内容模型。
2. 对比 Windows Form 设计文件 .Designer.cs 和 WPF 窗体设计文件 .xaml，两者存在哪些主要区别？
3. 画出样式 Style、设置器 Setter、触发器 Trigger 这几种 XAML 元素类型之间的关系图，并说说这种设计方式有哪些好处。
4. 扩展程序 P13_7，在蓝色线条的下方绘制一系列波浪线，模拟房子坐落在河边的绘图效果。要求将波浪线的几何对象作为资源使用。
5. 编写程序，在 WPF 窗体中绘制一个椭圆，且椭圆的尺寸会随着窗体的尺寸变化而自动缩放，而在打印时椭圆将占据整个纸张的大小。
6. 编写程序，在 WPF 窗体绘制一个会不断旋转和变色的彩灯。
7. Animatable 和 StoryBoard 类型在 WPF 动画中各自起着什么作用？这种设计方式对我们有哪些启示？
8. 在旅行社管理系统中，为 TourWindow、PackageWindow 和 LineWindow 定义共同的基类，并试着将 3 个窗体中通用的功能转移到基类中实现。

第 14 章
C# Web 应用程序设计

ASP.NET 是 .NET 平台上的 Web 应用开发模型。本章主要介绍 ASP .NET Web 应用程序界面的开发技术，包括 Web 窗体、Web 基本对象、HTML 控件和 Web 服务器控件。学习过程中要树立的观点是：Web 应用开发不是传统意义上的"做网页"，而是使用对象来设计和组装程序。

14.1 ASP .NET 技术概述

在 Internet 上，大多数 Web 网页都是基于 HTML（Hyper Text Markup Language，超文本标记语言）格式的，网络节点之间使用超文本传输协议 HTTP 进行数据传输。传统的 Web 网页编程技术主要是在 HTML 文档中嵌入脚本语言，其中微软的 ASP（Active Server Pages，动态服务器网页）技术就是在 HTML 中嵌入 VB Script 或 Java Script，从而支持面向对象的编程和脚本语言的解释执行。用户可从网络上的任意一台计算机连接到服务器，服务器程序根据用户浏览器的请求动态地生成 HTML 网页，并将结果发回给浏览器。

ASP.NET 是微软在 2000 年推出的 Web 应用开发模型，但它并不是 ASP 的简单升级，而是 .NET Framework 的一部分。它采用 C#、VB .NET 等编译式语言而非脚本语言，通过 .NET 类库提供的各种基础类型和控件来支持高效的应用开发。ASP.NET 应用程序经编译后运行在 CLR 之上，并利用早期绑定、实时编译、本地缓存等技术来提高程序性能。和 ASP 相比，ASP .NET 具有更高的开发和运行效率，且程序结构更为清晰。

与一般的 Windows 程序不同，ASP.NET 应用程序不仅包含已编译的程序集文件，还包括应用程序根目录中及其子目录中的其他各种文件，如一般网页文件（.html）、ASP .NET 页面文件（.aspx）、XML Web 服务文件（.asmx）、网站配置文件（.config）等。而且 ASP.NET 应用程序一般是在虚拟目录而非本地目录下浏览和运行。虚拟目录是对 Web 应用程序物理位置的映射，其中的文件来源不仅包括应用程序所在的目录，还可以是本地或网络上的其他位置；但在客户端浏览器看来，所有的文件都是来自于同一个虚拟目录。

在 ASP.NET 开发模型中，网页及其各种元素都是完全面向对象的，开发人员可以使用属性、方法和事件来处理各种网页元素。有关网页和控件的状态维护，以及服务器和浏览器之间的交互细节，基本上都由 ASP.NET 框架负责处理。开发人员主要关心的是如何设计业务对象、选择控件对象，以及处理各种对象之间的交互关系。

14.2 ASP .NET Web 窗体和基本对象

14.2.1 Web 窗体

每一个 ASP.NET 网页就是一个 Web 窗体。之所以叫做"窗体",是因为在 Visual Studio 中开发 Web 窗体与开发 Windows 窗体没有多大区别:在 Web 窗体的设计视图中,可从"工具箱"向窗体拖放各种 Web 控件,还可以通过属性窗口设置窗体和控件的属性,或是添加事件处理方法来实现各种程序逻辑。

在本书 2.3.5 小节简单地介绍了创建 ASP .NET 网站项目的方法。下面我们再通过一个项目示例来理解 Web 窗体的基本特点和功能。在 Visual Studio 中创建一个"ASP .NET 空 Web 应用程序"P14_1,向其中加入一个 Web 窗体 Default.aspx,打开代码文件 Default.aspx.cs 可看到如下的内容:

```
public partial class Default : System.Web.UI.Page
{
    protected void Page_Load(object sender, EventArgs e)
    {}
}
```

在.NET 类库中,System.Web.UI 命名空间中的 Page 类是对 Web 窗体的抽象。与 Windows 窗体类似,Web 窗体在启动时也会引发 Load 事件,上面的 Page_Load 就是表示该事件的处理方法。接下来向该方法中添加如下的代码:

```
this.Title = "Web 应用程序示例";
this.Response.Write("欢迎光临! ");
this.Response.Write("您的浏览器为: " + this.Request.Browser.Type);
```

编译运行程序,此时 Visual Studio 会打开一个 Internet Explorer 网页,网页的标题为"Web 应用程序示例",网页的内容中将包含文本"欢迎光临!"以及当前系统所使用的浏览器类型,如图 14-1 所示。

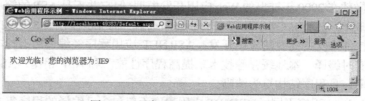

图 14-1 程序 P14_1 的输出结果示例

Page 的 Title 属性表示 Web 窗体的标题,上面还使用到了 Page 类的两个重要属性:
- Response——与窗体相关的 HttpResponse 对象,用于向浏览器输出内容。
- Request——与窗体相关的 HttpRequest 对象,用于从浏览器读取数据。

其中 HttpResponse 对象的 Write 方法用于将信息通过 HTTP 输出到浏览器。如果把浏览器看作是控制台,那么该方法就相当于 Console.Write 方法。HttpRequest 对象的 Browser 属性则能够返回当前用户的浏览器信息(程序运行在不同的浏览器上将得到不同的结果)。

用户通过客户端浏览器与服务器进行交互,那么一个网页可能会因为响应客户端的回发(如用户通过单击按钮来提交数据)而多次加载,而每次加载时都会引发一个 Load 事件。为此,Page 类提供了 IsPostBack 属性来判断页面是否为首次加载。例如我们可以在程序 P14_1 的主窗体上加入一个按钮控件,并将窗体的 Load 事件处理方法改写为如下内容:

```csharp
protected void Page_Load(object sender, EventArgs e)
{
    if (!IsPostBack){
        this.Title = "Web 应用程序示例";
        this.Response.Write("欢迎光临! ");
    }
    else
        this.Response.Write("您的浏览器为: " + this.Request.Browser.Type);
}
```

那么运行程序首次加载页面时,网页上显示的文字是"欢迎光临!",而在单击按钮之后将网页上显示浏览器的类型。

由于 Web 窗体自身的特点,Page 对象的生命周期会产生比 Windows 窗体更多的事件。其中 Init 事件在页面初始化时发生,它早于 Load 事件,并且只发生一次,因此控件的初始化代码可放在该事件的处理方法中完成。对应的,页面卸载时将发生 UnLoad 事件。此外,Init 事件前后分别发生 PreInit 和 InitComplete 事件,Load 事件前后分别发生 PreLoad 和 LoadComplete 事件。

在 Web 窗体的代码文件中我们看不到像 "this.Load +=new EventHandler(Default_Load)" 这样的代码。这是因为 ASP.NET 默认为 Web 窗体设置了 AutoEventWireup 属性,它表示 Web 窗体的事件将自动与事件处理方法相关联。也就是说,如果我们在 Default.aspx.cs 代码文件中定义了 Page_Init、Page_Load 这样的方法,ASP.NET 就会自动将它们作为 Web 窗体 Default 的 Init 和 Load 事件的处理方法。Default.aspx 文件的第一行 HTML 源代码为:

```
<%@ Page Language="C#" AutoEventWireup="true" CodeFile="Default.aspx.cs" Inherits="_Default" %>
```

其中 AutoEventWireup="true"就表示为 Web 窗体启动了自动事件关联的特性。如果想取消这一特性并手动添加代码,只要将其中的 true 改为 false 即可。

除了 Response 和 Request 之外,还可通过 Page 对象的 Server、Session 等属性来访问 ASP.NET 的其他内置对象,下面将依次介绍这些重要的 ASP.NET 基本对象。

14.2.2 请求和响应

和本地 Windows 应用程序不同,服务器上的 Web 应用程序不能直接访问客户端的内存数据,因此就要频繁地在服务器和客户端之间进行数据传递。System.Web 命名空间中的 HttpRequest 类型封装了客户端通过 HTTP 发送的请求信息,HttpResponse 则封装了来自 ASP.NET 服务器的 HTTP 响应信息。上面使用的 HttpRequest 的 Browser 属性,其类型为 HttpBrowserCapabilities,通过它可以获得发送请求的浏览器信息,并在需要时根据不同的浏览器来选择不同的操作。看下面的代码示例:

```csharp
HttpBrowserCapabilities browser = this.Request.Browser;
StringBuilder sb1 = new StringBuilder("名称:");
sb1.AppendLine(browser.Browser);
sb1.Append("版本:");
sb1.AppendLine(browser.Version);
sb1.Append(".NET Framework 版本:");
sb1.AppendLine(browser.ClrVersion.ToString());
sb1.Append("系统平台:");
sb1.AppendLine(browser.Platform);
sb1.Append("支持 Cookies:");
sb1.AppendLine(browser.Cookies ? "是" : "否");
this.Response.Write(sb1.ToString());
```

HttpRequest 的 UserHostAddress 和 UserHostName 属性分别返回客户端的主机 IP 地址和 DNS

名称；如果请求是通过代理发送，那么 UserAgent 属性将返回代理的相关信息。

除了客户端自身的一些信息外，HTTP 请求中还包含要求服务器返回的内容，其中 HttpRequest 的 Url 属性就表示的所请求网页的地址，而 Path 和 PhysicalPath 属性分别表示请求网页的虚拟路径和物理路径。假设网页 Default.aspx 位于 "C:\Inetpub\wwwroot\ws" 目录下，对应的虚拟目录为 "http://192.168.1.1/ws"，那么请求该网页的 Url 为 "http://192.168.1.1/ws/Default.aspx"，而 Path 和 PhysicalPath 属性值分别为 "/ws/Default.aspx" 和 "C:\Inetpub\wwwroot\ws\Default.aspx"。此外，使用 HttpRequest 对象的 MapPath 方法能够将虚拟路径转换为物理路径。

要通过不同的参数来请求窗体返回不同的结果，这在 Windows 窗体中可使用带参构造函数来实现，而 Web 窗体可以把参数封装在请求的 Url 字符串中，并通过 HttpRequest 的 QueryString 属性来读取相关参数。该属性的类型为 NameValueCollection，可通过其索引函数来查询指定的参数值。设当前虚拟目录为 "http://192.168.1.1/ws"，用户请求的 Url 为 "http://192.168.1.1/ws/Default.aspx?id=1&grade=2"，那么通过当前请求对象的 QueryString[0]或 QueryString["id"]将得到参数值 1，而通过 QueryString[1]或 QueryString["grade"]将得到 2。

下面的程序 P14_2 就演示了如何通过请求的参数来返回不同的学生对象信息：

```csharp
//程序 P14_2(Default.aspx.cs)
using System;
public partial class Default : System.Web.UI.Page
{
    protected void Page_Load(object sender, EventArgs e)
    {
        string id = this.Request.QueryString["id"];
        if (id == null)
            Response.Write("错误:未指定学生学号");
        else if (id == "1001")
            Response.Write(new Student("周军", true, new DateTime(1988,2,8)));
        else if (id == "1002")
            Response.Write(new Student("刘莉", false, new DateTime(1990,9,1)));
        else
            Response.Write("没有找到指定的学生");
    }
}

public class Student
{
    private string name;
    private bool gender;
    private DateTime birthday;

    public Student(string name, bool gender, DateTime birthday)
    {
        this.name = name;
        this.gender = gender;
        this.birthday = birthday;
    }

    public override string ToString()
    {
        return string.Format("姓名:{0} 性别:{1} 年龄:{2}", name, gender ? '男' : '女',
DateTime.Now.Year - birthday.Year);
    }
}
```

直接运行打开程序时，网页上将输出 "错误:未指定学生学号"；而在网页地址的 Default.aspx

之后加上"?id=1001",得到的输出就是"姓名:周军性别:男年龄:21"。

再看 HttpResponse。除了用于输出文本的 Write 方法外,它还可通过 BinaryWrite 方法来输出二进制内容,或是通过 WriteFile 方法来输出文件内容。此外,HttpResponse 的 End 方法可完成输出并停止当前页面的执行,例如下面的代码在输出"欢迎光临!"后将停止,而不会再输出浏览器的类型:

```
this.Response.Write("欢迎光临! ");
this.Response.End();  //停止输出
this.Response.Write("您的浏览器为:" + this.Request.Browser.Type);
```

HttpResponse 的 BufferOutput 属性表示对页面输出是否进行缓存,其值为 false 时每次输出都会直接发送到浏览器,为 true 时则会缓存输出内容,直至处理完整个页面后再一次性发送。此外,HttpResponse 的 Flush 方法将强制输出缓存内容,Clear 方法则会清空缓存。

HttpResponse 的输出内容还可以是图像、电子表格、音频和视频等多种多样的类型,它的 ContentType 属性就表示输出内容的类型,默认值"text/html"表示 HTML 文本。表 14-1 列出了一些常见的 HTTP 输出类型。

表 14-1　　　　　　　　　　ContentType 的常用属性值列表

取 值	类 型	取 值	类 型
application/acad	AutoCAD 程序文档	image/gif	gif 图像
application/msexcel	Microsoft Excel 程序文档	image/jpeg	jpeg 图像
application/mspowerpoint	Microsoft Powerpoint 程序文档	image/png	png 图像
application/msword	Microsoft Word 程序文档	image/tiff	tiff 图像
application/pdf	Adobe PDF 程序文档	text/html	html 文本
application/rft	Microsoft 写字板程序文档	text/plain	无格式文本
application/zip	Zip 压缩程序文档	text/richtext	多样式文本
audio/x-midi	midi 音频	text/xml	xml 文本
audio/x-mpeg	mpeg 音频	video/mpeg	mpeg 视频
audio/x-wav	wav 音频	video/x-msvideo	avi 视频

下面的程序就随机绘制了一幅验证码图像,并通过 HttpResponse 的 BinaryWrite 方法将其输出到网页上,程序的运行效果如图 14-2 所示。

```
//程序 P14_3(Default.aspx.cs)
using System;
using System.Drawing;
using System.Drawing.Imaging;
using System.IO;
public partial class Default : System.Web.UI.Page
{
    protected void Page_Load(object sender, EventArgs e)
    {
        Bitmap bmp = new Bitmap(80, 40);
        Graphics g = Graphics.FromImage(bmp);
        g.Clear(Color.White);
        g.DrawRectangle(Pens.Black, 0, 0, 79, 39);  //绘制边框
        Color[] colors = { Color.Black, Color.Red, Color.Blue, Color.DarkGreen, Color.Purple, Color.DarkGoldenrod, Color.Chocolate };
        string[] fonts = { "宋体", "楷体_GB2312", "隶书", "Arial", "Comic Sans MS", "Microsoft Sans Serif", "Times New Roman" };
```

图 14-2　程序 P14_3 的输出结果示例

```csharp
            Random rand = new Random();
            char[] chs = new char[] { (char)(65 + rand.Next(26)), (char)(65 + rand.Next(26)),
(char)(65 + rand.Next(26)) }; //随机生成验证字符
            int x, y;
            for (int i = 0; i < 3; i++) //用随机字体和颜色绘制验证码
            {
                Brush brush = new SolidBrush(colors[rand.Next(7)]);
                Font font = new Font(fonts[rand.Next(7)], 18, FontStyle.Bold);
                x = i * 20 + 2;
                y = 5 + rand.Next(5);
                g.RotateTransform(rand.Next(-10, 9));
                g.DrawString(chs[i].ToString(), font, brush, x, y);
            }
            Pen[] pens = { Pens.Gray, Pens.LightGray };
            for (int i = 0; i < 200; i++) //绘制随机噪点
            {
                x = rand.Next(bmp.Width - 1);
                y = rand.Next(bmp.Height - 1);
                g.DrawEllipse(pens[i % 2], x, y, 1, 1);
            }
            MemoryStream ms = new MemoryStream();
            bmp.Save(ms, ImageFormat.Jpeg);
            this.Response.Clear();
            this.Response.ContentType = "image/Jpeg";
            this.Response.BinaryWrite(ms.ToArray()); //输出图像
            g.Dispose();
            bmp.Dispose();
        }
    }
```

HttpResponse 还有一个常用方法 Redirect，它用于将浏览器重新定位到另一个网页上。比如网站的旧邮箱主页的地址是"http://192.168.1.1/freemail.aspx"，而升级后的新邮箱主页是"http://192.168.1.1/mail.aspx"，那么可以删除旧页面的其余内容，只在其 Load 事件处理方法中加入下面一行代码，这样用户在访问旧页面时就会被自动引导到新页面：

```csharp
this.Response.Redirect("http://192.168.1.1/mail.aspx");
```

如果要转到的新页面与当前页面属于同一虚拟目录，那么可以不写出完整的 Url 地址，而是用"~/"表示当前的虚拟目录，例如：

```csharp
this.Response.Redirect("~/mail.aspx");
```

14.2.3 服务器对象

通过当前 Web 窗体的 Server 属性可获得一个 HttpServerUtility 对象，它提供了服务器端的一些基本属性和方法。其中 MachineName 属性表示服务器的名称，而 ScriptTimeout 表示处理请求的超时时间（默认值为 90 秒），任何执行时间超过该属性值的网页都会被中止。那么在执行较长时间的任务时（如从数据库读取大量数据，或是上传大文件），应将服务器对象的 ScriptTimeout 属性设置为一个较大的值，例如：

```csharp
this.Server.ScriptTimeout = 300; //超时时间设为 300 秒
```

Web 服务器的一个基本功能是处理 HTML 文档，而 HttpServerUtility 的 HtmlEncode 和 HtmlDecode 分别提供了对 HTML 的标记文本进行编码和解码的功能。例如 HTML 标记 h2 表示二级标题，那么下面第 2 行代码将直接在网页上以二级标题格式输出"欢迎光临"；而在服务器对象对字符串进行编码后，第 4 行输出的将是编码后的字符串"<h2>欢迎光临</h2>"（如图 14-3 所示）：

```csharp
string s1 = "<h2>欢迎光临</h2>";
```

```
this.Response.Write(s1); //以二级标题格式输出"欢迎光临"
s1 = this.Server.HtmlEncode(s1);
this.Response.Write(s1); //以普通文本输出"<h2>欢迎光临</h2>"
```

图 14-3　HTML 编码和解码的不同输出效果

编码后的内容使用 HtmlDecode 方法进行解码，输出不带标记的字符串内容：

```
s1 = this.Server.HtmlDecode(s1);
this.Response.Write(s1); //重新二级标题格式输出"欢迎光临"
```

许多浏览器都要求网页的 Url 地址不能带有特殊字符，此时可利用服务器对象的 UrlEncode 方法和 UrlDecode 方法进行编码和解码。例如要以学生的姓名作为 Url 的参数字符串，又要防止有的浏览器不支持中文地址，那么可先对该参数值进行编码：

```
string name = "王小红";
string url = "http://192.168.1.1/student.aspx?name=";
Response.Redirect(url + Server.UrlEncode(name));
```

那么实际的 Url 将是 "http://192.168.1.1/student.aspx?name=%e5%91%a8%e5%86%9b"。而在学生页面 student.aspx 中，可使用下面的代码从 Url 中解码出学生姓名：

```
string name = Request.QueryString["name"];
name = Server.UrlDecode(name); //name = "王小红"
```

此外，服务器对象也定义了将请求的虚拟路径转换为物理路径的 MapPath 方法，以及将浏览器定位到新网页的 Transfer 方法。和 HttpResponse 的 Redirect 方法相比，Transfer 方法更节省服务器资源，但它不能转移到当前服务器以外的站点。

14.2.4　应用程序、会话、视图和缓存

Page 对象的 Application 属性表示当前的应用程序状态，其类型为 HttpApplicationState，它和 NameValueCollection（HttpRequest 的 QueryString 属性类型）一样都实现了 IEnumerable 和 ICollection 接口，也都是维护一组键/值对的集合。例如下面的页面在第一次被访问时会为程序创建一个名为 RequestCount 的状态变量，其值在每次访问页面时都会加 1：

```
protected void Page_Load(object sender, EventArgs e)
{
    if (Application["RequestCount"] == null)
        Application.Add("RequestCount", 1);
    else
        Application["RequestCount"] = (int)Application["RequestCount"] + 1;
    Response.Write("本页面访问次数已达: " + Application["RequestCount"]);
}
```

注意其中的值类型为 object，那么写入和读取时要进行装箱和拆箱转换。使用 HttpApplicationState 的索引函数对状态变量赋值，程序会自动检查指定的索引名是否存在，如不存在则创建一个该名称的变量，否则就修改原有的变量值。例如上面 if 分支中的代码可改写为如下语句：

```
Application["RequestCount"]= 1;
```

如果不再需要使用某些状态变量，那么可使用 Remove 方法将其删除；使用 RemoveAll 方法或 Clear 方法则可以删除程序中的所有状态变量。例如：

```
Application.Remove("RequestCount");
Application.RemoveAll();
```

Page 对象的 Session 属性则表示当前的会话状态,其类型为 System.Web.SessionState 命名空间中的 HttpSessionState。它的使用方法和 HttpApplicationState 非常类似。例如在网站的登录页面中成功登录后,可创建一个名为 UserName 的会话状态变量来记录用户名:

```
Session["UserName"] = "王小红";
```

那么在会话有效的范围内,其他页面就可以从会话中读取该状态变量,例如:

```
Response.Write(this.Session["UserName"] + ":欢迎光临!");
```

类似的,HttpSessionState 的 Remove 方法用于删除指定的状态变量,Clear 方法则用于清空整个会话。此外,通过其 Timeout 属性可读取或设置会话状态的超时时间(默认值为 20 分钟),例如:

```
Session.Timeout = 30; //超时时间设为 30 分钟
Response.Write(string.Format("{0}分钟内无操作将终止会话,Session.Timeout"));
```

 应用程序状态是指整个 ASP.NET 程序所共享的全局信息,该状态只有在服务器上的程序被关闭后才会失效。而会话则是指客户端与服务器之间的"对话",它从客户端浏览器第一次请求程序的某个页面开始,其状态在会话超时或浏览器关闭时就会失效。

Page 对象的 ViewState 属性表示当前的视图状态,它也可以用来保存特定的状态变量,只不过该状态只对当前页面有效。下面的示例程序用于在网页上输出二十四节气的内容(如图 14-4 所示),输出的行数保存在名为 iRows 的视图状态变量中;只要行数小于 4,每次单击按钮都会增加一个输出行(HTML 标记
 表示换行):

图 14-4 程序 P14_4 的输出结果示例

```
//程序 P14_4(Default.aspx.cs)
using System;
public partial class Default : System.Web.UI.Page
{
    protected void Page_Load(object sender, EventArgs e)
    {
        if(!IsPostBack)
            ViewState["iRows"] = 1;
        string[] seasons = {"立春", "雨水", "惊蛰", "春分", "清明", "谷雨", "立夏", "小满",
"芒种", "夏至", "小暑", "大暑", "立秋", "处暑", "白露", "秋分", "寒露", "霜降", "立冬", "小雪",
"大雪", "冬至", "小寒", "大寒"};
        int iRows = (int)ViewState["iRows"];
        for (int i = 0; i < iRows; i++){
            for (int j = 0; j < 24 / iRows; j++)
                Response.Write(seasons[i * iRows + j] + ' ');
            Response.Write("<br/>");
        }
    }
    protected void Button_Click(object sender, EventArgs e)
    {
        int iRows = (int)ViewState["iRows"];
        if (iRows < 4)
            ViewState["iRows"] = iRows + 1;
    }
}
```

上面介绍的三种状态对象都只适合存储少量信息,否则会占用较多的服务器资源。而对于需要频繁访问而不经常修改的数据,建议将它们保存在服务器缓存之中。缓存对象可通过 Page 的 Cache 属性获得。例如要把程序 P14_4 中的字符串数组保存在缓存中,那么字符串数组 seasons

的初始化代码可改为如下内容:
```
string[] seasons;
if (Cache["seasons"] == null)
    Cache["seasons"] = seasons = new string[]{"立春","雨水","惊蛰","春分","清明",
"谷雨","立夏","小满","芒种","夏至","小暑","大暑","立秋","处暑","白露","秋分","寒露",
"霜降","立冬","小雪","大雪","冬至","小寒","大寒"};
else
    seasons = (string[])Cache["seasons"];
```

14.3 HTML 控件

14.3.1 从 HTML 元素到 HTML 控件

HTML 是一种结构化的标记语言,ASP .NET 允许将所有的 HTML 元素都转换为对象,这样在服务器端就能通过对象的属性、方法和事件来进行编程,同时使网页的界面要素与程序代码相分离。

例如下面的 HTML 内容定义了 4 个元素:一个 div 段落,还有嵌套在段落中的 label 标签、br 换行符和 button 按钮,那么生成的页面中将看到蓝底黄字的文本以及下方的按钮,如图 14-5(a)所示:
```
<div style="background-color:Blue">
    <label style="color:Yellow;">HTML 元素</label>
    <br />
    <button>使用控件</button>
</div>
```

要将 HTML 元素转换为 HTML 控件很简单,只需要设置元素的 runat 属性值为 server,再设置一个 id 属性值来标识控件对象。例如下面的 HTML 内容就将 div、label 和 button 元素都转换为了 HTML 控件:
```
<div id="Div1" runat="server" style="background-color:Blue;">
    <label id="Label1" runat="server" style="color:Yellow;">HTML 元素</label>
    <br />
    <button id="Button1" runat="server" onserverclick="Button1_Click">使用控件</button>
</div>
```

这样就能够将控件作为对象来使用,特别是方便地处理各种控件事件。例如上面按钮 Button1 的属性 onserverclick="Button1_Click"就表示按钮的单击事件将在服务器端处理,且处理方法为 Button1_Click。此时就需要在对应的.aspx.cs 源文件中添加方法定义:
```
protected void Button1_Click(object sender, EventArgs e)
{
    Div1.Style["background-color"] = "Yellow";
    Label1.Style["color"] = "Blue";
    Label1.InnerText = "HTML 控件";
}
```

运行程序并在网页中单击按钮,输出的内容将如图 14-5(b)所示。此时通过 IE 浏览器的菜单命令"查看→源文件"可在记事本中打开网页的 HTML 内容,其中可以找到如下语句:
```
<div id="Div1" style="background-color:Yellow;">
    <label id="Label1" style="color:Blue;">HTML 控件</label>
    <br />
    <button id="Button1" onclick="doPostBack('Button1','')">使用控件</button>
</div>
```

也就是说,我们在事件处理方法中编写的 C#代码,经服务器运行后会生成标准的 HTML 内容,而后再发送到客户端。

(a)   (b)

图 14-5   HTML 元素输出示例

14.3.2   HtmlControl 类型

从前面的介绍中可以了解，ASP .NET 在执行网页时会检查 HTML 元素是否标记了属性 runat="server"，如果没有就将其作为普通内容发送，有则从.NET 类库中加载相应的控件类型，通过程序处理控件对象的属性和方法，再将程序执行的结果转换为 HTML 内容一并发送到客户端，最后由客户端浏览器处理并显示所有的 HTML 内容。

Visual Studio 工具箱的 HTML 选项卡下列出了一些基本的 HTML 控件，通过鼠标拖曳就可以将控件加入到窗体中；当然我们也可以通过编写 HTML 标记来创建 HTML 控件。.NET 类库中的各种 HTML 控件类型均位于 System.Web.UI.HtmlControls 命名空间中，它们的共同基类就是 HtmlControl。该类型提供了以下常用属性：

- ID——控件的 ID 号。
- Disable——控件是否可用。
- Visible——控件是否可见。
- Page——控件所在的 Web 窗体。
- Controls——控件的子控件集合。
- ViewState——控件的视图状态。
- Style——控件的样式集合。
- Attributes——控件的属性集合。

其中 Style 属性存储了控件的一系列样式名称和值的集合，在上一小节的代码中我们就看到了通过它设置控件背景色和前景色的例子，样式的值是通过字符串来表示的。表 14-2 列出了一些基本的 HTML 样式。

表 14-2   常用 HTML 样式

样式名称	含 义	取 值
Color	前景色	颜色名称或 RGB 值（如 Blue 或#008000）
background-color	背景色	颜色名称或 RGB 值
background-image	背景图像	图像的 Url 地址
font-family	字体族	字体名称
font-size	字号	large，medium，small，smaller，x-small 等，或字体的磅值（如 12pt）
font-style	是否斜体	normal 或 italic
font-weight	是否粗体	normal 或 bold
border-style	边框类型	none，solid，dashed，dotted 等
border-color	边框颜色	颜色名称或 RGB 值
border-width	边框粗细	Thin，medium，thick 等

此外，控件的任意样式值都可以设为"inherit"，此时控件会自动使用父控件的样式。例如一个 label 标签嵌套在 div 段落中，那么下面的代码表示它使用和段落相同的前景色；如果父控件也没有设置该样式值，那么使用的将是窗体的默认设置：

```
Label1.Style["color"] = "inherit";
```

如果在 Style 的索引函数中使用了不正确的样式名，或是设置了不正确的值，那么 ASP .NET 仍然会生成相应的 HTML 标记。只不过收到这些标记后，浏览器将会忽略其中的错误。

HtmlControl 的 Attributes 属性存储了控件对象的 HTML 属性名称和值的集合，例如我们也可改用下面的代码来设置控件的前景色和背景色：

```
Label1.Attributes["Style"] = "color:Blue; background-color:Yellow";
```

和 Style 类似，ASP .NET 也不会检查 Attributes 索引函数使用的属性名和属性值，而不合法的内容在浏览器显示时会被忽略。

HtmlControl 的直接派生类包括 HtmlImage、HtmlContainerControl、HtmlInputControl 等类型，其中 HtmlImage 用于显示图像，其 Src 表示图像的来源，Alt 属性则表示图像的说明文本，例如下面的 Html 语句将生成一个基于文件 asp.gif 的图像：

```
<image ID="Image1" runat="server" scr="~/asp.gif" alt="ASP"/>
```

HtmlContainerControl 对应于那些同时具有开始标记和结束标记的 HTML 元素，而它的派生类 HtmlGenericControl 提供了创建这里一般 HTML 控件的方法，使用时只需要将对应的 HTML 元素名传递给其构造函数即可，例如：

```
HtmlControl ctrl1 = new HtmlGenericControl("label"); //<label></label>
ctrl1 = new HtmlGenericControl("button"); //<button></button>
ctrl1 = new HtmlGenericControl("br"); //<br></br>
```

HtmlContainerControl 的 InnerText 属性和 InnerHtml 属性分别表示 HTML 开始/结束标记之间的文本和内容。具体地说，如果文本中带有 HTML 标记，那么将其赋值给 InnerText 属性时将会被正常编码，而赋值给 InnerHtml 属性时则按照 HTML 格式进行解析。

HtmlInputControl 将在 14.3.5 小节中介绍。

14.3.3 HtmlAnchor、HtmlTextArea 和 HtmlSelect 控件

HtmlAnchor 表示超链接控件，通过它能够以对象编程的方式来访问 HTML 中的<a>标记元素，例如下面的 HTML 语句将生成一个指向微软 MSDN 网站的超链接 Anc1：

```
<a id="Anc1" runat="server" href="http://msdn.microsoft.com">MSDN 帮助</a>
```

HtmlAnchor 的 HRef 属性表示超链接所指向的 Url 地址，而 Target 属性表示新页面内容的打开方式，它有如下一些特殊取值。

- _blank：在新窗口中打开。
- _self：在当前窗口框架中打开。
- _parent：在上一级父框架中打开。
- _top：在无框架的全窗口中打开。

例如下面的代码能够在页面启动时检查浏览器的用户语言是否支持中文，如是则将链接地址改为中文 MSDN 网站：

```
foreach (string s in Request.UserLanguages)
if (s == "zh-cn")
    Anc1.HRef = "http://www.microsoft.com/china/msdn";
```

HtmlTextArea 是一个文本框控件，它对应于 HTML 元素<textarea>。该控件支持多行输出，其 Cols 属性表示文本框的宽度（以字符为单位），Rows 属性则表示文本框的高度（以行为单位）。如果一行中输入的字符数超过了 Cols 属性值将自动换行，行数超过了 Rows 属性值则将自动显示

滚动条。例如下面的语句将生成一个大小为 10×3 的文本框：
```
<textarea id="TextArea1" runat="server" rows="3" cols="10"></textarea>
```
通过 HtmlTextArea 的 Value 属性可读取文本框中输入的内容，例如下面的代码将分行输出文本框中的内容：
```
string[] ss = TextArea1.Value.Split('\r');
foreach (string s in ss)
    Response.Write(s + "<br/>");
```
HtmlSelect 是一个选择控件，它对应于 HTML 元素<select>。该控件默认显示为一个组合框，各个选择项的内容通过嵌套的<option>元素来设置。例如下面的语句就生成了一个 HtmlSelect 控件，它在浏览器中的显示结果如图 14-6（a）所示：
```
<select id="Select1" runat="server">
    <option>周军</option>
    <option selected="selected">王小红</option>
    <option>方小白</option>
</select>
```
和 Windows 列表控件类似，HtmlSelect 的 Items 属性表示列表项的集合，SelectedIndex 属性则表示所选项的索引。例如下面的代码也可向 HtmlSelect 控件中加入一组列表项：
```
Select1.Items.Add("周军");
Select1.Items.Add("王小红");
Select1.Items.Add("方小白");
Select1.SelectedIndex = 1;
```
有趣的是，HtmlSelect 的 Size 属性表示控件的高度（以行为单位），默认值为 1；当该值大于 1 时，控件将显示成为一个列表框。另外，HtmlSelect 的 Multiple 属性表示控件是否支持多项选择，当该值为 true 时显示的同样也是列表框，而此时如果没有设置 Size 属性值则浏览器会自动调整控件的高度，如图 14-6（b）所示。

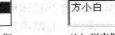

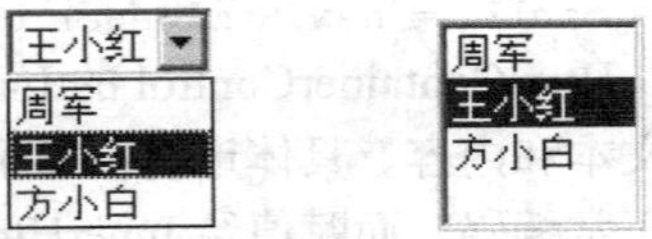

（a）组合框　　（b）列表框

图 14-6　HtmlSelect 控件的不同显示样式

HtmlSelect 没有提供 SelectedIndices 这样的属性。如果启用了多项选择，那么 SelectedIndex 属性值表示第一个选择的元素索引。要访问所有的选择项，就需要遍历控件的 Items 属性，其中每一项的类型为 ListItem，其 Selected 属性表示列表项是否被选择，Text 属性则表示列表项的文本。下面的代码演示了如何获取 HtmlSelect 中所选集合的内容：
```
StringBuilder sb1 = new StringBuilder();
foreach (ListItem item in Select1.Items)
    if (item.Selected)
            sb1.AppendLine(item.Text);
```
ListItem 还有一个 Value 属性，通过它将列表项与特定数据值相关联。例如下面的语句将在列表项文本中显示学生的姓名，并把学生的学号存放在列表项的 Value 属性中：
```
<select id="Select1" runat="server">
    <option value="1002">周军</option>
    <option value="1002">王小红</option>
    <option value="1003">方小白</option>
</select>
```
这样程序就能够根据列表项的 Value 属性值来执行特定的操作，例如：
```
if (Select1.SelectedIndex >= 0){
    string id = Select1.Items[Select1.SelectedIndex].Value;
    Response.Redirect("http://192.168.1.1/student.aspx?id=" + id);
}
```

14.3.4 HtmlTable 控件

表格是网页上极为常见的一种元素，HtmlTable 控件就用于实现这一目的。它对应的 HTML 标记为<table>，其中嵌套标记<tr>表示表格的行，<tr>中又可以嵌套单元格标记<td>。下面就定义了一个 HtmlTable 控件，其显示的表格内容如图 14-7 所示。 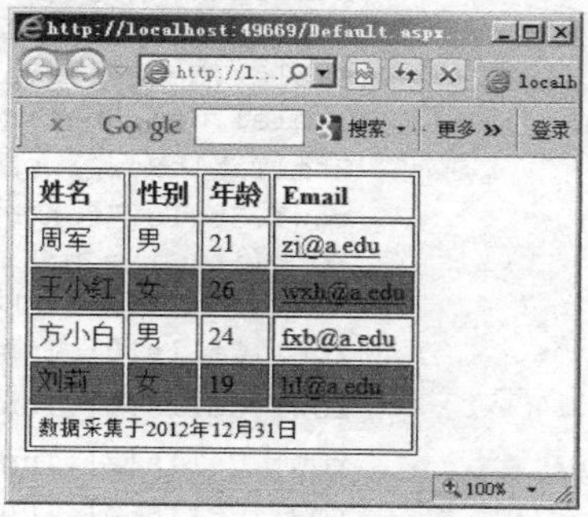

```
<table  id="Table1"  runat="server"  border="1"
cellpadding="4" cellspacing="2">
    <tr style="font-weight:bold">
        <td>姓名</td><td>性别</td><td>年龄</td><td>Email</td>
    </tr>
    <tr>
        <td>周军</td><td>男</td>
        <td>21</td><td><a href="mailto:zj@a.edu">zj@a.edu</a></td>
    </tr>
    <tr style="background-color:Gray">
        <td>王小红</td><td>女</td>
        <td>26</td><td><a href="mailto:wxh@a.edu">wxh@a.edu</a></td>
    </tr>
    <tr>
        <td>方小白</td><td>男</td>
        <td>24</td><td><a href="mailto:fxb@a.edu">fxb@a.edu</a></td>
    </tr>
    <tr style="background-color:Gray">
        <td>刘莉</td><td>女</td>
        <td>19</td><td><a href="mailto:lil@a.edu">lil@a.edu</a></td>
    </tr>
    <tr>
        <td colspan="4" style="font-size:small">数据采集于 2008 年 12 月 31 日</td>
    </tr>
</table>
```

图 14-7 HtmlTable 示例

HtmlTable 的 border 属性表示表格的边框宽度（不设置或值为 0 时无边框），cellpadding 属性表示单元格内容和边框之间的距离，cellspacing 属性则表示相邻单元格之间的距离，这三个属性都是以像素为单位。

要访问表格中的数据，那么可通过 HtmlTable 的 Rows 属性来获得表格的行集，其中每一行的类型为 HtmlTableRow，通过其 Cells 属性又可获得行中的单元格集合，其中每个单元格的类型为 HtmlTableCell。HtmlTable、HtmlTableRow 和 HtmlTableCell 也都是 HtmlContainerControl 的派生类。

下面的程序就通过编程方式创建了一个表格对象，其内容和图 14-7 中的表格是一致的：

```csharp
//程序 P14_5(Default.aspx.cs)
using System;
using System.Web.UI.HtmlControls;
public partial class _Default : System.Web.UI.Page
{
    protected void Page_Load(object sender, EventArgs e)
    {
        Student[] students = new Student[] {
            new Student("周军", true, new DateTime(1988, 5, 10), "zj@a.edu"),
            new Student("王小红", false, new DateTime(1983,2,8), "wxh@a.edu"),
            new Student("方小白", true, new DateTime(1985,12,1), "fxb@a.edu"),
```

```csharp
            new Student("刘莉", false, new DateTime(1990,9,15), "lil@a.edu")
        };
        HtmlTable Table1 = new HtmlTable();
        Table1.Border = 1;
        Table1.CellPadding = 4;
        Table1.CellSpacing = 2;
        HtmlTableRow row1 = new HtmlTableRow();   //标题行
        row1.Style["font-weight"] = "bold";
        for (int i = 0; i < 4; i++)
            row1.Cells.Add(new HtmlTableCell());
        row1.Cells[0].InnerText = "姓名";
        row1.Cells[1].InnerText = "性别";
        row1.Cells[2].InnerText = "年龄";
        row1.Cells[3].InnerText = "Email";
        Table1.Rows.Add(row1);
        for (int i = 0; i < students.Length; i++)   //各数据行
        {
            row1 = new HtmlTableRow();
            if (i % 2 == 1)
                row1.Style["background-color"] = "Gray";
            for (int j = 0; j < 4; j++)
                row1.Cells.Add(new HtmlTableCell());
            row1.Cells[0].InnerText = students[i].Name;
            row1.Cells[1].InnerText = students[i].Gender ? "男" : "女";
            row1.Cells[2].InnerText = students[i].Age.ToString();
            row1.Cells[3].InnerHtml = string.Format("<a href='mailto:{0}'>{0}</a>", students[i].Email);
            Table1.Rows.Add(row1);
        }
        row1 = new HtmlTableRow();   //脚注行
        HtmlTableCell cell1 = new HtmlTableCell();
        cell1.InnerText = "数据采集于2012年12月31日";
        cell1.Style["font-size"] = "small";
        cell1.ColSpan = 4;
        row1.Cells.Add(cell1);
        Table1.Rows.Add(row1);
        this.Controls.Add(Table1);
    }
}

public class Student
{
    private string name;
    public string Name
    {
        get { return name; }
    }

    private bool gender;
    public bool Gender
    {
        get { return gender; }
    }

    private DateTime birthday;
    public int Age
    {
        get { return DateTime.Now.Year - birthday.Year; }
```

```csharp
    }
    public string Email { get; set; }
    public Student(string name, bool gender, DateTime birthday, string email)
    {
        this.name = name;
        this.gender = gender;
        this.birthday = birthday;
        this.Email = email;
    }
}
```

该表格的最后一行只有一个单元格，但它占据了 4 个单元格的跨度，这是通过 HtmlTableCell 的 ColSpan 属性来设置的，而其 RowSpan 属性则可用于设置单元格所跨的行数。此外，表格中最后一列显示的是超链接内容。如果不使用单元格的 InnerHtml 属性，那么也可创建 HtmlAnchor 控件对象并将其作为单元格的子控件：

```csharp
HtmlAnchor Anc1 = new HtmlAnchor();
Anc1.HRef = "mailto:" + students[i].Email;
Anc1.InnerText = students[i].Email;
row1.Cells[3].Controls.Add(Anc1);
```

14.3.5　HtmlInputControl 控件

HtmlInputControl 控件对应标记为<input>的 HTML 元素，它有一个 type 属性，且随着属性值的不同会产生不同类型的派生控件：

- 文本框控件——类型为 HtmlInputText，对应 type="text"；该类型还有一个派生类 HtmlInputPassword，对应 type="password"，表示密码框。
- 按钮控件——类型为 HtmlInputButton，对应 type="button"；该类型还有两个派生类 HtmlInputSubmit 和 HtmlInputReset，对应 type="submit"和 type="reset"，分别表示提交按钮和重置按钮。
- 图像按钮控件——类型为 HtmlInputImage，对应 type="image"。
- 单选框控件——类型为 HtmlInputRadioButton，对应 type="radio"。
- 复选框控件——类型为 HtmlInputCheckBox，对应 type="checkbox"。
- 上传文件控件——类型为 HtmlInputFile，对应 type="file"。
- 隐藏控件——类型为 HtmlInputHidden，对应 type="hidden"。

和 HtmlContainerControl 不同，HtmlInputControl 对应于的 HTML 元素不需要在开始和结束标记之间放置内容，而是可以在一对尖括号中以反斜杠/收尾。下面的 HTML 内容中就定义了 4 个 HtmlInputControl 控件，其显示效果如图 14-8 所示。

```html
用户：<input type="text" id="Text1" runat="server"/><br />
密码：<input type="password" id="Text1" runat="server"/><br />
<input type="button" value="确认" id="Button1" runat="server"/>
<input type="reset" id="Reset1" runat="server"/>
```

图 14-8　HtmlInputControl 示例

和一般的<button>元素不同，<input type="button">是通过 Value 属性值来设置按钮上的文本，不过它们都是使用 ServerClick 事件来处理单击按钮的情况。而对于单选框和复选框，它们的 Value 属性表示与控件相关联的值，Checked 属性来表示控件是否被选中，选项发生改变时则会引发 SeverChanged 事件，不过该事件只有在 Web 窗体被发送回服务器时才会引发。也就是说，用户在网页上改变了单选框或复选框的选项后，还需要通过 HtmlInputButton 或 HtmlInputImage 等按钮进行回送，之后才会执行相应的事件处理方法。

下面的程序在窗体中放置了一组单选框和复选框控件、一个图像按钮和一个文本框（设计文件 Default.aspx），其中各单选框和复选框的 SeverChanged 事件共用一个处理方法（源代码文件 Default.aspx.cs）：

```
//程序 P14_6(Default.aspx)
<%@ Page Language="C#" AutoEventWireup="true" CodeFile="Default.aspx.cs" Inherits="_Default" %>
<html xmlns="http://www.w3.org/1999/xhtml">
<head runat="server">
  <title>HtmlInputControl 示例程序</title>
</head>
<body>
  <form id="form1" runat="server">
  <div>
    搜索领域(多选):<br/>
      <input type="checkbox" id="CheckBox1" runat="server" onserverchange="RC_Change"/>Windows 开发
      <input type="checkbox" id="CheckBox2" runat="server" onserverchange="RC_Change"/>网站开发
      <input type="checkbox" id="CheckBox3" runat="server" onserverchange="RC_Change"/>数据库开发
      <br />使用语言(单选):<br/>
      <input type="radio" id="Radio1" runat="server" onserverchange="RC_Change"/>C#
      <input type="radio" id="Radio2" runat="server" onserverchange="RC_Change" />VB
      <br/><input type="image" runat="server" src="search.bmp" alt="搜索"/>
      <textarea id="TextArea1" runat="server" cols="30" rows="5"></textarea>
  </div>
  </form>
</body>
</html>

(Default.aspx.cs)
using System;
using System.Text;
public partial class _Default : System.Web.UI.Page
{
    protected void Page_Load(object sender, EventArgs e)
    {}

    protected void RC_Change(object sender, EventArgs e)
    {
        StringBuilder sb1 = new StringBuilder();
        if (Radio1.Checked){
            if (CheckBox1.Checked)
                sb1.AppendLine("C# Windows Form 程序设计");
            if (CheckBox2.Checked)
                sb1.AppendLine("ASP .NET 2.0 深度解析");
            if (CheckBox3.Checked)
                sb1.AppendLine("ADO .NET 2.0 高级编程");
        }
        else{
```

```
            if (CheckBox1.Checked)
                sb1.AppendLine("Windows Forms Programming in VB .NET");
            if (CheckBox2.Checked)
                sb1.AppendLine("Essential ASP.NET with Examples in VB .NET");
            if (CheckBox3.Checked)
                sb1.AppendLine("ADO.NET Programming in VB .NET");
        }
        TextArea1.Value = sb1.ToString();
    }
}
```

编译运行程序，在 Web 窗体中改变单选框或复选框的选项，此时结果文本并不会出现在文本框中，而是需要单击图像按钮后才会显示，如图 14-9 所示。

HtmlInputFile 控件中包含一个文本框和一个"浏览"按钮，供用户选择文件以备上传。下一节将介绍 Web 服务器中的 FileUpload 控件，二者实现文件上传的方式基本相同。

图 14-9　程序 P14_7 的输出结果示例

14.4　Web 服务器控件

HTML 控件需要映射到已存在的 HTML 元素，这要求开发人员掌握一定的 HTML 知识。Web 服务器控件则是一组全新的.NET 控件，它侧重于 ASP .NET 开发模型，能够表现更为丰富的界面元素。这些控件类型在 System.Web.UI.WebControls 命名空间中定义。

14.4.1　标准窗体控件

Web 标准窗体控件包括 Label、TextBox、Button、CheckBox、ListBox 等控件，它们可以在 Visual Studio 工具箱的"标准"选项卡下找到。例如从中拖放一个 Button 控件到窗体上，再切换到 HTML 源视图可看到生成的如下内容：

```
<asp:Button ID="Button1" runat="server" Text="Button" />
```

Web 服务器控件也是通过 ID 属性来标识对象，并通过 runat="server"属性声明在服务器上工作。所有的 Web 服务器控件都以前缀"asp:"开头，而且它们有着共同的基类 WebControl。和 HtmlControl 类似，WebControl 也提供了 Visible、Page、Style、Attributes 等基本属性，不过许多控件样式可以通过更为直观的属性来设置，其中常用的有：

- ForeColor——前景色，类型为 Color 结构。
- BackColor——背景色，类型为 Color 结构。
- Font——字体，类型为 FontInfo 类。
- BorderStyle——边框样式，类型为 BorderStyle 枚举，枚举值包括 None（无边框）、Solid（实线）、Double（双线）、Dotted（虚线）、Dashed（点划线）等。
- BorderWidth——边框宽度，类型为 uint。
- BorderColor——边框颜色，类型为 Color 结构。
- ToolTip——提示文本，类型为 string。

例如对于一个按钮控件 Button1，下面两行代码的作用是等价的：

```
Button1.Style["background-color"] = "Yellow";
Button1.BackColor = Color.Yellow;
```

但第二行代码含义更为明确，不像第一行代码那样容易出错。此外，Web 服务器控件的编程方式也更贴近于 Windows 窗体控件，如 Label、TextBox、CheckBox 等控件都是通过 Text 属性来表示控件上的文本，RadioButton 和 CheckBox 的选项改变事件都是 CheckChanged。再如 TextBox 控件，它也是通过 ReadOnly 属性来表示文本框是否只读，通过 MaxLength 属性来限制输入字符的数量。下面着重介绍其中几个独具特色的 Web 窗体控件。

1. RadioButtonList 和 CheckBoxList 控件

考虑到单选框和复选框经常需要成组使用，Web 服务器控件中提供了 RadioButtonList 和 CheckBoxList，分别用于封装一组单选框和复选框。这两个控件有很多类似点，最主要的就是它们都通过 Items 属性来表示选项的集合，而每一项都是一个 ListItem 类型的对象。

例如下面的 HTML 语句就创建了一个包含 6 个选项的 RadioButtonList 控件，其显示效果如图 14-10（a）所示：

```
<asp:RadioButtonList ID="RadioButtonList1" runat="server" BorderStyle="Solid">
    <asp:ListItem Text="程序设计" Selected="True"></asp:ListItem>
    <asp:ListItem Text="网站开发"></asp:ListItem>
    <asp:ListItem Text="操作系统维护"></asp:ListItem>
    <asp:ListItem Text="数据库管理"></asp:ListItem>
    <asp:ListItem Text="图形图像"></asp:ListItem>
    <asp:ListItem Text="办公应用"></asp:ListItem>
</asp:RadioButtonList>
```

而下面的代码可在程序中动态地创建 RadioButtonList 控件：

```
RadioButtonList RadioButtonList1 = new RadioButtonList();
RadioButtonList1.Items.Add("程序设计");
RadioButtonList1.Items.Add("网站开发");
RadioButtonList1.Items.Add("操作系统维护");
RadioButtonList1.Items.Add("数据库管理");
RadioButtonList1.Items.Add("图形图像");
RadioButtonList1.Items.Add("办公应用");
RadioButtonList1.Items[0].Selected = true;
RadioButtonList1.BorderStyle = BorderStyle.Solid;
this.form1.Controls.Add(RadioButtonList1);
```

对于 RadioButtonList，通过其 SelectedIndex 属性能够读取或设置所选项的索引，当然这也可以通过对应 ListItem 的 Selected 属性来进行。不过对于 CheckBoxList，通过其 SelectedIndex 属性只能读取或设置第一个选项的索引；如果要选取多个项，那么只能通过各个 ListItem 对象来进行。RadioButtonList 和 CheckBoxList 也都提供了 SelectedItem 属性，不过该属性是只读的。

14.3.3 小节介绍的 HtmlSelect 控件，其列表项和 RadioButtonList 以及 CheckBoxList 的选项实际上是同一类型，即 System.Web.UI.WebControls 命名空间中的 ListItem。

RadioButtonList 和 CheckBoxList 中的选项默认都是垂直显示，此时其 RepeatDirection 属性值为 RepeatDirection.Vertical。如果将枚举取值改为 Horizontal，那么各个选项将水平显示，页面一行排列不下时将自动换行，如图 14-10（b）所示：

```
RadioButtonList1.RepeatDirection = RepeatDirection.Horizontal;
```

这两个控件还提供了一个 RepeatColumns 属性，通过它能够设置使各个选项分列显示，例如下面的代码就使 RadioButtonList1 中的单选框分两列显示，如图 14-10（c）所示：

```
RadioButtonList1.RepeatColumns = 2;
```

RadioButtonList 和 CheckBoxList 的选项改变也都是通过 SelectedIndexChanged 事件来处理。但无论是 RadioButtonList、CheckBoxList，还是单个的 RadioButton 和 CheckBox 控件，它们都和

HTML 控件中的单选框和复选框一样，选项改变事件只有在窗体被回送到服务器后才会引发。不过 Web 服务器控件中的这些选项控件提供了一个 AutoPostBack 属性，其默认值为 false；将其值改为 true 后，用户改变控件中的选项将直接导致窗体被回送，这样 SelectedIndexChanged 事件就会立即得到处理，而不必依靠其他按钮控件来进行回送。

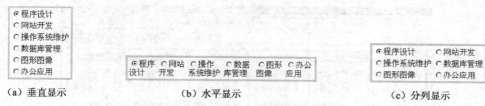

图 14-10 RadioButtonList 的不同排列效果

2. DropDownList、ListBox 和 BulletedList 控件

Web 服务器控件中的 DropDownList 和 ListBox 控件类似于 Windows 窗体中的组合框和列表框，只不过前者只能单选，而后者还支持多选。它们也都是通过 Items 属性来返回列表项的集合，通过 SelectedIndex 属性来读取或设置选项的索引，并在选项改变时引发 SelectedIndexChanged 事件。

ListBox 还有一个 SelectionMode 属性，其类型为 ListSelectionMode 枚举，默认值为 Single 时表示只能单选，取值为 Multiple 时则支持多选。

BulletedList 控件则只是静态地显示一个列表，不支持列表项的选择。但它能够在每个列表项之前显示一个项目符号，这是通过其 BulletStyle 属性来设置的。该属性类型为 BulletStyle 枚举，取不同的值则显示不同样式的项目符号，如 Numbered 为阿拉伯数字编号，UpperRoman 为大写罗马数字编号，Square 为正方形符号等。下面的程序就创建了 6 个显示不同项目符号的 BulletedList 控件，并将列表内容及对应的 BulletStyle 属性值显示在表格中，程序输出如图 14-11 所示：

```
//程序 P14_7(Default.aspx.cs)
using System;
using System.Web.UI.HtmlControls;
using System.Web.UI.WebControls;
public partial class _Default : System.Web.UI.Page
{
    protected void Page_Load(object sender, EventArgs e)
    {
        string[] ss = { "程序设计","网站开发","操作系统维护","数据库管理","图形图像","办公应用" };
        HtmlTableCell cell1, cell2;
        HtmlTableRow row1 = new HtmlTableRow();
        HtmlTableRow row2 = new HtmlTableRow();
        BulletedList[] bls = new BulletedList[6];
        for (int i = 0; i < bls.Length; i++){
            bls[i] = new BulletedList();
            foreach (string s in ss)
                bls[i].Items.Add(s);
            bls[i].BulletStyle = (BulletStyle)i;
            cell1 = new HtmlTableCell();
            cell1.InnerText = bls[i].BulletStyle.ToString();
            row1.Cells.Add(cell1);
            cell2 = new HtmlTableCell();
            cell2.Controls.Add(bls[i]);
            row2.Cells.Add(cell2);
        }
        HtmlTable table1 = new HtmlTable();
```

```
            table1.Border = 1;
            table1.Rows.Add(row1);
            table1.Rows.Add(row2);
            this.form1.Controls.Add(table1);
    }
}
```

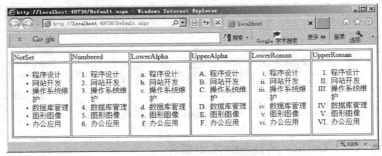

图 14-11　BulletedList 的不同项目符号

当使用各种数字编号时,通过 FirstBulletNumber 属性可设置第一个列表项的编号,例如下面的代码能够使 BulletedList 中列表项的编号从 0 开始:

```
BulletedList1.BulletStyle = BulletStyle.Numbered;
BulletedList1.FirstBulletNumber = 0;
```

除了显示普通文本外,BulletedList 还能够将列表项显示为超链接,这是通过将其 DisplayMode 属性值设为 HyperLink 或 LinkButton 来实现的,例如:

```
BulletedList1.Items.Add("http://www.microsoft.com");
BulletedList1.Items.Add("http://www.csdn.net");
BulletedList1.Items.Add("http://www.ict.ac.cn");
BulletedList1.DisplayMode = BulletedListDisplayMode.HyperLink;
```

在 BulletedList 中使用超链接时,如果列表项文字本身不是超链接内容,那么可通过 ListItem 的 Value 属性来设置链接的 Url 地址,例如:

```
BulletedList1.Items.Add("Microsoft");
BulletedList1.Items.Add("中国程序员网");
BulletedList1.Items[0].Value = "http://www.microsoft.com";
BulletedList1.Items[1].Value = "http://www.csdn.net";
BulletedList1.DisplayMode = BulletedListDisplayMode.LinkButton;
```

3. FileUpload 控件

FileUpload 控件用于上传文件,它的外观和 HtmlInputFile 一样,都是一个文本框和一个"浏览"按钮。实际使用时,要增加一个用于执行上传命令的按钮,有时还应再增加一个表示上传结果的文本标签,例如:

```
<asp:FileUpload runat="server" />
<asp:Button ID="Button1" runat="server" onclick="Button1_Click" Text="上传" />
<asp:Label ID="Label1" runat="server" ForeColor="Red"></asp:Label>
```

FileUpload 控件的 HasFile 属性表示控件中是否选择了文件,FileName 属性表示文件名(不包括路径),PostedFile 属性则用于访问文件对象,其类型为 HttpPostedFile,通过该对象的 FileName、ContentType 和 ContentLength 属性可分别获得文件的路径名、类型和大小,通过其 SaveAs 方法保存文件。例如下面的语句表示将文件上传并保存为 C 盘 temp 目录下的 a.txt 文件:

```
FileUpload1.PostedFile.SaveAs("C:\\temp\\a.txt");
```

实际应用中往往还要考虑文件的大小和类型限制、访问权限等因素。下面的程序演示了一个用于上传 Word 文档的网页,网站虚拟目录下的 Files 子目录专门用于保存上传的文件,文件大小限制为 1MB,且每个文件会从 1 开始自动编号:

```
//程序P14_8(Default.aspx.cs)
using System;
public partial class Default : System.Web.UI.Page
{
    protected void Page_Load(object sender, EventArgs e)
    {
        if (Application["fid"] == null)
            Application["fid"] = 1;
    }
    protected void Button1_Click(object sender, EventArgs e)
    {
        if (!FileUpload1.HasFile)
            Label1.Text = "请选择要上传的文件!";
        else if (FileUpload1.PostedFile.ContentType != "application/msword")
            Label1.Text = "文件类型必须为doc!";
        else if (FileUpload1.PostedFile.ContentLength > 1048576)
            Label1.Text = "文件大小不能超过1MB!";
        else
        {
          try
          {
              int id = (int)Application["fid"];
              FileUpload1.PostedFile.SaveAs(Server.MapPath("~/Files/") + id + ".doc");
              Application["fid"] = id + 1;
              Label1.Text = "成功上传文件" + FileUpload1.FileName;
          }
          catch (Exception exp)
          {
              Label1.Text = "上传失败:" + exp.ToString();
          }
        }
    }
}
```

4. Image 及其派生控件

Image 类似于 Windows 窗体中的 PictureBox 控件，它通过 ImageUrl 属性来指定实际图像的 Url 地址，通过 AlternateText 属性来指定说明文本，还可以通过 Width 和 Height 属性来设置要显示的图像大小，例如：

```
<asp:Image id="Image1" runat="server" ImageUrl="pic/computer.jpg"AlternateText="计算机" Width="32"Height="32"/>
```

Image 控件本身不提供任何事件，但它的派生类 ImageButton 在显示图像的基础上增加了按钮的功能，即能够通过 Click 事件来响应用户单击。如果单击的作用是引导到某个 Url，那么可设置其 PostBackUrl 属性值为 Url 地址，而不必再编写 Click 事件处理方法，例如：

```
<asp:ImageButton id="Button1" runat="server" ImageUrl="pic/computer.jpg"AlternateText="计算机"PostBackUrl="http://www.lenovo.com.cn"/>
```

Image 还有一个派生类 ImageMap，它上面的图像可以被划分为多个"热点区域"，每个区域都是一个 HotSpot 对象。在单击图像的不同区域时，ImageMap 的 Click 事件参数能够识别出当前的区域，这样就可以执行不同的操作，或是根据 HotSpot 的 NavigateUrl 属性来引导到不同的 Url 地址。

ImageMap 上可以有三种不同类型的 HotSpot 区域：
- RectangleHotSpot——矩形区域，通过矩形的 4 个顶点来指定。
- PolygonHotSpot——多边形区域，通过多边形的一组顶点来指定。

- CircleHotSpot——圆形区域，通过圆心和半径来指定。

例如在下面的 HTML 语句中，ImageMap 控件上显示了一个电脑图片，图片上的显示器、机箱、键盘和鼠标部分各自构成了一个 HotSpot 区域，并设置了各自的 NavigateUrl 和 AlternateText 属性，那么这些区域时就能够链接到相应的产品网页上，如图 14-12 所示：

```
<asp:ImageMap ID="ImageMap1" runat="server"ImageUrl="~/pic/computer.jpg">
    <asp:RectangleHotSpot Left="275" Top="5" Right="430" Bottom="290"NavigateUrl=
"http://www.lenovo.com.cn" AlternateText="联想电脑"/>
    <asp:PolygonHotSpot
Coordinates="128,53,103,209,263,229,274,32"NavigateUrl="http://lcd.zol.com.cn"
AlternateText="显示器"/>
    <asp:PolygonHotSpot
Coordinates="5,250,69,233,215,270,150,308"NavigateUrl="http://keyboard.zol.com.cn/"
AlternateText="键盘"/>
    <asp:CircleHotSpot X="240" Y="290" Radius="35"NavigateUrl="http://mouse.zol.com.
cn"AlternateText="鼠标"/>
</asp:ImageMap>
```

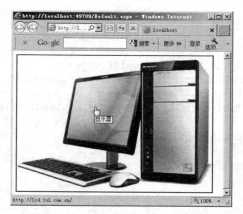

图 14-12　ImageMap 控件的显示效果示例

14.4.2　验证控件

在上传文件的示例程序 P14_8 中，Web 窗体上放置了一个 Label 控件，当用户没有选择文件或文件类型错误时就给出提示。这种验证用户输入的方式需要将窗体发送回服务器，并将验证失败的消息再传回客户端，而且需要手动编写不少的验证代码。

ASP .NET 提供了一组专门的验证控件，其特点是设置方便，且在客户端进行验证，验证失败的话窗体不会被回送到服务器。这些控件都可以在 Visual Studio 工具箱的"验证"选项卡下找到，而且它们有一个共同的基类 BaseValidator，其 ControlToValidate 属性表示要验证输入的控件 ID，ErrorMessage 属性则表示验证失败所显示的提示信息。

1. RequiredFieldValidator 控件

RequiredFieldValidator 要求目标控件的输入不能为空，否则就会验证失败。例如可以将该控件拖放到程序 P14_8 的窗体中，设置其 ControlToValidate 属性值为 FileUpLoad 对象，ErrorMessage 属性值为"未选择上传文件"，那么再运行程序，直接按下"上传"按钮将会显示红色的提示信息，如图 14-13 所示。

图 14-13　验证控件示例

2. RangeValidator 控件

RangeValidator 验证控件的输入值是否在指定的范围内，其 Type 属性表示值的类型，取值包括 String、Integer（整数）、Double（实数）、Date（时间）和 Currency（货币）。它的 MinimumValue 和 MaximumValue 属性则共同用于设置值的范围。

例如 Web 控件中不包含数值框，如果要限制用户在文本框中只能输入数值，那么可以把一个 RangeValidator 控件关联到文本框，设置验证控件的 Type 属性为 Integer，MinimumValue 和 MaximumValue 属性值分别为 0 和 100，这样用户在文本框中输入非整数或 0～100 之外的整数都会验证失败：

```
<asp:RangeValidator ID="RangeValidator1" runat="server" ErrorMessage="必须输入 0~100 之间的一个整数" ControlToValidate="TextBox1" Type="Integer"MinimumValue="0" MaximumValue="100">
</asp:RangeValidator>
```

不过 RangeValidator 并不检查控件的输入是否为空；要真正确保输入的是整数，它还需要和 RequiredFieldValidator 配合使用。

3. CompareValidator 控件

CompareValidator 用于对控件的输入进行比较式验证，比较的方式有两种：

- 和某个常数值进行比较，此时通过其 ValueToCompare 属性来设置要比较的值。
- 和另一个控件的输入内容进行比较，此时通过其 ControlToCompare 属性来设置要比较的控件。

CompareValidator 控件同样也使用 Type 属性来设置要比较的值的类型；它还通过 Operator 属性来设置比较操作，取值包括 Equal（相等）、NotEqual（不等）、GreaterThan（大于）、GreaterThanEqual（大于等于）、LessThan（小于）、LessThanEqual（小于等于）。例如下面的 CompareValidator1 用于验证两个文本框中输入的内容是否相等：

```
<asp:CompareValidatorID="CompareValidator1" runat="server"ErrorMessage="两次输入的密码不一致" ControlToValidate="TextBox1"ControlToCompare="TextBox2"Type="String"Operator ="Equal">
</asp:RangeValidator>
```

4. RegularExpressionValidator 控件

RegularExpressionValidator 使用正则表达式来验证控件的输入内容，表达式内容通过其 ValidationExpression 属性来设置，例如 "[A-Z]" 就表示输入的字符必须为大写字母。正则表达式有着很强的表达能力，能够实现各种复杂的验证功能。例如下面的验证控件就要求文本框中输入的必须是一个合法的 E-mail 地址：

```
<asp:RegularExpressionValidatorID="RegularExpressionValidator1"runat="server"ErrorMessage="Email 地址有误" ControlToValidate="TextBox1"ValidationExpression="[A-Za-z0-9_\-\.]{1,}@[A-Za-z0-9_\-\.]{3,}">
</asp:RegularExpressionValidator>
```

14.5 案例研究——旅游信息查询网站

本节将带领读者开发旅行社网站 TravelWeb。网站静态内容（如旅游常识、公司简介等）主要由网站管理员进行维护，我们在这里重点关心那些与业务对象密切相关且内容随着对象信息变化而变化的网页，主要介绍几个具有代表性的页面的开发过程。

14.5.1 网站母版页

在旅行社管理系统解决方案中新建一个 ASP .NET 应用程序项目 TravelWeb，并为其添加对类库项目 TravelLib 的引用。和空项目不同，Visual Studio 会为 ASP .NET 应用程序项目创建一个网站框架，其中一项很重要的内容就是母版页 Site.Master。母版页是 ASP .NET 2.0 引进的一项重要革新，它能够为网站定义一个标准的外观和行为模式，其他网页可以共享一个母版页，从而简化设计、提高一致性。

打开 Site.Master 的 HTML 视图，可以看到母版页主框架 div.page 包含三个嵌套的 div 元素：header、main 和 clear。Header 块中又嵌套了一个 clear hideSkiplink 段落，其中定义了一个名为 NavigationMenu 的菜单控件。修改其中菜单项的内容，并在菜单下方再增加一个 div 块来支持线路搜索：

```
<div class="clear hideSkiplink">
    <asp:Menu ID="NavigationMenu" runat="server" CssClass="menu" EnableViewState="false" IncludeStyleBlock="false" Orientation="Horizontal">
        <Items>
            <asp:MenuItem NavigateUrl="~/Default.aspx" Text="主页"/>
            <asp:MenuItem NavigateUrl="~/TourLines.aspx" Text="线路大全"/>
            <asp:MenuItem NavigateUrl="~/ViewScenes.aspx" Text="景点大观"/>
            <asp:MenuItem NavigateUrl="~/ViewMap.aspx" Text="地图导航"/>
            <asp:MenuItem NavigateUrl="~/TravelKnowledge.aspx" Text="旅行百科"/>
            <asp:MenuItem NavigateUrl="~/User.aspx" Text="会员服务"/>
            <asp:MenuItem NavigateUrl="~/About.aspx" Text="关于我们"/>
        </Items>
    </asp:Menu>
    <div style="padding: 3px">
        <asp:TextBox runat="server" ID="TbLine" Text="搜索旅游线路" />
        <asp:ImageButton runat="server" ID="BtnSearch" ImageUrl="Image/Search.ico" AlternateText="搜索" onclick="BtnSearch_Click"/>
    </div>
</div>
```

main 块中包含一个 ContentPlaceHolder 控件，使用该母版页的其他页面内容都会放置在这里：

```
<div class="main">
    <asp:ContentPlaceHolder ID="MainContent" runat="server"/>
</div>
```

clear 块的内容为空，这里我们加入下列内容，以便在母版页底部提供两个导航链接：

```
<div class="clear">
    <table>
    <tr>
        <td><a href="~/Sitemap.aspx">网站地图</a></td>
        <td><a href="~/Admin.aspx">管理员入口</a></td>
    </tr>
    </table>
</div>
```

最后定位到母版页页脚 div.footer 部分，在其中加入下列内容：

```
<div class="footer">
    京ICP备05001号<br/>
    CopyRight &copy;2008<b>碧水丹山旅行社</b><br/>
    联系电话：010-88888888 传真：010-88889999<br/>
    联系地址：中国·北京·西长安街北里1001号邮编：100101<br />
</div>
```

此时在 Visual Studio 中看到的母版页设计视图如图 14-14 所示，使用该母版页的页面内容都会出现在 MainContent 控件处。

图 14-14　母版页设计视图

14.5.2　网站首页与线路浏览

接下来切换到主页 Default.aspx，可以看到其 HTML 代码的顶部指令中包含了下面这项内容，它表示该 aspx 页面使用 Site.master 作为母版页：

```
MasterPageFile="~/Site.master"
```

使用了母版页后，网页的主要内容不是放在 body 元素中，而是放在 Content 控件中，该控件通过 ContentPlaceHolderID 来和母版页中的容器相关联：

```
<asp:Content ID="BodyContent" runat="server" ContentPlaceHolderID="MainContent">
    //在这里放置页面内容…
</asp:Content>
```

旅行社网站首页需要展示一组特色旅游线路上的组团方案，那么在上添加如下内容的一个表格：

```
<table id="Table1" runat="server">
  <tr>
  <td colspan="2" style="background-color:#94CF6B">华北大地</td>
  </tr><tr>
  <td><img src="Image/Scene/1_small.jpg" alt=""/></td>
  <td><asp:BulletedList ID="BulletedList1" runat="server" DisplayMode="HyperLink">
</asp:BulletedList></td>
  </tr><tr>
  <td colspan="2" style="color:White; background-color:Green">关外风情</td>
  </tr><tr>
  <td><img src="Image/Scene/212_small.jpg" alt=""/></td>
  <td><asp:BulletedList ID="BulletedList2" runat="server" DisplayMode="HyperLink">
</asp:BulletedList></td>
  </tr><tr>
  <td colspan="2" style="background-color:#94CF6B">江南水乡</td>
  </tr><tr>
  <td><img src="Image/Scene/103_small.jpg" alt=""/></td>
  <td><asp:BulletedList ID="BulletedList3" runat="server" DisplayMode="HyperLink">
</asp:BulletedList></td>
  </tr><tr>
  <td colspan="2" style="color:White; background-color:Green">南国揽胜</td>
  </tr><tr>
  <td><img src="Image/Scene/249_small.jpg" alt=""/></td>
  <td><asp:BulletedList ID="BulletedList4" runat="server" DisplayMode="HyperLink">
</asp:BulletedList></td>
  </tr><tr>
  <td colspan="2" style="background-color:#94CF6B">巴山蜀水</td>
  </tr><tr>
  <td><img src="Image/Scene/28_small.jpg" alt=""/></td>
  <td><asp:BulletedList ID="BulletedList5" runat="server" DisplayMode="HyperLink">
```

```
</asp:BulletedList></td>
    </tr><tr>
      <td colspan="2" style="color:White; background-color:Green">塞外风光</td>
    </tr><tr>
      <td><img src="Image/Scene/301_small.jpg" alt=""/></td>
      <td><asp:BulletedList ID="BulletedList6" runat="server" DisplayMode="HyperLink">
</asp:BulletedList></td>
    </tr>
</table>
```

该表格分为 6 个部分，每部分包含一个图片和一个 BulletedList 控件。程序要做的工作是把各个地区的旅行团方案信息分别读取到对应的 BulletedList 中，这可以通过如下代码来实现：

```
protected void Page_Load(object sender, EventArgs e)
{
    BulletedList[] bls = { BulletedList1, BulletedList2, BulletedList3, BulletedList4, BulletedList5, BulletedList6 };
    for (int i = 0; i < 6; i++){
        foreach (Line l in Line.Get((Area)i)){
            if (l.Packages.Count > 0){
                Package p = l.Packages[0];
                ListItem item = new ListItem(p.ToString(), "PackageInfo.aspx?id="+ p.Id);
                bls[i].Items.Add(item);
            }
            if (bls[i].Items.Count >= 5)
                break;
        }
    }
}
```

其中 Line 的静态方法 Get 用于获取指定地区的线路集合。编译运行程序，可以看到如图 14-15 所示的页面内容。这里各个 BulletedList 控件中的列表项都显示为超链接，指向的 Url 为 PackageInfo.aspx 页面，并以相应 Package 对象的 Id 作为参数封装在请求的 Url 字符串。

首页只显示具有代表性的组团方案。接下来创建"线路大全"导航菜单所对应的 Web 页面 TourLines.aspx（注意新建页面时选择使用 Site.master 作为母版页）。在页面的 MainContent 控件内放置的 HTML 控件代码如下，其中左侧的 GridView 控件用于显示所有线路信息，右侧的 Table 控件用于显示所选线路的组团方案信息：

图 14-15　旅行社首页 Default.aspx

```
<table>
  <tr>
    <td>
      <asp:GridView ID="GridView1" runat="server" AutoGenerateColumns="false" Font-Size="Medium">
      <Columns>
        <asp:BoundField HeaderText="编号" DataField="Id" />
        <asp:BoundField HeaderText="名称" DataField="Name" />
        <asp:BoundField HeaderText="天数" DataField="Days" />
        <asp:BoundField HeaderText="晚数" DataField="Nights" />
        <asp:BoundField HeaderText="备注" DataField="Remark" />
```

```
                <asp:HyperLinkField Text="查看旅行团方案" DataNavigateUrlFormatString= "TourLines.
aspx?id={0}"
DataNavigateUrlFields="Id" />
              </Columns>
          </asp:GridView>
      </td>
      <td style="vertical-align: top">
        <asp:Table ID="Table1" runat="server" BackColor="LightBlue" >
          <asp:TableHeaderRow ID="HeaderRow" runat="server">
            <asp:TableHeaderCell ID="HeaderCell" runat="server" Font-Size="Medium" />
          </asp:TableHeaderRow>
        </asp:Table>
      </td>
    </tr>
</table>
```

该页面定义了一个 lines 数组来存放所有线路信息。在页面初始化时，线路信息被读取到数组中，并与 GridView 控件进行绑定。此外，GridView 控件包含一个 HyperLinkField 列，单击其中的字段会重新打开当前页面，且页面地址上增加了一个当前线路的 id 作为查询字符串；如果其值不为空，那么调用 LoadPackages 方法来显示组团方案信息：

```
Line[] lines;
protected void Page_Load(object sender, EventArgs e)
{
    lines = Line.GetAll();
    GridView1.DataSource = lines;
    GridView1.DataBind();
    if(Request.QueryString["id"] != null)
    {
        int id;
        if (int.TryParse(Request.QueryString["id"], out id))
            this.LoadPackages(id);
    }
}

void LoadPackages(int id)
{
    Line line = null;
    foreach (Line l in lines)
    {
        if (l.Id == id)
        {
            line = l;
            break;
        }
    }
    Package[] packages = line.Packages.ToArray();
    HeaderCell.Text = line.Name;
    foreach (Package p in packages)
    {
        HyperLink link = new HyperLink() { Text = p.Name };
        link.NavigateUrl = "PackageInfo.aspx?id=" + p.Id;
        link.Target = "_blank";
        TableCell cell = new TableCell();
        cell.Controls.Add(link);
        TableRow row = new TableRow();
        row.Cells.Add(cell);
        Table1.Rows.Add(row);
    }
}
```

重新编译运行程序，打开的"线路大全"页面，内容如图 14-16 所示。注意页面右侧表格中的组团方案同样是以超链接形式显示的，链接的目标也是 PackageInfo.aspx 页面。下面就来实现该 Web 页面。

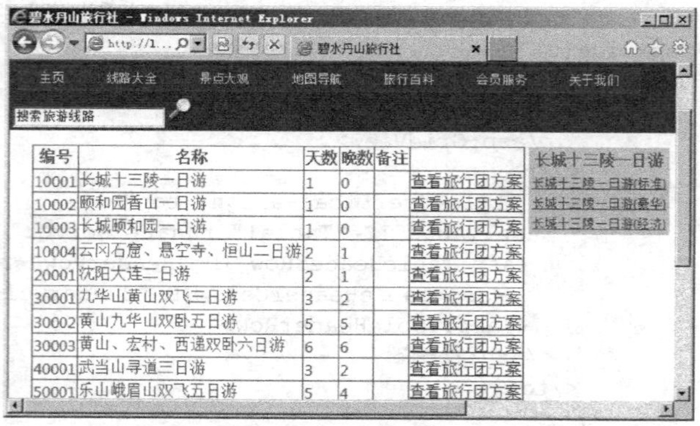

14.5.3 旅行团方案页面

向网站中加入一个 Web 窗体 PackageInfo.aspx，它也是使用表格来显示有关组团方案的信息，不过设计时表格的内容为空：

图 14-16 旅行社首页 Default.aspx

```
<table id="Table1" runat="server" border="1" style="font-size:small">
  <tr><td align="center" colspan="2"></td></tr>
  <tr><td></td><td></td></tr>
  <tr><td></td><td></td></tr>
  <tr><td></td><td></td></tr>
  <tr><td colspan="2"></td></tr>
</table>
```

该窗体在载入时先读出参数字符串中的方案 Id 号，根据该值来获得 Package 对象，将基本信息显示在表格的各单元格中；接下来再获取线路上的景点集合，将每个景点的信息加入表格的一个新行中：

```csharp
protected void Page_Load(object sender, EventArgs e)
{
    int id;
    if (Request.QueryString["id"] == null || !int.TryParse(Request.QueryString["id"], out id)){
        Response.Write("未指定组团方案编号");
        return;
    }
    Package package = Package.Get(id);
    Line line = package.Line;
    //显示线路及组团方案基本信息
    this.Title = "碧水丹山旅行社——" + package.ToString();
    Table1.Rows[0].Cells[0].InnerHtml = string.Format("<h2>{0}</h2>", line.Name);
    Table1.Rows[1].Cells[0].InnerText = "价格:" + package.Price;
    Table1.Rows[1].Cells[1].InnerText = "人数:" + package.Number;
    Table1.Rows[2].Cells[0].InnerText = "餐饮:" + package.Dinner;
    Table1.Rows[2].Cells[1].InnerText = "住宿:" + package.Hotel;
    Table1.Rows[3].Cells[0].InnerText = "类型:" + package.Type;
    Table1.Rows[3].Cells[1].InnerText = string.Format("时间:{0}天{1}晚", line.Days, line.Nights);
    Table1.Rows[4].Cells[0].InnerHtml = package.Introduction;
    //显示线路上各景点信息
    HtmlTableRow row1;
    HtmlTableCell cell1, cell2;
    ImageButton button1;
    foreach (Scene s in line.Scenes){
        button1 = new ImageButton();
        button1.ImageUrl = string.Format("Image/Scene/{0}_small.jpg", s.Id);
        button1.AlternateText = s.Name;
        button1.PostBackUrl = "SceneInfo.aspx?id=" + s.Id;
```

```
            button1.Width = 120;
            button1.Height = 80;
            cell1 = new HtmlTableCell();
            cell1.Controls.Add(button1);
            cell2 = new HtmlTableCell();
            cell2.InnerHtml = s.Introduction;
            row1 = new HtmlTableRow();
            row1.Cells.Add(cell1);
            row1.Cells.Add(cell2);
            Table1.Rows.Add(row1);
        }
    }
```

在表格的每个新行中,左侧的单元格是使用 ImageButton 控件来显示景点图片(景点的缩略图位于"Image/Scene"子目录下,且图片文件名为景点的 Id 号加上"_small.jpg");如果单击图片,那么将引导到景点的详细信息页面 SceneInfo.aspx。旅行团方案页面的输出效果如图 14-17 所示。

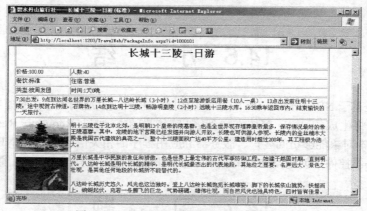

图 14-17　旅行团方案页面 PackageInfo.aspx

14.5.4　景点信息页面

景点信息页面 SceneInfo.aspx 的内容也很简单,其最上方显示景点的全景图片(文件名为景点的 Id 号),接下来是景点的星级和介绍,下方则通过一个表格来显示景点的交通、住宿和票价信息,页面输出效果如图 14-18 所示。对应的启动事件处理方法如下:

```
protected void Page_Load(object sender, EventArgs e)
{
    int id;
    if (Request.QueryString["id"] == null || !int.TryParse(Request.QueryString["id"], out id)){
        Response.Write("未指定景点编号");
        return;
    }
    Scene scene = Scene.Get(id);
    this.Title = scene.Name;
    for (byte i = 0; i < scene.Star; i++)
        Label1.Text += "<img src=\"Image/star.gif\">";
    Label2.Text = scene.Introduction;
    Table1.Rows[0].Cells[1].Text = scene.TrafficInfo;
    Table1.Rows[1].Cells[1].Text = scene.LodgeInfo;
    Table1.Rows[2].Cells[1].Text = scene.Price.ToString();
    Table1.Rows[3].Cells[1].Text = scene.OffSeasonPrice.ToString();
    Image1.ImageUrl = string.Format("Image/Scene/{0}.jpg", id);
}
```

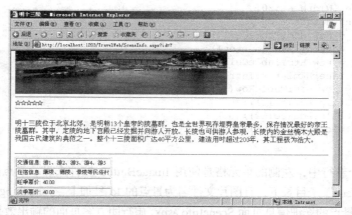

图 14-18　景点信息页面 SceneInfo.aspx

14.6　小　　结

ASP .NET 是基于 .NET Framework 的 Web 应用开发模型，其应用程序在服务器端编译和运行，并根据客户请求将生成的 Web 窗体发送到客户端浏览器。通过 Request、Response、Server、Application 等基本对象能够访问客户、服务器和应用程序的相关信息。ASP .NET 提供了 HTML 控件和 Web 服务器控件来创建 Web 窗体界面，使用时要注意它们与 Windows 窗体控件的区别。

14.7　习　　题

1. 简述 Web 窗体和 Windows 窗体在使用上的主要不同之处。
2. 怎样使用户能够通过网页浏览和编辑服务器上的图片，而又不暴露图片在服务器上的物理路径？
3. 有哪些技术能够在不同的网页之间传递数据，它们各有哪些优缺点？
4. 为程序 P14_5 增加表格排序的功能。
5. 哪些 Web 服务器控件提供了 AutoPostBack 属性？使用该属性时要注意哪些问题？
6. 将程序 P14_3 的验证码图片与 14.4.2 小节中介绍的验证控件功能结合起来，验证用户是否正确地输入了验证码图片中的文本。
7. 设计一个供学生进行学籍注册的 Web 窗体，在其中使用验证控件来规范输入。
8. 说说母版页的设计原理及其优点。
9. 一个网站的首页是访问最为频繁的，因此需要更高的执行效率。试修改旅行社网站首页的执行代码，对所显示的旅行团方案信息进行缓存。
10. 绘制出 HTML 控件和 Web 服务器控件的类继承示意图。

第 15 章
对象持久性——数据库存取和 LINQ 查询

目前绝大多数的企业应用程序都使用数据库来进行信息存取，而数据库系统中最为成熟、应用最为广泛的就是关系数据库。本章将介绍如何使用 ADO .NET 技术来访问关系数据库，如何使用 LINQ 技术来查询对象数据，从而对应用程序中的各种对象进行有效的管理和维护。

15.1 关系数据库概述

15.1.1 关系表和对象

关系模型是建立在关系代数基础上的数据模型，它由关系数据结构、关系操作和完整性约束共同组成。关系数据库使用关系模型来描述数据。简单地说，一个关系表就是一个二维表格，表中的列叫做字段，表中的行叫做记录或元组。表 15-1 定义了 Province 关系，它具有 id、name、capital、municipal 和 introduction 这 5 个字段，并且包含了 5 条记录；对于第一条记录而言，其 id 和 name 的字段值分别为 1 和北京。

表 15-1   关系表 Province

id	name	capital	municipal	introduction
1	北京	北京	true	北京是世界闻名的历史古城，是全国的政治、经济和文化中心
2	天津	天津	true	天津市是中国北方最大的沿海开放城市，地处华北平原东北部
3	河北	石家庄	false	河北省东临渤海、中部环绕北京、天津两大直辖市
4	山西	太原	false	山西位于太行山以西、黄河以东，是中华民族的发祥地之一
5	内蒙古	呼和浩特	false	

如果把一条记录类比于一个对象，则记录的字段类似于对象的字段。很自然地，人们会想到将对象存储在关系表中，但有以下几点情况需要注意：

- 关系数据库支持的字段类型是有限的，它们在一般情况下只能描述程序设计语言中的值类型（如字符串、整数、实数、布尔、时间等）。
- 关系表是严格的二维结构，不允许在表格中嵌套表格，因此不能在记录的某个字段中存储集合型数据（如数组和列表）。
- 关系表可以指定某些字段不能为空，这叫做表的完整性约束。例如对于 Province 表，其

id 和 name 字段都不应为空。
- 关系表可以使用一个或多个字段作为主键，此时表格中记录的主键不能为空，且任意两条记录的主键值不能相等。例如 Province 表中的 id 或 name 都可以作为主键。

考虑到对象类型与关系表的差异，要利用关系数据库来实现对象的持久性，在将类型转换为关系的过程中可综合采用以下设计策略：

（1）首先找出需要进行存取的对象类型，为每一个类型设计一个关系表，将类型中需要存取的每个简单值字段定义为表格中的一个字段。

（2）在构造对象时必须进行初始化的字段，在表格中应将其定义为非空字段。

（3）如果对象没有明显的标识字段，那么可在表格中增加一个 id 字段作为主键，使用递增的整数来标识记录。

（4）如果对象的某些字段类型不是简单值类型，但表述较为简单，那么可以将其在表格中存为字符串字段，并在存取时通过 ToString() 和 Parse(string) 等方法进行转换。

（5）如果对象之间存在包含（聚合）关系，那么可在被包含类型的表格（子表）中增加一个字段，该字段值与父表的主键值相对应，以此来表明子记录被哪个父记录所包含（这在关系模型中称为外键）。例如一个省包含一组城市，那么可在表格 City 中加入一个 Province 字段来表示城市所属的省份，如表 15-2 所示。

表 15-2　　　　　　　　　　　　　　　关系表 City

id	name	province	fullname	introduction
1	石家庄	3	石家庄市	石家庄市是河北省省会，现有国家级重点文物保护单位 18 处
2	秦皇岛	3	秦皇岛市	秦皇岛地处河北省东北部，南濒渤海，北依燕山
3	太原	4	太原市	太原市是一座具有 2400 多年历史的古城
4	临汾	4	临汾地区	临汾古为帝尧之都，因地处汾水之滨而得名

（6）如果对象之间是一般的一对多的关联关系，那么同样可采用外键的处理方式，将"一"端表格的主键加入"多"端表格的字段中。如每个旅行团有一个导游，但导游可能带过多个旅行团，那么应为表格 Tour 定义一个 Guide 字段，用以存储旅行团所对应的导游的 id。

（7）如果对象之间存在一对一的关联关系，那么可将其中任意一个关系表的主键加到另一个表的字段中。

（8）如果对象之间存在多对多的关系，那么可增加一个新的关系表，并将原来两个表格的主键都作为新表的字段。如一条旅游线路上有多个景点，一个景点又可能属于多条线路，那么可以新建一个 LineScene 表来记录线路与景点之间的所有对应关系，如表 15-3～表 15-5 所示：

表 15-3　关系表 Line

id	name
1	长城十三陵一日游
2	长城颐和园一日游

表 15-4　关系表 Scene

id	name	price	City
1	长城	40	1
2	十三陵	30	1
3	颐和园	40	1

表 15-5　关系表 LineScene

line	scene
1	1
1	2
2	1
2	3

（9）如果对象之间存在继承关系，那么基类和派生类可以分别定义各自的关系表，也可以共用一个关系表。对于后者，表格中应包含派生类的所有字段；如果基类对象没有相应的字段，那么记录的字段值为空；为了方便，有时还可以在表中增加一个字段，以表明对象的实际类型。例

如在表 15-6 的关系表 Staff 中，前两条记录对应的不是导游对象，因此其 guideCard 和 guideDue 字段值都为空。

表 15-6　　　　　　　　　　　　　　　关系表 Staff

id	name	gender	birthday	idcard	guideCard	guideDue	type
1001	程学兵	true	1972-2-20	256301720220100200			1
1002	马蕾	false	1977-10-20	127309771020300115			2
1003	张文强	true	1980-9-7	300201800907410230	D1000001801	2010-1-31	3
1004	吴小艳	false	1981-11-5	290102811105367002	D1000003201	2011-5-31	3

15.1.2　关系数据库语言 SQL

SQL（Structured Query Language）是面向关系数据库的结构化查询语言。和过程化的程序设计语言不同，SQL 语言是描述性的语言，即开发人员只说明需要什么样的数据，而不必说明使用什么样的过程来获取数据。SQL 语言可分为数据定义语言（DDL）、数据操纵语言（DML）和数据控制语言（DCL）3 部分。下面将介绍最基本的一些 SQL 语句，更详细的 SQL 语法和功能请参考有关的基础书籍。

1. 数据定义

SQL 数据定义语句包括 CREATE、DROP 和 ALTER 语句，它们分别用于创建、取消（删除）和修改基本数据对象。例如下面的语句表示创建一个名为 Travel 的数据库：

```
CREATE DATABASE [Travel]
```

下面的语句则在 Travel 的数据库中创建了一个名为 Province 的表格，注意表名后的括号中依次指定了表格各字段的名称和数据类型，对字符串（nvarchar）类型的字段还指定了长度限制，此外 PRIMARY KEY 关键字用于指定表格的主键，NOT NULL 关键字则用于指定非空字段：

```
USE [Travel]
CREATE TABLE [Province](
    [id] int PRIMARY KEY,
    [name] nvarchar(20) NOT NULL,
    [capital] nvarchar(20),
    [municipal] bit,
    [introduction] nvarchar(1024)
)
```

上面的 USE 语句用于范围限定，例如使用了一次 USE [Travel]语句后，之后的 SQL 语句都默认在 Travel 数据库的范围内执行。

SQL 语句中的自定义名称，如表名、字段名、变量名等，不能与系统的保留关键字相冲突，而将这些名称用中括号括起来就可以避免冲突。此外，数据库系统中一般都提供了查询工具，在其中输入 SQL 语句就可以执行并查看结果。例如对于 Microsoft SQL Server，在其管理工具 Management Studio 中就可以打开查询窗口。

对应的，DROP TABLE 语句和 DROP DATABASE 语句分别用于删除表格和数据库，例如：

```
DROP TABLE [Province]
DROP DATABASE [Travel]
```

ALTER TABLE 语句则用于修改现有表格的结构，包括增加、修改和删除其中的字段，这分别通过 ADD、ALTER COLUMN 和 DROP COLUMN 子句来实现，例如：

```
ALTER TABLE [Province]    ADD [remark] nvarchar(1024)
ALTER TABLE [Province]    ALTER COLUMN [introduction] nvarchar(256)
ALTER TABLE [Province]    DROP COLUMN [capital]
```

2. 数据查询

SQL 语言中的 SELECT 语句用于数据查询，例如下面的语句表示查询 Province 表格中的所有记录：

```
SELECT * FROM [Province]
```

其中的*为通配符，表示返回表格的所有字段，而 FROM 关键字后面跟的是表格的名称。如果只需要返回指定的字段，那么可以在 SELECT 关键字之后列出这些字段名，例如：

```
SELECT [name], [introduction] FROM [Province]
```

对于查询字段，还可以通过关键字 AS 来指定字段的别名，例如：

```
SELECT [name] AS [名称], [introduction] AS [intro] FROM [Province]
```

SELECT 语句还可对记录进行筛选，具体的筛选条件由 WHERE 子句指定，例如下面的语句表示查询 Staff 表格中的所有男职员记录：

```
SELECT * FROM [Staff] WHERE [gender]=1
```

布尔值在 SQL Server 等数据库中使用 0 和 1 来表示，而在另一些数据库（如 Microsoft Access）中则使用关键字 true 和 false 来表示，对于后者上面的查询子句应改为"WHERE [gender]=true"。

如果条件字段为字符串类型，那么可使用 LIKE 关键字加通配符%来进行查询，例如下面第一条 SQL 语句用于查询所有以"岛"结尾的城市名，第二条语句则用于查询所有姓张的职员记录（注意 SQL 语句中的字符串是包含在一对单引号中的）：

```
SELECT [name] FROM [City] WHERE [name] LIKE '%岛'
SELECT * FROM [Staff] WHERE [name] LIKE '张%'
```

通过 AND 或 OR 关键字还可以对多个筛选条件进行组合，例如：

```
SELECT [name], [introduction] FROM [Province] WHERE [id]>1 AND [id]<5
```

再加上 ORDER BY 子句，还能对查询返回的结果进行排序，例如下面的语句会将职员记录按生日的升序进行排列：

```
SELECT * FROM [Staff] WHERE [gender]=1 ORDER BY [birthday]
```

如果希望返回的结果按降序排列，那么可在上面语句的最后加上关键字 DESC。

SELECT 语句还能一次性对多个表进行查询，此时可以在 FROM 关键字之后列出这些表名，并通过 WHERE 子句来指定表的连接条件，例如：

```
SELECT [City].[name], [Province].[name] AS [pName], [City].[introduction]
FROM [Province],[City]
WHERE [City].[Province] = [Province].[ID]
```

该语句表示查询各省及其城市的名称，以及城市的简介，查询的结果集内容如表 15-7 所示（注意如果查询语句中的列名存在歧义，那么应在前面加上表格名称作为前缀）。

表 15-7　　　　　　　　　　　多表连接查询结果

name	pName	introduction
石家庄	河北	石家庄市是河北省省会，现有国家级重点文物保护单位 18 处
秦皇岛	河北	秦皇岛地处河北省东北部，南濒渤海，北依燕山
太原	山西	太原市是一座具有 2400 多年历史的古城
临汾	山西	临汾古为帝尧之都，因地处汾水之滨而得名

此外，在 SQL 语句中还可以使用+、-、*、/等基本运算符，SIN、COS、SQRT 等初等数学函数，以及下面这些聚合函数：

- COUNT——对记录数量进行统计。

- SUM——对数值型的字段进行求和。
- AVG——对数值型的字段计算平均值。
- MAX——取数值型字段的最大值。
- MIN——取数值型字段的最小值。

例如下面的语句就用于统计职员的数量：
```
SELECT COUNT(*) FROM [Staff]
```
而接下来的这条语句可计算出北京市所有景点的平均票价：
```
SELECT AVG([price]) FROM [Scene],[City]
WHERE [Scene].[City] = [City].[ID] AND [City].[name]='北京'
```

3. 数据更新

数据更新包括插入、修改和删除。其中 INSERT 语句用于在表格中插入记录，表格的名称放在 INTO 关键字之后，而记录的各个字段值放在 VALUES 关键字之后，例如：
```
INSERT INTO [Province]
VALUES (6, '辽宁', '沈阳', 0, '辽宁省位于东北地区南部辽东半岛')
```
对于空字段值，SQL 语句中使用关键字 NULL 来表示。如果空字段值较多，那么还可在 INSERT 语句中只指定取值不为空的字段名，并使其与 VALUES 子句中的字段值一一对应，例如：
```
INSERT INTO [City] ([id], [name], [province], [introduction])
VALUES (6, '沈阳', 6, NULL)
INSERT INTO [City] ([id], [name], [province])
VALUES (7, '大连', 6)
```
当然，在插入记录时，非空字段不能取空值，否则插入就会失败。

DELETE 语句则用于删除表格中的记录，表格名放在 FROM 关键字之后，例如下面的语句将删除 Staff 表中的所有记录：
```
DELETE FROM [Staff]
```
和 SELECT 语句类似，DELETE 语句中的 FROM 子句后也可跟 WHERE 子句，这样就可以指定要删除的记录，例如：
```
DELETE FROM [City]WHERE [province] = 2
```
修改表格中的现有记录则应当使用 UPDATE 语句，它通过 SET 子句来指定修改值，例如下面的语句将 Scene 表中所有记录的 Price 字段值加 10：
```
UPDATE [Scene] SET [Price]=[Price]+10
```
同样，UPDATE 语句也可使用 WHERE 子句来指定筛选条件，从而只修改满足条件的记录，例如：
```
UPDATE [Scene] SET [Price] = [Price]+10 WHERE [City]=1
```

4. 视图管理

视图也是一种关系表，但它在数据库中没有实际的物理存储，而是在其他物理表的基础上进行查询而得到的结果，因此也叫做虚拟表。SQL 语言中的 CREAT VIEW 语句用于创建视图，它在 AS 关键字之后指定用于查询的 SELECT 语句，例如：
```
CREATE VIEW [VCity] AS
    SELECT [City].[name], [Province].[name] AS pName, [City].[introduction]
    FROM [Province],[City]
    WHERE [City].[Province] = [Province].[ID]
```
在数据库中创建了视图后，如果基础表的数据发生了变化，那么视图中所看到的内容也会自动更新。单独的 SELECT 语句也可以对视图进行查询，而且在视图的基础上还可以再创建视图。视图的典型应用情况有以下三种：
- 子集视图，即视图在一个基础表上创建，其字段集是基本表字段集的子集，这常用于基础表字段数量较多的情况。
- 连接视图，即视图是对两个或两个以上基础表进行连接查询的结果，这能够方便用户看到直观的数据内容。
- 带虚拟列的视图，即视图中可以包含基础表中没有的字段，这些字段值是根据基础表的其他字段进行计算而得到的。

与一般表格类似，ALTER VIEW 语句用于修改视图定义，DROP VIEW 语句则用于删除视图。

15.2　ADO.NET 数据访问模型

ADO .NET 是.NET 平台上的数据访问编程模型，它所提供的类型可分为连接类型和非连接类型两大部分。连接类型的对象针对具体的数据库平台，它们负责维护程序与数据库的连接，并在程序对象和物理数据存储区之间交换数据。非连接类型的对象并不依赖于数据连接，也就是说它们不需要连接到真正的数据库就可以工作。

15.2.1　非连接类型

1. 数据表格 DataTable

ADO .NET 非连接类型中最核心的就是 DataTable 类，每一个 DataTable 对象就是一个二维关系表格。对应的，DataRow 和 DataColumn 中两个类分别表示表格中的行和列，它们都在 System.Data 命名空间中定义。

可以使用无参构造函数来创建 DataTable 对象，但更多的时候是在创建对象时指定表格的名称，例如下面的代码就创建了一个名为 Province 的表格对象：

```
DataTable table1 = new DataTable("Province");
```

DataTable 的 TableName 属性表示表格的名称，其 Columns 和 Rows 属性分别表示表格的列集合和行集合（类型分别为 DataColumnCollection 和 DataRowCollection，它们都支持 ICollection 接口）。下面的代码演示了向表格 table1 中加入列、设置主键、以及添加记录，其中 DataTable 的 PrimaryKey 属性表示表格的主键，而 DataColumn 的 AllowDBNull 属性表示该列字段是否可以取空值（非主键列默认为 true）：

```
table1.Columns.Add("id", typeof(int));
table1.Columns.Add("name", typeof(string));
table1.Columns.Add("capital", typeof(string));
table1.Columns.Add("municipal", typeof(bool));
table1.Columns.Add("introduction", typeof(string));
table1.PrimaryKey = new DataColumn[] { table1.Columns[0] };
table1.Columns[1].AllowDBNull = false;
table1.Rows.Add(1, "北京", "北京", true, "北京是全国的政治、经济和文化中心。");
table1.Rows.Add(3, "河北", "石家庄", false, "");
```

DataColumnCollection 的 Add 方法有多种重载形式来加入新列，上面代码中采用的方式是直接指定列名和数据类型；如果要显式地使用 DataColumn 对象，那么可采用如下形式的代码：

```
DataColumn col1 = new DataColumn("id", typeof(int));
table1.Columns.Add(col1);
table1.PrimaryKey = new DataColumn[] { col1 };
```

DataRowCollection.Add 方法也有两种重载形式，参数类型分别为 object[]数组和 DataRow 对象。但 DataRow 没有提供公有构造函数，因此不能直接构造对象，而必须使用 DataTable 对象的 NewRow 方法来得到新行，再通过索引函数来访问各个字段值，例如：

```
DataRow row1 = table1.NewRow();
row1[0] = 6;
row1[1] = "辽宁";
row1[2] = "沈阳";
```

```
table1.Rows.Add(row1);
```
DataRow 的索引函数使用的参数既可以是整数（列的下标），也可以是字符串（列的名称）。例如上面设置 row1 字段值的代码可以改为：
```
row1["id"] = 6;
row1["name"] = "辽宁";
row1["capital"] = "沈阳";
```
DataRow 的索引函数类型为 object，因此在对字段赋值时实际上进行了装箱转换，而要从中读取值则需要拆箱转换，例如：
```
DataRow row2 = table1.Rows[1];
Province p = new Province((string)row2["name"]);
p.Municipal = (bool)row2["municipal"];
p.Introduction = (string)row2["introduction"];
```
当然，如果不能保证所有的字段值都不为空，那么在读取可空字段的值之前要进行判断，例如上面最后两行代码换成如下内容会更安全：
```
if(row2["municipal"] != DBNull.Value)
    p.Municipal = (bool)row2["municipal"];
if (row2["introduction"] != DBNull.Value)
    p.Introduction = (string)row2["introduction"];
```

注意　　对字段值是否为空的判断不是和 null 进行比较，而是和 DBNull.Value 进行比较，后者专门用于表示数据字段的空值。如果 DataRow 对象的某个字段没有被赋值，甚至被赋予了空值 null，那么通过索引函数访问到的字段值仍然是 DBNull.Value。

2. 数据视图 DataView

DataView 表示在 DataTable 基础上创建的视图对象，其 RowFilter 属性表示查询的筛选条件，Sort 属性则表示对查询结果的排序方式。通过 DataTable 的 DefaultView 属性可得到表格的默认视图，也可以基于指定表格来创建 DataView 对象，例如：
```
DataView view1 = new DataView(table1);
view1.RowFilter = "municipal=0";
view1.Sort = "name";
```
由此得到的 DataView 对象相当于如下 SQL 查询语句所返回的记录集：
```
SELECT* FROM [Province] WHERE [municipal]=0 ORDER BY [name]
```
DataView 对象的 AddRow 和 Delete 方法分别用于在视图中增加和删除行，其中 AddRow 方法返回一个 DataRowView 对象，但对该对象的操作与 DataRow 对象是基本相同的，例如：
```
DataRowView row1 = view1.NewRow();
row1[0] = 7;
row1[1] = "吉林";
row1[2] = "长春";
table1.Rows.Add(row1);
```
由于 DataTable 和 DataView 对象本身的数据都在内存中，因此 DataView 对象可以直接转换为 DataTable，这是通过其 ToTable 方法来进行的，例如：
```
DataTable table2 = view1.ToTable();
```
DataView 类实现了 IList 接口，DataTable 类则实现了 IListSource 接口，因此它们都可以绑定到 Windows 窗体的 DataGridView 控件、WPF 的 DataGrid 控件，以及 ASP.NET 的 GridView 控件。在下面的程序中，窗体 Form1 的 GetDataTable 方法返回一个表格对象，字段 dataView 则是基于该表格的一个数据视图：
```
//程序 P15_1(Form1.cs)
using System;
using System.Data;
using System.Text;
```

```csharp
using System.Windows.Forms;
namespace P15_1
{
    public partial class Form1 : Form
    {
        private DataView dataView;

        public Form1()
        {
            InitializeComponent();
        }
        private void Form1_Load(object sender, EventArgs e)
        {
            dataView = new DataView(this.GetDataTable());
            dataGridView1.DataSource = dataView;
            cmbFilterField.SelectedIndex = cmbOperator.SelectedIndex = cmbOrderField.SelectedIndex = 0;
        }
        private void button1_Click(object sender, EventArgs e)
        {
            StringBuilder sb1 = new StringBuilder(cmbFilterField.Text);
            if (cmbOperator.SelectedIndex == 0)
                sb1.Append(">");
            else if (cmbOperator.SelectedIndex == 1)
                sb1.Append("=");
            else
                sb1.Append("<");
            sb1.Append(numericUpDown1.Value);
            dataView.RowFilter = sb1.ToString();
            dataView.Sort = cmbOrderField.Text;
        }
        public DataTable GetDataTable()
        {
            DataTable table1 = new DataTable("Staff");
            table1.Columns.Add("id", typeof(int));
            table1.Columns.Add("姓名", typeof(string));
            table1.Columns.Add("性别", typeof(bool));
            table1.Columns.Add("年龄", typeof(byte));
            table1.Columns.Add("工资", typeof(decimal));
            table1.PrimaryKey = new DataColumn[] { table1.Columns[0] };
            table1.Columns[1].AllowDBNull = false;
            table1.Columns[2].AllowDBNull = false;
            table1.Rows.Add(1, "程学兵", true, 37, 3500);
            table1.Rows.Add(2, "张文强", true, 29, 2500);
            table1.Rows.Add(3, "马秋萍", false, 30, 3000);
            table1.Rows.Add(4, "何艳", false, 24, 1800);
            table1.Rows.Add(5, "薛冰", false, 26, 2400);
            return table1;
        }
    }
}
```

该窗体在启动时将数据视图绑定到 DataGridView 控件，那么网格中将显示原有表格的内容。而在用户选择了不同的筛选条件和排序依据后，视图数据发生变化，网格中显示的内容也随之变化，如图 15-1 所示。

图 15-1 程序 P15_1 的输出结果示例

3. 数据集 DataSet

DataSet 对象由一组 DataTable 组成,并能够维护表格之间的关系,因此可以把它看成是一个内存中的小型数据库。DataSet 的 Tables 属性(类型为 DataTable Collection)表示表格的集合。例如下面的代码先创建了一个表格 dtProvince,而后将它加入数据集 ds1 的表格集合中:

```
DataTable dtProvince = new DataTable("Province");
dtProvince.Columns.Add("id", typeof(int));
dtProvince.Columns.Add("name", typeof(string));
dtProvince.Columns.Add("introduction", typeof(string));
DataSet ds1 = new DataSet();
ds1.Tables.Add(dtProvince);
```

除了 Tables 属性,DataSet 还有一个 Relations 属性,它表示数据集中表格的关系集合,集合元素的类型为 DataRelation。例如下面的代码又向数据集 ds1 中加入了一个表格 dtCity,并将 dtProvince 的 id 列与表格 dtCity 的 Province 列关联起来:

```
DataTable dtCity = new DataTable("City");
dtCity.Columns.Add("id", typeof(int));
dtCity.Columns.Add("name", typeof(string));
dtCity.Columns.Add("province", typeof(int));
ds1.Tables.Add(dtCity);
ds1.Relations.Add(dtProvince.Columns["id"], dtCity.Columns["province"]);
```

如果要显式地使用 DataRelation 对象,那么上面最后一行代码可改为:

```
DataRelation rel = new DataRelation("Relation1", dtProvince.Columns["id"], dtCity.Columns["province"]);
ds1.Relations.Add(rel);
```

4. 类型化数据集和数据表格

DataSet 和 DataTable 本身都是非类型化,例如可以把一个字符串赋值给表格中的一个整数字段,程序仍然能够通过编译,但运行时会发生错误:

```
DataRow row1 = dtCity.NewRow();
row1["id"] = "abc"; //类型不匹配,但能通过编译
row1["name"] = "上海";
row1["province"] = "上海";  //类型不匹配,但能通过编译
```

一种解决办法是针对每个特定的表格架构定义一个 DataRow 的派生类,在其中指定每个列的具体类型,这样就能防止类型不匹配的字段赋值;进一步可以定义类型化数据表格,它是 DataTable 的派生类,且由类型化的列组成;一组类型化的表格又可以组成类型化的数据集。

Visual Studio 中提供了相应的工具来帮助我们创建类型化的数据集。通过菜单命令"项目→添加新项",在如图 15-2 所示的"添加新项"窗口中选择"数据集",输入数据集的名称并单击"确定"按钮,这样就生成了一个类型化的数据集框架。

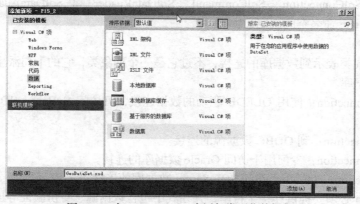

图 15-2 在 Visual Studio 中添加类型化数据集

打开类型化数据集（后缀名为 xsd）的设计视图，那么从工具箱中可以直接拖曳 DataTable 组件到视图上，之后可对表格的各列进行可视化的编辑，保存时就会生成类型化的表格框架；通过拖曳表格之间的相关列，还可以生成类型化的数据关系，如图 15-3 所示。

如果数据库已经存在，那么还可以在 Visual Studio 的服务器资源管理器（如图 15-4 所示）中打开指定的数据连接，选取指定的数据库表格并将其直接拖放到数据集编辑器上，这样 Visual Studio 就会根据数据库表格的结构来自动生成类型化的 DataTable 和 DataSet。

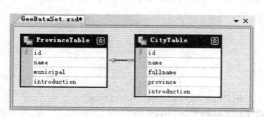
图 15-3　Visual Studio 中的类型化数据集编辑器

图 15-4　Visual Studio 服务器资源管理器中的数据连接

使用类型化的数据对象，不仅可以减少拆箱和装箱转换的次数，还能够使程序在编译时就检查出错误的类型转换。此外，在数据集对象中访问表格、行和列都可以通过名称直接进行，而不必再使用集合的索引函数。看下面的代码示例：

```
GeoDataSet ds1 = new GeoDataSet(); //创建类型化的数据集
GeoDataSet.ProvinceDataTable table1 = ds1.Province; //使用类型化数据表格
GeoDataSet.ProvinceRow row1 = table1.NewProvinceRow(); //使用类型化数据行
row1.id = 1; //必须赋值为整数
row1.name = "上海"; //必须赋值为字符串
```

15.2.2　连接类型

ADO .NET 中的基本连接类型都在 System.Data.Common 命名空间中定义，其中包括数据连接类 DbConnection、数据命令类 DbCommand、数据阅读器类 DbDataReader、数据适配器类 DbDataAdapter 等；这些类型针对不同的数据库还有不同的派生类，例如用于 SQL Server 数据访问的就分别是 SqlConnection、SqlCommand、SqlDataReader 和 SqlDataAdapter，它们都在 System.Data.SqlClient 命名空间中定义。

1. 数据连接

DbConnection 类表示到数据库的连接，不过它是一个抽象类，它的下列派生类分别用于连接到指定类型的数据库。

- OleDbConnection：使用 OLEDB 驱动的数据连接，可访问 SyBase、SQL Server、Access 等数据库。
- OdbcConnection：到 ODBC 数据源的连接。
- OracleConnection：专门用于访问 Oracle 数据库的连接。
- SqlConnection：专门用于访问 SQL Server 数据库的连接。

DbConnection 必须要提供连接字符串，并在其中包含数据源、身份验证等连接信息。连接字符串可通过 ConnectionString 属性来设置，也可以在派生类的构造函数中指定。例如下面的代码创建了两个连接对象，前者通过 Windows 集成身份验证来访问本地服务器上的 Travel 数据库，后者则用于访问 C 盘上的 Access 数据库 demo.mdb：

```
DbConnection conn1 = new SqlConnection();
conn1.ConnectionString = "Server=localhost;Database=Travel;IntegratedSecurity=true";
DbConnection conn2 = new OleDbConnection("Provider=Microsoft.Jet.OLEDB.4.0;Data Source=C:\\demo.mdb");
```

不同类型数据库的连接字符串格式也不尽相同，其中一些基本信息还可以通过连接对象的属性来访问。以 SQL Server 数据库为例，SqlConnection 的 DataSource 和 Database 属性就分别表示服务器实例名和数据库的名称。

创建了 DbConnection 对象之后，还需要使用 Open 方法打开连接后才能使用；当不再需要连接时，应使用 Close 方法将其关闭。通过 DbConnection 的 State 属性判断连接的当前状态，其类型为 ConnectionState 枚举，取值包括 Open（打开）、Closed（关闭）、Broken（中断）等。当然，由于身份验证失败、服务器故障、网络故障等异常情况都可能导致数据连接打开失败，因此通常应进行异常处理：

```
try
{
    conn1.Open();
    //进行数据访问……
}
catch
{
    //处理异常情况
}
finally
{
    if (conn1 != null && conn1.State == ConnectionState.Open)
        conn1.Close();
}
```

2. 数据命令

DbCommand 用于封装 SQL 语句并执行相应的操作。可通过 DbConnection 对象的 CreateCommand 方法来返回一个数据命令对象，并通过其 CommandText 属性来设置命令的内容，例如：

```
DbCommand cmd1 = conn1.CreateCommand();
cmd1.CommandText = "CREATE TABLE [Scene] ([id] int,[name] nvarchar(20), [price] money)";
```

对于 SqlCommand 这样的具体类型，也可以在其构造函数中同时指定命令文本和连接对象，例如：

```
SqlCommand cmd2 =new SqlCommand("DELETE FROM [Scene]", conn1);
```

根据命令的不同，DbCommand 的执行方式也各不相同。对于不要求返回结果的一般 SQL 语句（如数据定义语句和数据更新语句），可使用 DbCommand 的 ExecuteNonQuery 方法来执行命令，方法的返回值表示受影响的记录数量，例如：

```
SqlCommandcmd1 =new SqlCommand("DELETE FROM [Scene]", conn1);
int i = cmd1.ExecuteNonQuery();
Console.WriteLine("成功删除{0}条语句", i);
```

而执行查询命令的方式有两种，一是使用 DbCommand 的 ExecuteScalar 方法，该方法将返回结果集中第一行第一列的值（类型为 object），这主要适用于从表格中检索单个值的情况，例如：

```
cmd1.CommandText ="SELECT AVG(price) FROM [Scene]";
decimal x = (decimal)cmd1.ExecuteScalar();
Console.WriteLine("景点的平均价格为{0}元", x);
```

另一种方式就是使用 DbCommand 的 ExecuteReader 方法，该方法返回一个 DataReader 对象以读取记录集，下面就对 DataReader 进行介绍。

3. 数据阅读器

DbDataReader 用于对记录集执行前向和只读的访问。该对象维护一个游标，其开始位置在第一条记录之前；只要还未到达记录集的最后一行，那么调用其 Read 方法就会返回布尔值 true，同时读取下一条记录，并将游标位置推进一行；否则其 Read 方法就返回 false。每次读取了记录后，DbDataReader 就可以像 DataRow 那样通过索引函数来访问指定的字段值，下面的代码就通过循环语句来使 DataReader 对象遍历整个记录集：

```
SqlCommand cmd1 = new SqlCommand("SELECT * FROM [Scene]", conn1);
SqlDataReader reader1 =cmd1.ExecuteReader();
while (reader1.Read()){
    Console.WriteLine("名称:" + reader1["name"]);
    Console.WriteLine("价格:" + reader1["price"]);
}
```

当然，如果只需要访问记录集中的第一条记录，或是确定查询命令最多返回一条记录，那么对 DbDataReader 对象只需调用一次 Read 方法即可，例如：

```
SqlCommand cmd1 = new SqlCommand("SELECT * FROM [Scene] WHERE id=0", conn1);
SqlDataReader reader1 =cmd1.ExecuteReader();
if (reader1.Read())
    Console.WriteLine("{0}的价格为{1}", reader1["name"], reader1["price"]);
```

使用完 DbDataReader 对象后，应通过其 Close 方法来关闭阅读器，否则在同一个数据连接下就不能执行其他的数据命令。而通过 DbDataReader 的 IsClosed 属性则可以判断阅读器是否已经关闭，例如：

```
if (reader1 != null && !reader1.IsClosed)
    reader1.Close();
```

此外，数据表格类 DataTable 专门提供了一个 Load 方法，方法的参数就是一个数据阅读器对象，它能够将记录集的全部内容一次载入到表格中，例如：

```
DataTable table1 = new DataTable();
table1.Load(reader1);
```

也就是说，DataTable 的 Load 方法相当于使用 DbDataReader 对象依次读取所有记录，并将各条记录作为新行加入到表格中，最后关闭数据阅读器（不需要再调用 DbDataReader 的 Close 方法）。

4. 数据适配器

数据阅读器的特点是访问效率高，但只能单向逐次阅读记录，而不能一次跳过多条记录，也不能回到之前已读过的记录。数据适配器 DbDataAdapter 则能够将记录集一次性填充到数据集或数据表格中，跟踪表格中数据的变化，并将结果保存到数据库。这是因为一个数据适配器对象实际上封装了多个数据命令对象，这些命令可分别通过 DbDataAdapter 的下列属性来进行访问。

- SelectCommand：查询命令，用于获取记录集。
- InsertCommand：插入命令，用于插入新记录。
- DeleteCommand：删除命令，用于删除现有记录。
- UpdateCommand：修改命令，用于修改现有记录。

以 SqlDataAdapter 为例，其构造函数中可以指定查询命令，而后通过其 Fill 方法将查询结果填充到表格中，例如：

```
SqlCommand cmd1 = new SqlCommand("SELECT * FROM [Scene]", conn1);
SqlDataAdapter adapter1 = new SqlDataAdapter(cmd1);
DataTable table1 = new DataTable("Scene");
adapter1.Fill(table1);
```

数据适配器对象的 Fill 方法还能向数据集中的指定表格填充数据，例如：
```
DataTable table1 = new DataTable("Scene");
DataSet ds1 = new DataSet();
ds1.Tables.Add(table1);
adapter1.Fill(ds1, "Scene");
```
可以手动创建数据适配器对象的其余数据命令，但最简单的方法是使用 SqlCommandBuilder 来自动生成这些命令。设置了 SqlDataAdapter 对象的 SelectCommand 命令后，只要创建一个关联到该数据适配器的 SqlCommandBuilder 对象，适配器就能自动生成用于插入、删除和修改数据命令，例如：
```
SqlCommandBuilder builder1 = new SqlCommandBuilder(adapter1);
```
这样在修改表格数据之后，只需要调用 SqlDataAdapter 的 Update 方法，数据适配器就会检查这些修改，并自动执行所需的 UpdateCommand、InsertCommand 或 DeleteCommand 命令来保存修改，例如：
```
adapter1.Update(ds1, "Scene");
```
也就是说，数据适配器在物理数据源和 DataTable/DataSet 之间架设了一座桥梁，它能够方便地保持内存数据与物理数据的同步，同时大大降低了编程的工作量。如果要同时维护多个表格，那么可为每个表格创建一个数据适配器，并通过这些适配器对象来统一操作数据集，例如：
```
SqlCommand cmd1 = new SqlCommand("SELECT * FROM [Province]", conn1);
SqlCommand cmd2 = new SqlCommand("SELECT * FROM [City]", conn1);
SqlDataAdapter adapter1 = new SqlDataAdapter(cmd1);
SqlDataAdapter adapter2 = new SqlDataAdapter(cmd2);
DataSet ds1 = new DataSet();
adapter1.Fill(ds1, "Province");
adapter2.Fill(ds1, "City");
SqlCommandBuilder builder1 = new SqlCommandBuilder(adapter1);
SqlCommandBuilder builder2 = new SqlCommandBuilder(adapter2);
//修改数据集中的数据
//…
adapter1.Update(ds1, "Province");
adapter2.Update(ds1, "City");
```
下面演示一个简单的数据库管理应用程序的创建过程。新建一个 WPF 应用程序项目 P15_2，将本书配套源文件中的 Access 数据库 Geo.mdb 复制并添加到项目中，此时 Visual Studio 将自动打开如图 15-5（a）所示的"数据源配置向导"，单击"下一步"，在图 15-5（b）所示的界面中选择所有的表，单击"完成"按钮，这样项目中就会自动创建一个 GeoDataSet.xsd 数据集文件，其中包含类型化的数据表 ProvinceDataTable、CityDataTable 和 SceneDataTable，以及针对每个数据表格的数据适配器。

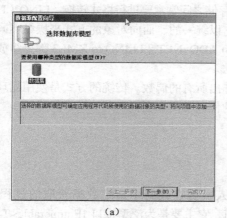

(a)   (b)

图 15-5　程序 P15_2 的示例输出结果

回到 MainWindow 的设计视图，向窗体上加入一个 ComboBox 控件、一个 Button 控件和一个 DataGrid 控件。然后切换到代码视图，为 MainWindow 类添加以下字段成员：

```
GeoDataSet geoDataSet = new GeoDataSet();
ProvinceTableAdapter pAdapter = new ProvinceTableAdapter();
CityTableAdapter cAdapter = new CityTableAdapter();
SceneTableAdapter sAdapter = new SceneTableAdapter();
```

在窗体的 Load 事件处理方法中加入下列代码，就能够将表格的名称列入组合框中：

```
private void Window_Loaded(object sender, RoutedEventArgs e)
{
    foreach (DataTable dt in geoDataSet.Tables)
        comboBox1.Items.Add(dt.TableName);
    pAdapter.Fill(geoDataSet.Province);
    cAdapter.Fill(geoDataSet.City);
    sAdapter.Fill(geoDataSet.Scene);
}
```

用户在组合框中选择一个表名时，对应的表格数据就可以显示在 DataGrid 控件中：

```
private void comboBox1_SelectedIndexChanged(object sender, EventArgs e)
{
    dataGrid1.ItemsSource = geoDataSet.Tables[comboBox1.SelectedIndex].DefaultView;
}
```

用户还可在数据网格控件中修改数据，并通过"保存"按钮来保存修改结果：

```
private void button1_Click(object sender, EventArgs e)
{
    pAdapter.Update(geoDataSet.Province);
    cAdapter.Update(geoDataSet.City);
    sAdapter.Update(geoDataSet.Scene);
}
```

程序 P15_2 的输出效果如图 15-6 所示。

图 15-6　程序 P15_2 的示例输出结果

15.3　LINQ 对象数据查询

应用程序中需要处理各种各样的对象，而对象数据的物理存储可以是一般文件、XML 文档、关系数据库等不同格式，因此在内存对象与其物理存储之间常常要进行格式转换。LINQ（Language Integrated Query）是一种数据抽象的编程模型，它以统一的、面向对象的方式来操纵不同格式的数据内容。LINQ 技术由 LINQ to Objects、LINQ to ADO .NET 和 LINQ to XML3 部分组成，下面以 LINQ to Objects 为例介绍 LINQ 查询的基本语法。

先来看一个简单的例子：从一个整数数组中选出所有的偶数。传统的方式是使用循环语句遍历整个数组，并从中选出满足条件的元素。这样的代码示例如下：

```
int[] a = { 2, 11, 9, 6, 8 };
List<int> a1 = new List<int>();
foreach (int x in a)
    if (x % 2 == 0)
        a1.Add(x);
```

上述代码属于"过程式"的语法，即开发人员要在代码中指定数据获取的过程。.NET Framework 3.5 新增了一个 System.LINQ 命名空间，它主要是为泛型接口 IEnumerable<T>定义了一组扩展方法。数组类型支持 IEnumerable<T>接口，因此可使用其中的 Where 方法来实现与上述

第 15 章 对象持久性——数据库存取和 LINQ 查询

代码相同的功能：

```
IEnumerable<int> a1 = a.Where(x => x % 2 == 0);
```

注意传递给 Where 方法的参数是一个 Lambda 表达式。这是因为该方法的参数类型实际上是泛型委托 Func<T,bool>，它表示筛选条件。Where 方法的作用就是选出可枚举集合 Ienumerable<T>中所有满足条件的 T 型元素，并将它们作为一个新集合返回。这种代码属于"声明式"的语法，即开发人员在代码中声明需要获取什么样的数据，而数据获取的具体过程由后台实现。很显然，"声明式"编程的抽象程度和编程效率都比"过程式"编程更高。

不过，Lambda 表达式对于某些开发人员而言可能难以阅读和理解。.NET 提供了另外一种声明式的语法——LINQ 查询表达式，采用它来实现上述功能的代码如下：

```
var a1 = from x in a
         where x % 2 == 0
         select x;
```

上述代码的等号右侧就是一个完整的 LINQ 查询表达式，它以 from 关键字开头，其后的 x 为局部变量，表示数据源中的每个元素，in 关键字后面指定的是数据源，where 关键字用于指定查询条件，最后的 select 关键字用于指定返回的结果项。注意整个语句结束的分号是在 select 子句之后，各子句间的分行只是为了阅读方便。根据需要也可将整个查询表达式放在一对括号中。这种查询语法和 SQL 语句非常相似，但表达式要求以 select 子句或 groupby 子句结尾。

通过 Visual Studio 的智能感知，可知上述隐式定义的变量 a1 的类型为 IEnumerable<int>。通过 foreach 等循环语句可遍历查询结果中的每个元素，例如：

```
foreach (int x in a1)
    Console.WriteLine(x);
```

提示 一般情况下，当程序执行到 LINQ 查询表达式的代码位置时，表达式背后的查询过程并不会立即执行；只有当使用循环语句访问到结果集合时，查询表达式才会被真正执行，这叫做"延迟求值"。如果希望查询表达式被立即执行，那么可对结果集合调用 IEnumerable<T> 中定义的 ToArray、ToList 等方法，将其由可枚举接口类型转换为具体的集合类型。

对于内存中的对象集合，只要集合类型支持 IEnumerable<T>接口，就可以使用 LINQ 查询表达式对其进行各种查询，这就是 LINQ to Objects。在接下来的例子中，我们将 Geo.mdb 数据库的 3 个数据表分别读取到了 provinces、cities 和 scenes 三个列表中（具体代码见配属源程序 P15_3），并通过它们来介绍更多的 LINQ 查询操作。

先看 select 子句，它实际对应的是 IEnumerable<T>的 Select 方法，传递给方法的参数是一个 Func<T,R>型委托变量。在查询表达式中，如果 select 关键字后的变量与 from 子句中的相同，则表达式返回类型也是 IEnumerable<T>。通过 select 子句可将源数据中的 T 型变量"投影"为一个 R 型变量。例如下面的表达式将查询列表 scenes 中的所有景点，并返回景点的"所在城市名"+"景点名"所构成的集合。查询结果显示在 ASP.NET 网页的 GridView 控件中的效果如图 15-7（a）所示。

```
var result = from s in scenes
             select s.City.Name + s.Name;
GridView1.DataSource = result;
GridView1.DataBind();
```

where 子句和 IEnumerable<T>.Where 方法上面已经介绍过，它可以通过与操作符"&&"以及或操作符"||"来组合多个筛选条件。在 select 子句之前还可以放上 orderby 子句来对查询结果进行排序，例如下面的表达式将查询指定省份的所有景点，并按景点所在的城市名进行排序：

```
var result = from s in scenes
             where s.City.Province.Name == DropDownList1.SelectedValue
```

```
            orderby s.City.Name
            select new { Name = s.City.Name + s.Name };
```
上面的 select 子句中使用了匿名对象创建表达式，查询结果如图 15-7（b）所示，结果中每个元素都是该匿名类型的一个对象，对象中只包含一个名为 Name 的字段。

orderby 子句实际对应的是 IEnumerable<T>的 OrderBy 方法，方法的参数类型是 Func<R,K>，其中 R 与 select 子句的结果类型相一致，而 K 表示用于比较的关键项类型。orderby 子句中可以使用多个排序关键项，这相当于 OrderBy 方法的嵌套调用；如果要逆序排列，那么可在 orderby 子句最后加上一个 descending 关键字，这实际上对应的是 IEnumerable<T>的 OrderByDescending<T,K>方法，例如：

```
var result = from s in scenes
             where s.City.Province.Name == DropDownList1.SelectedValue
             orderby s.City.Name, s.Name descending
             select new { Name = s.City.Name + s.Name };
```

如果要在多个集合之间进行联接查询，就要使用到 join 子句和 on 子句，前者指定要联接的集合，后者指定联接条件。下面的表达式就对列表 scenes 和 cities 进行了联接查询，其结果如图 15-7（c）所示。

```
var result = from s in scenes
             join c in cities
             on s.City equals c
             select new { Prov = c.Province.Name, s.Name };
```

查询表达式中还可以使用 groupby 子句来进行分组查询，其对应的是 IEnumerable<T>的 GroupBy<T,K>方法。该方法的参数类型是 Func<T,K>，其中 T 为源数据中的元素类型，K 为用于分组的关键项类型。方法的返回类型为 IEnumerable<IGrouping<K,T>>，也就是说，返回集合中的每一项都是一个 IGrouping<K,T>类型的元素，此元素本身又是一个集合，因为 IGrouping<K,T>继承了 IEnumerable<T>接口。

例如下面的表达式就将查询到的景点按城市进行分组，并将其中第一组的结果绑定到 GridView 控件：

```
var result = from s in scenes
             where s.City.Province.Name == DropDownList1.SelectedValue
             group s by s.City;
IGrouping<City, Scene>[] sGroups = result.ToArray();
Label1.Text = sGroups[0].Key.Name;
GridView1.DataSource = sGroups[0];
GridView1.DataBind();
```

如果通过窗体对象的字段保存查询结果，那么还可以通过不同的索引项来访问不同的分组，例如：

```
if (++current == sGroups.Length)
    current = 0;
Label1.Text = sGroups[current].Key.Name;
GridView1.DataSource = sGroups[current];
GridView1.DataBind();
```

代码的输出结果如图 15-7（d）和图 15-7（e）所示。

IEnumerable<T>的扩展方法还支持基本的集合操作和聚合操作，这些方法可直接应用于查询表达式的结果。其中 Distinct 方法用于删除集合中的重复项，例如下面的表达式将得到所有景点的价格的降序排列，且不含重复项：

```
var result = (from s in scenes
              order by s.Price descending
              select s.Price).Distinct();
```

Union、Intersect 和 Except 这 3 个方法都是作用于两个相同类型的集合，它们分别对应集合运算中的并集、交集和差集。下面的代码演示了它们的用法：
```
var r1 = from s in scenes
         where s.Price< 100
         select s;
var r2 = from s in scenes
         where s.Star >= 4
         select s;
var result1 = r1.Union(r2);
var result2 = r1.Intersect(r2);
```

(a)

(b)

(c)

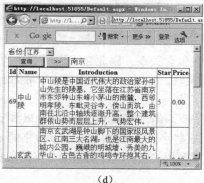
(d)

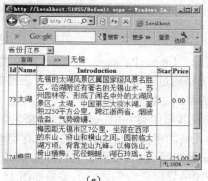
(e)

图 15-7　LINQ 查询表达式执行结果示例

此外，Count 方法用于获得集合的大小；对于数值型集合（如整数集合和小数集合），还可使用 Sum、Average、Min、Max 这些方法来计算集合的和、平均值、最小值和最大值，例如下面的表达式用于获取上海市所有景点价格的平均值：
```
decimal tp = (from s in scenes
             wheres.City.Name == "上海"
             select s.Price).Average();
```
LINQ 查询还支持嵌套查询，这样内部表达式中可以访问外部表达式中的变量。例如在下面的语句中，内部表达式会针对外部表达式中的每个城市 c 查询其景点集合，这样整个表达式将返回所有景点数量大于 5 的城市集合：
```
from c in cities
where (from s in scenes
       where s.City.Equals(c)
       select s).Count() >= 5
select c;
```

15.4　案例研究——旅行社管理系统的数据库解决方案

15.4.1　数据表格设计

要使旅行社管理系统运行在数据库平台上，首先是对系统中需要进行持久性存储的对象类型建立数据表。先考虑 Province、City 和 Scene 类，它们在数据库中都有对应的物理表，每个表格都增加一个 id 字段作为主键；类的简单值字段都作为表格的字段（枚举类型存储为对应的整数值），而且 City 表格应当通过一个 province 字段来与 Province 表格的主键相关联，Scene 表格则通过一个 city 字段与 City 表格的主键相关联，如图 15-8 所示。

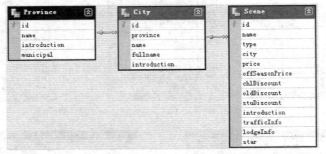

图 15-8　Province、City 和 Scene 表格

类似地，Line、Package 和 Tour 对象也依次存在包含关系，因此在它们所对应的表格中，Package 表格通过一个 line 字段与 Line 表格的主键相关联，而 Tour 表格通过一个 package 字段与 Package 表格的主键相关联。

Staff 及其派生类共用一个 Staff 表格，派生类的特殊字段（如 Guide 类的 guideId 和 guideDue）也都作为表格的字段，其他类型的对象存储时不占用这些字段。此外，Package 表格的 agent 字段以及 Tour 表格的 guide 字段，它们都与 Staff 表格的主键相关联，以标明组团方案的负责业务员，以及旅行团的导游。它们之间的关系如图 15-9 所示。

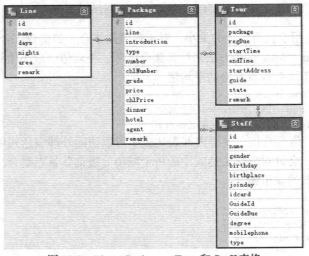

图 15-9　Line、Package、Tour 和 Staff 表格

Customer 类也需要建立一个单独的表格；为了表示旅行团和游客之间的关系，还应当专门建立一个 TourCustomer 表格，其中的 tour 字段与 Tour 表格的主键相关联，customer 字段则与 Customer 表格的主键相关联，如图 15-10 所示。

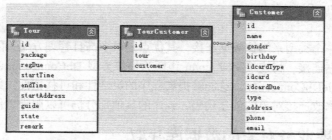

图 15-10　Tour 表格、Customer 表格及其关联表

类似的，针对 Line 表格和 Scene 表格也应当建立一个关系表 LineScene，其 line 字段与 Line 表格的主键相关联，scene 字段则与 Scene 表格的主键相关联，如图 15-11 所示。

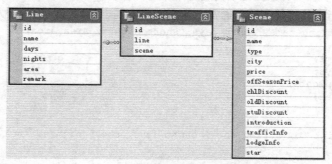

图 15-11　Line 表格、Scene 表格及其关联表

15.4.2　数据库连接管理

创建完数据库表格之后，就应当在程序中建立并维护与数据库的连接。接下来回到旅行社业务类库项目 TravelLib 中，在其中新建一个静态类 TravelData，它通过一个包含数据连接字段 _conn，数据连接在构造函数中初始化，并通过静态方法来 Connect 和 Disconnect 来打开和关闭数据连接：

```
public static class TravelData
{
    private static DbConnection _conn;
    public static DbConnectionConnection
    {
        get { return _conn; }
    }

    static TravelData()
    {
        _conn = new SqlConnection("server=(local); Database=Travel; Integrated Security=true; MultipleActiveResultSets=true");
    }

    public static void Connect()
    {
        if (_conn.State != ConnectionState.Open)
            _conn.Open();
```

```csharp
    }
    public static void Disconnect()
    {
        if (_conn.State != ConnectionState.Closed)
            _conn.Close();
    }
}
```

在程序中使用一个类型来维护数据连接，其优点是效率高且管理方便，因为这样能够避免程序中使用过多的数据连接。将连接字段_conn声明为抽象类型DbConnection，而非某个具体类型SqlConnection或OleDbConnection，这能够提高程序的灵活性和可移植性：如果数据库平台发生了变化，那么只要修改连接对象的创建代码，程序的其他部分仍可以照常工作。

15.4.3 实现业务对象的数据库存取

下面为系统的各个业务类型增加数据库存取的功能。首先考虑Line类，它的主要字段在数据表格Line中都有对应的字段，其中_id对应表格的主键。这时可考虑为其定义两个构造函数，需要指定_id字段的用于从现有数据记录构造对象,不需要指定_id字段的则用于在程序中新建对象；在构造函数中需要指定的其他字段，通常在数据表格中属于必填字段（不可为空）：

```csharp
public Line(string name, short days, short nights, Area area)
{
    _name = name;
    _days = days;
    _nights = nights;
    _area = area;
}
private Line(int id, string name, short days, short nights, Area area)
{
    _id = id;
    _name = name;
    _days = days;
    _nights = nights;
    _area = area;
}
```

接下来修改其静态方法GetAll，在其中使用数据命令和数据阅读器来读出各条记录，并返回一个Line[]数组：

```csharp
public static Line[] GetAll() //查询所有线路
{
    List<Line> lines = new List<Line>();
    DbCommand cmd = TravelData.Connection.CreateCommand();
    cmd.CommandText = "SELECT * FROM [Line]";
    DbDataReader reader = cmd.ExecuteReader();
    while (reader.Read())
    {
        Line l = new Line((int)reader["id"], (string)reader["name"], (Area)(byte)reader["area"], (short)reader["days"], (short)reader["nights"]);
        if (reader["remark"] != DBNull.Value)
            l.Remark = (string)reader["remark"];
        lines.Add(l);
    }
    reader.Close();
    return lines.ToArray();
}
```

数据命令和数据阅读器使用的也都是抽象类型DbCommand和DbDataReader，显然这比使用具体类型更加灵活。对于数据表格中非必填的且没有默认值的字段，读取时要注意判断其是否为

第15章 对象持久性——数据库存取和LINQ查询

空值。此外，对于_area这样的枚举类型字段，其一般是以整数值存储在表格中，那么在读取时要进行两次类型转换：先拆箱转换为整数，再将整数转换为枚举。

Line类原有的静态方法Get仍然能够工作，不过程序有时候只需要查询一条线路信息，此时通过GetAll方法读取所有线路再进行查找就会大大降低效率，那么可使用带筛选条件的数据命令来从数据库中读取单条记录：

```
public static Line Get(int id) //查询指定线路
{
    Line line = null;
    DbCommand cmd = TravelData.Connection.CreateCommand();
    cmd.CommandText = "SELECT * FROM [Line] Where [id]=" + id;
    DbDataReader reader = cmd.ExecuteReader();
    if (reader.Read())
    {
        line = new Line(id, (string)reader["name"], (Area)(byte)reader["area"], (short)reader["days"], (short)reader["nights"]);
        if (reader["remark"] != DBNull.Value)
            line.Remark = (string)reader["remark"];
    }
    reader.Close();
    return line;
}
```

再来考虑对象数据的保存，由于新对象是使用公有构造函数创建，_id字段值默认为0，那么应使用INSERT语句将其插入数据库表格中；而对于从数据库中读取的Line对象，修改后应使用UPDATE语句来覆盖数据库表格中的原有记录。此外，Line对象的修改也会影响到LineScene表格，那么可先删除所有与当前线路相关的LineScene记录（如果存在），再根据Line对象的Scenes属性来向LineScene表格加入新记录。那么保存线路的Save方法代码如下：

```
public void Save()
{
    DbCommand cmd = TravelData.Connection.CreateCommand();
    if (_id <= 0) //未设置id, 新记录
    {
        _id = Line.NewId(_area);
        cmd.CommandText = string.Format("INSERT INTO [Line] ([id],[name],[days],[nights],[area],[remark]) Values({0},'{1}',{2},{3},{4},'{5}')", _id, _name, Days, Nights, (byte)_area, Remark);
    }
    else //修改现有记录
    {
        cmd.CommandText = string.Format("UPDATE [Line] SET [name]='{0}',[days]={1},[nights]=2,[area]=3,[remark]='{4}' WHERE [id]={5}", _name, Days, Nights, (byte)_area, Remark, _id);
        cmd.ExecuteNonQuery();
        cmd.CommandText = "DELETE FROM [LineScene] WHERE [lineId]=" + _id;
    }
    cmd.ExecuteNonQuery();
    int nId = 0;
    cmd.CommandText = "SELECT MAX(id) FROM [LineScene]";
    object o1 = cmd.ExecuteScalar();
    if (o1 != null && o1 != DBNull.Value)
        nId = (int)o1;
    foreach (Scene s in _scenes)
    {
        cmd.CommandText = string.Format("INSERT INTO [LineScene] ([id],[lineId],[sceneId]) VALUES({0},{1},{2})", ++nId, _id, s.Id);
        cmd.ExecuteNonQuery();
    }
}
```

369

在增加新记录时还要考虑主键值的设置。如果数据表格中设置主键是自动编号，那么在 INSERT 语句中就不用指定主键值。而如果主键值要按照专门的业务规则来编号，就需要在程序中设置其值。Line 的 id 编号属于后一种情况，因此上面的代码是调用了 Line 的静态方法 NewId 来获取新的主键值（按地区分类编号，如华北地区的旅行线路编号依次为 10001，10002，…，华北地区的线路编号从 20001 开始，华东地区从 30001 开始，等等）：

```csharp
private static int NewId(Area area)
{
    DbCommand cmd = TravelData.Connection.CreateCommand();
    cmd.CommandText = "SELECT MAX(id) FROM [Line] Where [area]=" + (int)area;
    object o = cmd.ExecuteScalar();
    return (o != DBNull.Value) ? (int)o + 1 : (int)(area + 1) * 10000 + 1;
}
```

注意前面 GetAll 方法和 Get 方法都只是读出了线路的基本信息，而没有读出线路景点集合的信息，这样能够避免占用过多的数据库资源。Line 类可以提供一个单独的成员方法 GetScenes 来将景点信息读取到 _scenes 字段中，而其 Scenes 属性的 get 访问器也可以做相应修改，使 Line 对象只有在 Scenes 属性被首次访问时才去从数据库中读取景点记录：

```csharp
private List<Scene> _scenes;
public List<Scene> Scenes //景点集合
{
    get{
        if (_scenes == null)
            GetScenes();
        return _scenes;
    }
}

private void GetScenes()
{
    DbCommand cmd = TravelData.Connection.CreateCommand();
    cmd.CommandText = "SELECT [Scene].* FROM [Scene],[LineScene] Where [LineScene].[sceneId]=[Scene].[id] AND [LineScene].[lineId]=" + _id;
    DbDataReader reader = cmd.ExecuteReader();
    _scenes = new List<Scene>();
    List<int> cityIds = new List<int>();
    while (reader.Read())
    {
        Scene s = new Scene((int)reader["id"], (string)reader["name"]);
        if (reader["star"] != DBNull.Value)
            s.Star = (byte)reader["star"];
        s.Price = (decimal)reader["price"];
        s.OffSeasonPrice = (decimal)reader["offSeasonPrice"];
        s.ChlDiscount = (decimal)reader["chlDiscount"];
        s.OldDiscount = (decimal)reader["oldDiscount"];
        s.StuDiscount = (decimal)reader["stuDiscount"];
        if (reader["introduction"] != DBNull.Value)
            s.Introduction = (string)reader["introduction"];
        if (reader["trafficInfo"] != DBNull.Value)
            s.TrafficInfo = (string)reader["trafficInfo"];
        if (reader["lodgeInfo"] != DBNull.Value)
            s.LodgeInfo = (string)reader["lodgeInfo"];
        if (reader["restInfo"] != DBNull.Value)
            s.LodgeInfo = (string)reader["restInfo"];
        if (reader["remark"] != DBNull.Value)
            s.LodgeInfo = (string)reader["remark"];
        _scenes.Add(s);
        cityIds.Add((int)reader["cityId"]);
    }
```

```
        reader.Close();
        for (int i = 0; i < _scenes.Count; i++)
            _scenes[i].City = City.Get(cityIds[i]);
    }
```

以上通过 Line 类说明了实现业务对象数据存取的基本方法,并介绍了在存储过程中如何处理与其他对象的关联关系。其他的各个业务类也都可以采用类似的技术来进行数据库存取,这里不再一一列出相关的详细代码。下面只给出 Staff 类的数据存取方法,使用时要注意对不同类型的职员对象用不同的处理方式:

```
    public static Staff Get(int id) //根据编号查询指定职员
    {
        Staff staff = null;
        DbCommand cmd = TravelData.Connection.CreateCommand();
        cmd.CommandText = "SELECT * FROM [Staff] Where [id]=" + id;
        DbDataReader reader = cmd.ExecuteReader();
        if (reader.Read())
        {
            switch ((byte)reader["type"])
            {
                case 0:
                    staff = new Manager(id, (string)reader["name"], (bool)reader["gender"],
(DateTime)reader["birthday"], (DateTime)reader["joinday"]);
                    break;
                case 1:
                    staff = new Director(id, (string)reader["name"], (bool)reader["gender"],
(DateTime)reader["birthday"], (DateTime)reader["joinday"]);
                    break;
                case 2:
                    staff = new Agent(id, (string)reader["name"], (bool)reader["gender"],
(DateTime)reader["birthday"], (DateTime)reader["joinday"]);
                    break;
                case 3:
                    staff = new Guide(id, (string)reader["name"], (bool)reader["gender"],
(DateTime)reader["birthday"], (DateTime)reader["joinday"]);
                    break;
                default:
                    staff = new Staff(id, (string)reader["name"], (bool)reader["gender"],
(DateTime)reader["birthday"], (DateTime)reader["joinday"]);
                    break;
            }
            if (reader["birthplace"] != DBNull.Value)
                staff.Birthplace = (string)reader["birthplace"];
            if (reader["idcard"] != DBNull.Value)
                staff.IdCard = (string)reader["idcard"];
            if (reader["degree"] != DBNull.Value)
                staff.Degree = (Degree)(byte)reader["degree"];
            if (reader["phone"] != DBNull.Value)
                staff.Phone = (string)reader["phone"];
            if (reader["email"] != DBNull.Value)
                staff.Email = (string)reader["email"];
            if (staff is Guide)
            {
                ((Guide)staff).GuideCard = (string)reader["cerID"];
                ((Guide)staff).GuideDue = (DateTime)reader["cerDue"];
            }
        }
        reader.Close();
        return staff;
    }

    public static Staff[] GetAll() //查询所有职员
```

```csharp
        {
            List<Staff> staffs = new List<Staff>();
            DbCommand cmd = TravelData.Connection.CreateCommand();
            cmd.CommandText = "SELECT * FROM [Staff]";
            DbDataReader reader = cmd.ExecuteReader();
            Staff s;
            while (reader.Read())
            {
                switch ((byte)reader["type"])
                {
                    case 0:
                        s = new Manager((int)reader["id"], (string)reader["name"], (bool)reader["gender"], (DateTime)reader["birthday"], (DateTime)reader["joinday"]);
                        break;
                    case 1:
                        s = new Director((int)reader["id"], (string)reader["name"], (bool)reader["gender"], (DateTime)reader["birthday"], (DateTime)reader["joinday"]);
                        break;
                    case 2:
                        s = new Agent((int)reader["id"], (string)reader["name"], (bool)reader["gender"], (DateTime)reader["birthday"], (DateTime)reader["joinday"]);
                        break;
                    case 3:
                        s = new Guide((int)reader["id"], (string)reader["name"], (bool)reader["gender"], (DateTime)reader["birthday"], (DateTime)reader["joinday"]);
                        break;
                    default:
                        s = new Staff((int)reader["id"], (string)reader["name"], (bool)reader["gender"], (DateTime)reader["birthday"], (DateTime)reader["joinday"]);
                        break;
                }
                if (reader["birthplace"] != DBNull.Value)
                    s.Birthplace = (string)reader["birthplace"];
                if (reader["idcard"] != DBNull.Value)
                    s.IdCard = (string)reader["idcard"];
                if (reader["degree"] != DBNull.Value)
                    s.Degree = (Degree)(byte)reader["degree"];
                if (reader["phone"] != DBNull.Value)
                    s.Phone = (string)reader["phone"];
                if (reader["email"] != DBNull.Value)
                    s.Email = (string)reader["email"];
                if (s is Guide)
                {
                    ((Guide)s).GuideCard = (string)reader["cerID"];
                    ((Guide)s).GuideDue = (DateTime)reader["cerDue"];
                }
                staffs.Add(s);
            }
            reader.Close();
            return staffs.ToArray();
        }
```

15.4.4 终端数据访问

由于在业务类中良好地封装了数据访问功能，那么旅行社内部管理子系统和旅行社网站的程序项目无须做较大改动，而只要在访问数据时打开数据连接就可以了。不过 Windows 应用程序和网站程序在性能上会有不同的考虑。旅行社内部管理子系统通常运行在本地局域网上，需要频繁地进行数据访问，那么可以在程序主窗体启动时打开数据连接，并在退出主窗体时关闭数据连接，而建立连接和访问数据的代码还应当具有异常处理的功能：

```csharp
public MainWindow()
{
    try
    {
        TravelData.Connect();
        _users = User.GetAll();
        TravelWin.LoginForm form = new TravelWin.LoginForm(_users);
        if (form.ShowDialog() != System.Windows.Forms.DialogResult.OK) {
            this.Close();
            return;
        }
        _lines = new List<Line>(Line.GetAll());
        if (File.Exists("agency.txt"))
            _agency = TravelAgency.Load("agency.txt");
    }
    catch (Exception ex)
    {
        MessageBox.Show("程序初始化错误: " + ex.Message);
        this.Close();
    }
    InitializeComponent();
}

private void Window_Closing(object sender, System.ComponentModel.CancelEventArgs e)
{
    TravelData.Disconnect();
}
```

旅行社网站的用户数量较多，过多的数据连接会严重消耗服务器资源。对于需要进行数据访问的 Web 窗体，通常可以在启动事件 Load 处理方法的第一行代码中打开数据连接，并在最后一行代码关闭数据连接，其他用于访问数据的控件事件处理方法也是如此。

此外，对于不经常更新的数据内容，使用 ASP .NET 缓存也能起到很好的性能提升效果。例如在 TourLines.aspx 页面中，可以使用 Cache 对象缓存所有线路信息；页面启动时，如果缓存为空则从数据库中读取数据，否则就直接从缓存中取数据：

```csharp
Line[] lines;

protected void Page_Load(object sender, EventArgs e)
{
    if (Cache["lines"] == null)
    {
        try
        {
            TravelData.Connect();
            Cache["lines"] = Line.GetAll();
            TravelData.Disconnect();
        }
        catch
        {
            return;
        }
    }
    lines = (Line[])Cache["lines"];
    ……
}
```

再来看线路搜索功能，单击母版页中的搜索按钮后，搜索关键字将作为查询字符串传递给 TourLines.aspx 页面：

```csharp
protected void BtnSearch_Click(object sender, ImageClickEventArgs e)
{
    if (TbLine.Text != "")
```

```
        Response.Redirect("TourLines.aspx?find=" + TbLine.Text);
}
```
对 TourLines.aspx.cs 的代码只需作简单的修改，在页面启动时检查键值为 find 的查询字符串是否存在，是则通过 LINQ 查询表达式查询结果，并显示在 GridView 控件上：
```
protected void Page_Load(object sender, EventArgs e)
{
    ......
    if (Request.QueryString["id"] != null)
    {
        ......
    else if (Request.QueryString["find"] != null)
    {
        string find = Request.QueryString["find"];
        GridView1.DataSource = lines.Where(l => l.Name.Contains(find));
    }
    GridView1.DataBind();
}
```

15.5 小　　结

数据库为应用程序的对象信息提供了更为安全可靠的存储平台。通过 ADO .NET 技术来进行数据库存取，需要使用 DbConnection 或其派生对象建立数据连接，通过 DbCommand 对象执行数据命令，并通过数据阅读器或数据适配器来返回查询结果。熟练地掌握这些数据对象的使用方法，就能够为自己的业务类型增加数据库存取的功能。

15.6 习　　题

1. 阅读一本 SQL Server 的基础使用教程，学习如何在 SQL Server 中建立和维护数据库、数据表格和数据视图。

2. 修改程序 P15_1，使其对数据的筛选可以是多个条件的组合，而且能够进行升序和降序排列。

3. 在 Access 或 SQL Server 数据库中创建一个 Student 表格，在其中存储学生的学号、姓名、性别、生日、班级、地址等字段；再创建一个 Score 表格，在其中存储学生的数学、英语、计算机等科目的考试成绩。通过 SQL 语句来执行下列查询：

（1）查询所有科目都及格的学生信息。

（2）查询总分以及各个单科分数最高的学生姓名和分数。

（3）查询每个班级中总分以及各个单科分数最高的学生姓名和分数。

4. 使用数据命令对象来完成上第 3 题所描述的数据查询功能，并将结果显示在 Windows 窗体上。

5. 画出下列类型之间的关系图：DataSet、DataTable、DataColumn、DataRow、DataTableCollection、DataColumnCollection、DataRowCollection。

6. 为旅行社管理系统中的 Package、Tour、Scene 等类型增加数据库存取的功能。

7. 基于数据库来实现旅行社管理系统的用户和登录管理功能。

8. 为旅行社网站增加"景点大观"页面 ViewScenes.aspx，页面左侧显示所有省份的列表；单击某一省份，其右侧将显示省份的所有城市列表；单击某一城市，其右侧将显示城市的所有景点列表。

第 16 章
Silverlight 客户端应用程序

Silverlight 是一项跨平台的 Web 客户端开发技术，使用它能够在浏览器中设计出具有丰富多媒体效果的交互式网络应用程序，其编程模型和 WPF 也有着很大的相似性。本章将对 Silverlight 程序的基本架构和开发方法进行讲解，并在案例研究中介绍必应地图 Silverlight 控件的用法。

16.1 Silverlight 应用开发基础

传统的 HTML 网页以文本内容为主，只支持一般的嵌入式图像，对多媒体功能的支持十分薄弱。随着人们对网络应用体验的要求越来越高，富 Internet 应用程序（Rich Internet Application, RIA）技术近年来得到了广泛的关注。

Silverlight 就是.NET 平台上的 RIA 解决方案，其开发框架在很大程度上借鉴了 WPF 技术，支持矢量图形、动画、3D、音频和视频等绚丽多彩的网络用户界面。最显著的一点就是 Silverlight 也是基于 XAML 文档来设计程序界面，其主要界面元素的集合可以看成是 WPF 类库的一个子集。因此，对于具有 WPF 经验的开发人员而言，学习 Silverlight 应用程序开发是一件非常轻松的事情。

和基于 WPF 的 Windows 应用程序不同，Silverlight 应用并不需要在浏览器的客户端安装.NET Framework。事实上，Silverlight 程序是由浏览器端的一个轻量级插件来运行的。如果客户端未安装 Silverlight 插件，那么浏览器在首次访问 Silverlight 应用时会被提示要求下载插件。不过整个插件只有几兆大小，下载和安装都非常便捷。

Silverlight 目前的最新版本是 Silverlight 5。使用 Visual Studio 开发 Silverlight 应用程序，需要先安装 Microsoft Silverlight SDK。此外，Microsoft Silverlight Toolkit 中还提供了许多工具和控件支持。它们都可以在 Microsoft 网站上免费下载。

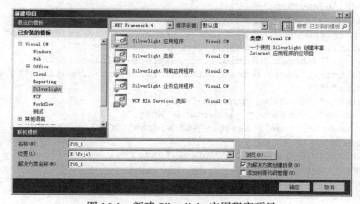

图 16-1 新建 Silverlight 应用程序项目

配置好了安装环境后，在 Visual Studio 中新建项目，在图 16-1 所示的对话框中，选择 Silverlight 节点下的"Silverlight 应用程序"模板，就可以创建 Silverlight 应用程序项目。

Silverlight 程序运行在浏览器的网页上，因此它需要有一个承载的 Web 程序。创建 Silverlight 应用程序时，Visual Studio 会询问是否创建这样一个 Web 程序项目，如图 16-2 所示。勾选其中的"在新网站中承载 Silverlight 应用程序"，Visual Studio 就会自动创建一个宿主 ASP.NET 程序；否则，Visual Studio 会生成测试页来运行 Silverlight 程序。

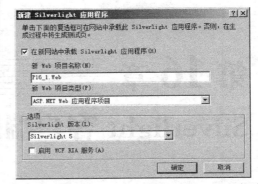

图 16-2　选择承载 Silverlight 应用程序的 Web 程序项目

16.2　Silverlight 程序架构

创建了 Silverlight 应用程序项目后，在 Visual Studio 的解决方案资源管理器中可以看到，项目中包含一个 App.xaml 文件和一个 MainPage.xaml 文件（每个 xaml 文件又有一个对应的 xaml.cs 源代码文件），前者用于描述整个程序的配置信息，后者则定义了 Silverlight 程序的主界面，这和 WPF 程序十分类似。例如也可以在 xaml 文件的 Resources 节点下定义各种资源，并在程序的其他位置方便地进行重用。

但要注意，Silverlight 程序中的 MainPage 并不是一个窗体或者页面，而是一个 Silverlight 用户控件。打开该 xaml 文件，可以看到其根元素是 UserControl。在该元素之下可以像 WPF 那样嵌套 Grid 元素，进而对各种界面元素进行布局和设置。

 Silverlight 支持的元素集合可看作是 WPF 元素的一个子集。例如，Silverlight 5 支持的布局型容器控件有 Grid、Canvas 和 StackPanel，但不支持 WrapPanel、DockPanel 和 UniformGrid。

这里模仿 WPF 程序 P13_9，将 UserControl 下的 Grid 元素替换成为一个 Canvas 元素，在其中放置两个 Ellipse 对象，并通过故事板动画来移动这两个椭圆：

```
//程序 P16_1 (MainPage.xaml)
<UserControl x:Class="P16_1.MainPage"
  xmlns="http://schemas.microsoft.com/winfx/2006/xaml/presentation"
  xmlns:x="http://schemas.microsoft.com/winfx/2006/xaml"
  xmlns:d="http://schemas.microsoft.com/expression/blend/2008"
  xmlns:mc="http://schemas.openxmlformats.org/markup-compatibility/2006"
  mc:Ignorable="d"
  d:DesignHeight="300" d:DesignWidth="400">
  <Canvas x:Name="LayoutRoot" Background="White">
    <Ellipse Name="ellipse1" Height="40" Width="40" Fill="Red">
      <Ellipse.Triggers>
        <EventTriggerRoutedEvent="Control.Loaded">
          <EventTrigger.Actions>
            <BeginStoryboard>
              <Storyboard>
                <DoubleAnimation Storyboard.TargetName="ellipse1"
```

```
Storyboard.TargetProperty="(Canvas.Left)" From="0" To="{Binding ElementName=LayoutRoot,
Path=ActualWidth}" BeginTime="0:0:0.5" Duration="0:0:1" RepeatBehavior="Forever" />
              <DoubleAnimation Storyboard.TargetName="ellipse1"
Storyboard.TargetProperty="(Canvas.Top)" From="0" To="{Binding ElementName=LayoutRoot,
Path=ActualHeight}" BeginTime="0:0:0.5" Duration="0:0:1" RepeatBehavior="Forever" />
            </Storyboard>
          </BeginStoryboard>
        </EventTrigger.Actions>
      </EventTrigger>
    </Ellipse.Triggers>
  </Ellipse>
  <Ellipse Name="ellipse2" Height="40" Width="40" Canvas.Top="260" Fill="Blue">
    <Ellipse.Triggers>
      <EventTrigger RoutedEvent="Control.Loaded">
        <EventTrigger.Actions>
          <BeginStoryboard>
            <Storyboard>
              <DoubleAnimation Storyboard.TargetName="ellipse2"
Storyboard.TargetProperty="(Canvas.Left)" From="0" To="{Binding ElementName=LayoutRoot,
Path=ActualWidth}" Duration="0:0:1" RepeatBehavior="Forever" />
              <DoubleAnimation Storyboard.TargetName="ellipse2"
Storyboard.TargetProperty="(Canvas.Top)" From="{Binding ElementName=LayoutRoot,
Path=ActualHeight}" To="0" Duration="0:0:1" RepeatBehavior="Forever" />
            </Storyboard>
          </BeginStoryboard>
        </EventTrigger.Actions>
      </EventTrigger>
    </Ellipse.Triggers>
  </Ellipse>
</Canvas>
</UserControl>
```

编译运行程序,Visual Studio 会打开浏览器页面,可以看到页面中两个椭圆形的小球在不停地移动,如图 16-3 所示。

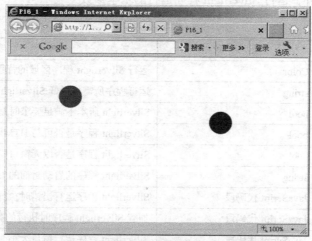

图 16-3 在网页中移动圆球的 Silverlight 应用程序

回到 Visual Studio 项目中,打开 Web 项目中的测试页面文件 P16_1TestPage.aspx,可以看到该文件的前半部分定义了一段 JavaScript 脚本,这主要是用于处理 Silverlight 程序的运行时错误;后半部分的 html 表单内容则定义如下:

```
<form id="form1" runat="server" style="height:100%">
  <div id="silverlightControlHost">
```

```
            <object                               data="data:application/x-silverlight-2,"
type="application/x-silverlight-2" width="100%" height="100%">
                <param name="source" value="ClientBin/P16_1.xap"/>
                <param name="onError" value="onSilverlightError" />
                <param name="background" value="white" />
                <param name="minRuntimeVersion" value="5.0.61118.0" />
                <param name="autoUpgrade" value="true" />
                <a
href="http://go.microsoft.com/fwlink/?LinkID=149156&v=5.0.61118.0" style="text- decoration:none">
                    <imgsrc="http://go.microsoft.com/fwlink/?LinkId=161376"  alt="  获 取   Microsoft
Silverlight" style="border-style:none"/>
                </a>
            </object>
            <iframe   id="_sl_historyFrame"   style="visibility:hidden;height:0px;width:0px;
border: 0px"></iframe>
        </div>
    </form>
```

这里的 object 元素就用于加载 Silverlight 插件；如果未安装该插件，通过其中<a href>元素指定的链接位置可以下载插件。object 元素中的各个 param 元素指定了 Silverlight 程序的运行参数，其中第一个名为 source 的 param 元素指定了 Silverlight 程序名，其值为 "ClientBin/P16_1.xap"。在解决方案资源管理器中，也可以看到 Web 项目中包含一个 ClientBin 文件夹，其中存放了一个 P16_1.xap 文件。每一个 Silverlight 程序被编译为一个 dll 文件，它和程序所使用的其他资源（如图片）和验证文件等都会被打包为一个 xap 文件并发送给承载的 Web 程序。Silverlight 插件负责从 xap 文件中提取相关内容、加载资源，并在浏览器端执行程序。

Silverlight 程序的 xap 文件较大时，用户首次访问的速度会很慢。因此，在使用的程序集较多时，可只将程序启动时必需的程序集放入 xap 文件，并在解决方案资源管理器中设置其他程序集的"复制本地"属性为 False，那么这些程序集会在使用过程中按需下载。

除了 source 参数外，运行 Silverlight 程序运行的基本参数见表 16-1。

表 16-1　　　　　　　　　　Silverlight 程序运行的基本参数

参 数 名	类　　型	说　　明
background	Color	运行 Silverlight 程序区域的背景色
minRuntimeVersion	string	运行程序所需的最低 Silverlight 版本
autoUpgrade	bool	Silverlight 版本不满足要求时是否自动升级
enableHtmlAccess	bool	Silverlight 程序是否可与 HTML 对象进行交互
windowless	bool	Silverlight 程序是否以无窗口模式运行
splashScreenSource	string	Silverlight 程序的启动封面的 URL 位置
onError	JavaScript 代码段	Silverlight 程序运行出错时执行的脚本代码
onLoad	JavaScript 代码段	加载 Silverlight 程序时执行的脚本代码
onResize	JavaScript 代码段	Silverlight 程序运行窗口大小改变时执行的脚本代码

回到 Silverlight 程序项目 P16_1，打开 app.xaml.cs 文件，其中定义的 App 类型从 System.Windows.Application 类继承，它封装了当前的 Silverlight 应用程序对象，通过其 Startup 和 Exit 事件处理方法能够在程序启动和退出时执行相应的操作。

Application 类提供了一些非常有用的属性来操纵 Silverlight 应用程序。通过其 RootVisual 属性可访问 Silverlight 用户控件的根容器，通过 Resources 属性可访问程序中定义的各种资源对象。

特别的，Application 的静态属性 Current 返回当前 Silverlight 程序对象。

要在 Silverlight 程序中与承载的 Web 项目进行交互，就要用到 Application 类的 Host 属性，其类型为 SilverlightHost，表示 Silverlight 程序的宿主对象，通过它可以与 Web 页面中的 HTML 对象进行交互。进一步，通过 SilverlightHost 的 Content 属性可获得 Web 页面中运行 Silverlight 程序的内容区域。例如可以在程序 P16_1 的主界面中放上一个按钮，在按钮的单击事件中修改 Content 的 IsFullScreen 属性，其值为 true 时程序将以全屏模式运行（如图 16-4 所示）：

```
private void button1_Click(object sender, RoutedEventArgs e)
{
    Content content = Application.Current.Host.Content;
    content.IsFullScreen= !content.IsFullScreen;
}
```

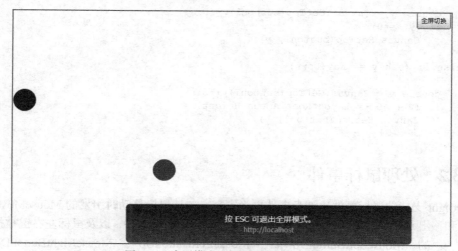

图 16-4　全屏模式运行 Silverlight 应用程序

再如，希望按钮始终位于浏览器窗口的右上角，那么可为 Content 对象的 Resized 事件关联一个处理方法，通过如下代码设置按钮控件在画布中的坐标位置：

```
void Content_Resized(object sender, EventArgs e)
{
    Canvas.SetLeft(button1, content.ActualWidth - button1.Width);
}
```

16.3　处理键盘和鼠标事件

16.3.1　处理键盘事件

Silverlight 应用程序中的用户界面元素都支持各种基本的键盘事件，事件模型和处理方式和 WPF 基本相同。其中最常用的就是 KeyDown 和 KeyUp 事件。

在示例程序 P16_2 中，主界面的画布上放置了一个按钮控件。这里希望实现的功能是：当用户按下键盘上的方向键时，按钮会向相应的方向移动。那么可以为按钮添加 KeyDown 事件处理方法，通过事件参数类 KeyEventArgs 的 Key 属性来判断按下的是哪一个键，并据此执行相应的操作：

```
void button1_KeyDown(object sender, KeyEventArgs e)
{
```

```csharp
        if (e.Key == Key.Left)
        {
            double x = Canvas.GetLeft(button1) - 5;
            if (x >= 0)
                Canvas.SetLeft(button1, x);
        }
        else if (e.Key == Key.Right)
        {
            double x = Canvas.GetLeft(button1) + 5;
            if (x + 80 <= LayoutRoot.ActualWidth)
                Canvas.SetLeft(button1, x);
        }
        if (e.Key == Key.Up)
        {
            double y = Canvas.GetTop(button1) - 5;
            if (y >= 0)
                Canvas.SetTop(button1, y);
        }
        else if (e.Key == Key.Down)
        {
            double y = Canvas.GetTop(button1) + 5;
            if (y + 40 <= LayoutRoot.ActualHeight)
                Canvas.SetTop(button1, y);
        }
    }
```

16.3.2 处理鼠标事件

Silverlight 支持的鼠标事件比键盘事件更为丰富,包括鼠标移动时引发的 MouseMove 事件,鼠标进入和移出元素位置时引发的 MouseEnter 和 MouseLeave 事件,以及鼠标左右键被按下和松开时引发的 MouseLeftButtonDown、MouseLeftButtonUp、MouseRightButtonDown 和 MouseRightButtonUp 事件。

在交互式 Web 程序中,界面元素常常需要支持鼠标拖放的功能。这里希望为程序 P16_2 的按钮控件增加拖放功能,那么首先应当将按钮的 ClickMode 属性值设为 Hover(该属性取默认值 Press,按钮的 Click 事件会屏蔽 MouseLeftButtonDown 等事件),而后为 MainPage 类添加如下两个字段,记录拖放状态及鼠标位置:

```csharp
bool mouseMoving = false;
Point mousePosition;
```

接下来为按钮添加 MouseEnter 和 MouseLeave 事件处理方法;当按钮不处于拖放状态时,光标移到按钮上将变为手形,移出后则恢复默认形状:

```csharp
private void button1_MouseEnter(object sender, MouseEventArgs e)
{
    if (!mouseMoving)
        button1.Cursor = Cursors.Hand;
}
private void button1_MouseLeave(object sender, MouseEventArgs e)
{
    if (!mouseMoving)
        button1.Cursor = Cursors.None;
}
```

在按钮上单击鼠标左键后,鼠标进入可拖曳状态,程序同时记录下当前的鼠标位置:

```csharp
private void button1_MouseLeftButtonDown(object sender, MouseButtonEventArgs e)
{
    mouseMoving = true;
```

```
mousePosition = e.GetPosition(null);
button1.CaptureMouse();
}
```
此后，在移动鼠标的过程中，程序通过计算鼠标的位置来更新按钮的位置：
```
private void button1_MouseMove(object sender, MouseEventArgs e)
{
    if (mouseMoving)
    {
        double dx = e.GetPosition(null).X - mousePosition.X;
        doubledy = e.GetPosition(null).Y - mousePosition.Y;
        Canvas.SetLeft(button1, dx + (double)Canvas.GetLeft(button1));
        Canvas.SetTop(button1, dy + (double)Canvas.GetTop(button1));
        mousePosition = e.GetPosition(null);
    }
}
```
松开鼠标左键后则结束拖放操作：
```
private void button1_MouseLeftButtonUp(object sender, MouseButtonEventArgs e)
{
    mouseMoving = false;
    button1.ReleaseMouseCapture();
    mousePosition.X = mousePosition.Y = 0;
}
```
编译运行程序，在按钮获得焦点后，通过上下左右方向键可在网页上移动按钮，通过鼠标左键则可以直接拖放按钮，如图 16-5 所示。

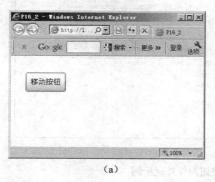
(a)

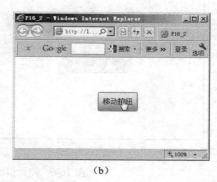
(b)

图 16-5 移动和拖放按钮控件

16.4 模板和自定义控件

16.4.1 使用控件模板

模板则提供了一种强大的重用技术。例如在下面定义的按钮控件就使用了控件模板（ControlTemplate）来设置外观，包括通过 Border 元素设置边框风格，以及通过 ContentPresenter 设置内容的排列风格，按钮的外观效果如图 16-6（a）所示。
```
<Button Content="模板控件" Width="80" Height="40">
  <Button.Template>
    <ControlTemplateTargetType="Button">
      <Border        Background="Yellow"BorderBrush="LightBlue"        BorderThickness="3" CornerRadius="10">
        <ContentPresenter VerticalAlignment="Center"HorizontalAlignment="Center"/>
```

```
      </Border>
    </ControlTemplate>
  </Button.Template>
</Button>
```

当然,为了方便重用,更常见的做法是将控件模板定义为资源,并通过特性 Key 为其指定键值:

```
<UserControl.Resources>
  <ControlTemplate TargetType="Button" x:Key="Temp1" >
    <Border BorderBrush="LightBlue" BorderThickness="3" CornerRadius="10" Background="Yellow">
      <ContentPresenter VerticalAlignment="Center" HorizontalAlignment="Center"/>
    </Border>
  </ControlTemplate>
</UserControl.Resources>
```

这样,不同的控件就可以通过 StaticResource 表达式来指定要使用的模板对象。例如,下面代码所设置的按钮的外观效果如图 16-6(b)所示。

```
<Button Content="进入简单模式" Width="100" Height="40" Template="{StaticResource Temp1}" />
<Button Content="进入双人模式" Width="150" Height="40" Foreground="Blue" Template="{StaticResource Temp1}" />
<Button Content="进入群组联网模式" Width="200" Height="40" FontSize="16" Template="{StaticResource Temp1}" />
```

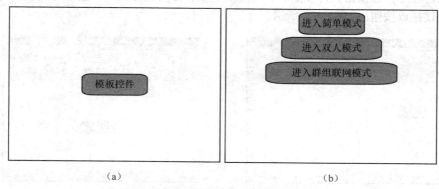

图 16-6　使用模板的按钮控件外观示例

 样式 Style 可以应用于 WPF 或 Silverlight 的所有元素(UIElement 的派生类),而模板 Template 只能应用于控件(Control 的派生类),因此也叫控件模板。

模板还可以定义为样式的一部分,例如:

```
<Style TargetType="Button" x:Key="Style1">
  <Setter Property="Foreground" Value="Blue" />
  <Setter Property="Template">
    <Setter.Value>
      <ControlTemplate TargetType="Button">
        <Border BorderBrush="LightBlue" BorderThickness="3" CornerRadius="10" Background="Yellow" Margin="1">
          <ContentPresenter VerticalAlignment="Center" HorizontalAlignment="Center"/>
        </Border>
      </ControlTemplate>
    </Setter.Value>
  </Setter>
</Style>
```

这样就可以通过样式来为控件应用预定义的模板，例如：
```
<Button Content="模板控件" Width="100" Height="40" Style="{StaticResource Style1}" />
```

16.4.2 创建自定义控件

Sliverlight 的自定义控件属于"无外观的控件"，其外观样式由控件模板定义。新建一个 Sliverlight 应用程序项目 P16_3，向项目中添加新项，在如图 16-7 的对话框中选择"模板化控件"，指定控件名称为 SpriteControl，单击"添加"按钮，Visual Studio 就会创建相应的控件定义。

图 16-7 向 Silverlight 项目添加模板化控件

上面创建自定义控件的类文件是 SpriteControl.cs，从代码中可以看到该类继承了 Control 类，其构造函数中通过 DefaultStyleKey 指定了控件所使用的样式：

```
public class SpriteControl : Control
{
    public SpriteControl()
    {
        this.DefaultStyleKey = typeof(SpriteControl);
    }
}
```

在 Sliverlight 程序项目中，所有自定义控件的样式统一放在 Themes 目录下的 Generic.xaml 文件中定义。该文件实际上是一个 xaml 资源字典，其中包含一组样式定义，每个样式通过 TargetType 特性关联到对应的自定义控件：

```
<Style TargetType="local:SpriteControl">
  <Setter Property="Template">
    <Setter.Value>
      <ControlTemplate TargetType="local:SpriteControl">
        <Border Background="{TemplateBinding Background}"BorderBrush="{TemplateBindingBorderBrush}" />
      </ControlTemplate>
    </Setter.Value>
  </Setter>
</Style>
```

上述类型和样式定义都是 Visual Studio 为自定义控件自动生成的，开发人员可根据需要对其进行修改和扩充。这里希望 SpriteControl 类能够在画布中移动，那么可为其添加如下的属性和方法定义，其中 X 和 Y 分别表示控件的横坐标合纵坐标，Vx 和 Vy 表示水平和垂直方向的移动速度，Rect 表示移动的范围，Move 方法用于在指定范围内移动控件（到达边界后自动折返），Intersect 则用于判断两个控件的位置是否相交：

```csharp
public double X
{
    get { return Canvas.GetLeft(this); }
    set { Canvas.SetLeft(this, value); }
}

public double Y
{
    get { return Canvas.GetTop(this); }
    set { Canvas.SetTop(this, value); }
}

public virtual RectRect
{
    get { return new Rect(X, Y, this.Width, this.Height); }
}

public double Vx { get; set; }
public double Vy { get; set; }
public virtual void Move(Rect border)
{
    double x = X + Vx;
    if (x >= border.Left&& x + this.Width<= border.Right)
        X = x;
    else
        Vx = -Vx;
    double y = Y + Vy;
    if (y >= border.Top&& x + this.Height<= border.Bottom)
        Y = y;
    else
        Vy = -Vy;
}

public virtual bool Intersect(SpriteControl s)
{
    Rectrect = this.Rect;
    rect.Intersect(s.Rect);
    return !rect.IsEmpty;
}
```

接下来我们在 SpriteControl 的基础上定义飞船控件 Boat 和推车控件 Cart。向项目中添加这样两个模板化控件，将控件的基类由 Control 改为 SpriteControl：

```csharp
public class Boat : SpriteControl
{
    public Boat()
    {
        this.DefaultStyleKey = typeof(Boat);
    }
}

public class Cart : SpriteControl
{
    public Cart()
    {
        this.DefaultStyleKey = typeof(Cart);
    }
}
```

切换到 Generic.xaml 文件，修改控件模板定义，使用指定的图片来设置控件外观：

```xml
<Style TargetType="local:Boat">
  <Setter Property="Template">
    <Setter.Value>
      <ControlTemplate TargetType="local:Boat">
        <Image Source="Image/boat.png" Stretch="Uniform"></Image>
      </ControlTemplate>
    </Setter.Value>
  </Setter>
</Style>

<Style TargetType="local:Boat">
  <Setter Property="Template">
    <Setter.Value>
      <ControlTemplate TargetType="local:Boat">
        <Image Source="Image/cart.png" Stretch="Uniform"></Image>
      </ControlTemplate>
    </Setter.Value>
  </Setter>
</Style>
```

这样就可以在 Silverlight 程序界面上使用上面定义好的控件了。值得注意的是，如果要在 XAML 代码文件中添加自定义控件，那么应在根元素 UserControl 中通过 xmlns 特性引入当前命名空间：

```xml
<UserControl x:Class="P16_3.MainPage" xmlns:local="clr-namespace:P16_3"…>
```

之后才能通过命名空间引用访问所定义的控件类型，例如：

```xml
<Canvas Name="canvas1">
  <local:Boat x:Name="boat" Canvas.Left="330" Canvas.Top="5" Width="150" Height="120" />
  <local:Cart x:Name="cart" Canvas.Left="0" Canvas.Top="220" Width="100" Height="100" />
</Canvas>
```

切换到 MainPage.xaml.cs 文件，按照下面的代码创建一个定时器控件 timer，通过其 Tick 事件处理方法使飞船控件在画布上自由移动，并在用户按下左右方向键时移动推车控件：

```csharp
public partial class MainPage : UserControl
{
    System.Windows.Threading.DispatcherTimer timer;
    Rect border;
    Random rand = new Random();

    publicMainPage()
    {
        InitializeComponent();
        boat.Vx = 2;
        border = new Rect(0, 0, canvas1.ActualWidth, canvas1.ActualHeight);
        timer = new DispatcherTimer();
        timer.Interval = TimeSpan.FromMilliseconds(40);
        timer.Tick += new EventHandler(timer_Tick);
        timer.Start();
    }

    private void UserControl_Loaded(object sender, RoutedEventArgs e)
    {
        App.Current.RootVisual.KeyDown += new KeyEventHandler(RootVisual_KeyDown);
        App.Current.RootVisual.KeyUp += new KeyEventHandler(RootVisual_KeyUp);
    }

    voidRootVisual_KeyDown(object sender, KeyEventArgs e)
    {
        if (e.Key == Key.Left)
            cart.Vx = -5;
```

```
        else if (e.Key == Key.Right)
            cart.Vx = 5;
    }

    voidRootVisual_KeyUp(object sender, KeyEventArgs e)
    {
        cart.Vx = 0;
    }

    voidtimer_Tick(object sender, EventArgs e)
    {
        boat.Move(border);
        if (cart.Vx != 0)
            cart.Move(border);
    }
}
```

编译运行程序，可以在网页中看到自定义控件的外观和动画效果。继续扩展该程序，加入更多的自定义控件，增加更丰富的控制方式和界面效果，就能够实现一个具有较强交互性和趣味性的网页游戏。在完整的程序 P16_3 中，各种自定义的卡通动物从飞船上不断下落，用户需要操纵推车来接住这些动物，程序运行效果如图 16-8 所示（详细的实现代码可在配书源程序中找到）。

图 16-8　程序 P16_3 运行效果示例

16.5　案例研究——使用必应地图服务

必应地图（Bing Maps）是基于 Silverlight 的在线地图控件，利用它能够在自己的 Web 程序中方便地集成地图应用，或进行相关的二次开发。本节将在旅行社网站中嵌入必应地图，帮助用户方便地浏览各地的旅游景点。

16.5.1　开发前的准备工作

基于必应地图开发 Silverlight 应用程序的第一步就是下载并安装 Bing Maps Silverlight Control SDK，下载地址为 http://www.microsoft.com/en-us/download/details.aspx?id=2949。

安装完毕后，要在自己开发的程序项目中使用地图控件，还需要有一个开发者许可的密钥 Key。获取 Key 的方式也很简单，访问 Bing Maps 账户中心网址 https://www.bingmapsportal.com/，在其中使用 Windows Live ID（MSN）账号登录（如没有账号可先申请一个），点击网页左上角"My Account"下的"Create or view keys"链接，按要求输入基本使用信息，即可创建一个 Key，如图 16-9 所示（一个 Windows Live ID 最多可创建 5 个 Key）。

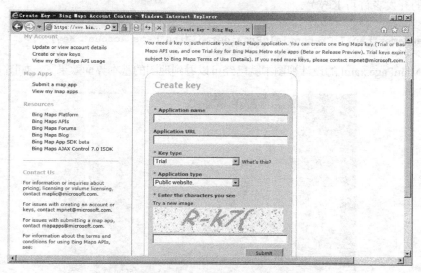

图 16-9　访问 Bing Maps 账户中心

16.5.2　创建程序并添加必应地图控件

打开旅行社管理系统的 Visual Studio 解决方案，在其中新建一个 Silverlight 应用程序项目 TravelSL，在如图 16-10 所示的对话框中选择"在解决方案中的新网站或现有的 TravelWeb 网站上承载 Silverlight 应用程序"复选框，且取消选择"添加引用该应用程序的测试页"，单击"确定"按钮完成项目创建。

接下来向项目添加必应地图控件的相关程序集引用。在如图 16-11 所示的"添加引用"对话框中，定位到 Bing Maps Silverlight Control SDK 的安装目录，选择 Microsoft.Maps.MapControl.dll 和 Microsoft.Maps.MapControl.Common.dll 这两个文件，单击"确定"按钮即可。

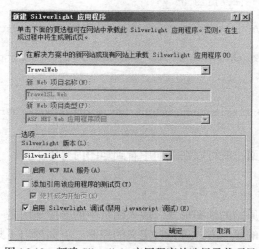

图 16-10　新建 Silverlight 应用程序并选择承载项目

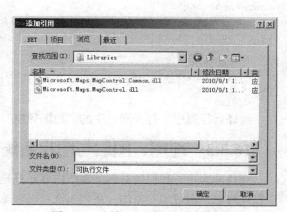

图 16-11　添加 Bing Maps 程序集引用

打开 MainPage.xaml 文件，在根元素 UserControl 中也通过 xmlns 指令引入命名空间：

```
xmlns:map="clr-namespace:Microsoft.Maps.MapControl;assembly=Microsoft.Maps.MapControl"
```

再通过如下的代码就在网格中放入了一个必应地图控件（中括号里的内容需要替换成前一小节中所创建的开发者许可 Key）：

```
<Grid x:Name="LayoutRoot" Background="White">
<map:Map Name="map1" CredentialsProvider="[Bing Maps Account Key]" />
</Grid>
```

此时在 MainPage.xaml 的设计视图中就已经可以看到地图的基本界面效果了，如图 16-12 所示。

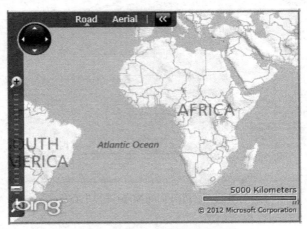

图 16-12　必应地图控件加载效果

下面将包含地图内容的 Silverlight 用户控件显示在 Web 页面中。转到 Web 程序项目 TravelWeb，向其中加入一个使用母版页的 Web 窗体 ViewMap.aspx。然后切换到该页面的代码视图，在其 MainContent 控件中嵌入以下内容（也可以从自动生成的 Silverlight 程序测试页中复制代码）：

```
<div id="silverlightControlHost">
    <object data="data:application/x-silverlight-2," type="application/x-silverlight-2" width="100%" height="100%">
        <param name="source" value="ClientBin/TravelSL.xap"/>
        <param name="background" value="white" />
        <param name="minRuntimeVersion" value="5.0.61118.0" />
        <param name="autoUpgrade" value="true" />
        <a href="http://go.microsoft.com/fwlink/?LinkID=149156&v=5.0.61118.0" style="text-decoration:none">
            <img src="http://go.microsoft.com/fwlink/?LinkId=161376" alt="获取 Microsoft Silverlight" style="border-style:none"/>
        </a>
    </object>
    <iframe id="_sl_historyFrame" style="visibility:hidden;height:0px;width:0px;border:0px"></iframe>
</div>
```

编译运行程序，打开网站中的"地图导航"页面，就可以在页面中看到地图内容了。

16.5.3　地图、图层与图片系统

到目前为止，我们所看到的地图内容还只是最原始的地图形式。嵌入网页中的必应地图是一个控件对象，通过该对象的一系列属性可以对地图的显示方式进行自定义的控制。下面列出了其中常用的一些属性。

- ZoomLevel：地图放大级数，有效值为1~16。
- Center：地图初始显示时的中心位置，以二维坐标（经纬度）格式显示。错误的经纬度数值会引发异常。
- NavigationVisibility：是否在地图上显示导航工具栏。
- Mode：地图显示模式，包括路况模式Road和卫星模式Aerial。

提示　　必应地图新推出的Beta版本还支持两种新的模式：鸟瞰模式BirdseyeMode和街道模式StreetsideMode。不过这需要下载支持扩展模式的Bing Maps Silverlight控件，以及支持扩展模式的组件 Microsoft.Maps.MapControl.ExtendedModes.dll。详情可访问Microsoft Connect必应地图站点http://connect.microsoft.com/silverlightmapmodesbeta。

当然，上述属性即可也在XMAL代码中直接设置，也可以在C#代码中进行动态控制。例如北京的位置坐标大约是（39.909，116.397），那么采用如下代码来设置地图控件，则地图将以北京为中心、放大4倍显示：

```
<map:Map Name="map1" Center="39.909,116.397" ZoomLevel="4" CredentialsProvider=…/>
```

实际应用中，我们常常需要在基础地图之上显示各种图形、标记等对象，这些对象是通过图层来进行组织的。Microsoft.Maps.MapControl名称空间下的MapLayer类对图层进行了封装。例如下面的XAML代码就在地图中放置了一个图层对象layer1：

```
<map:Map Name="map1" CredentialsProvider=…>
    <map:MapLayer x:Name="layer1" />
</map>
```

地图的主要功能之一就是定位，比如在指定的位置放上一个图钉标记。必应地图的图钉功能是通过Pushpin类来提供的。下面的C#代码就在地图上放置了一个图钉，并将其加入到图层layer1的指定位置，地图及图钉显示效果如图16-13所示：

```
Pushpin pin = new Pushpin();
layer1.AddChild(pin, new Location(39.909, 116.397));
```

图16-13　地图及图层上的图钉标记

和其他主要的在线电子地图一样，必应地图也提供自己的图片（Tile）系统，通过内置的算法对一组预处理好的图片进行无缝拼接，并与基础地图进行映射。图片被放置在一种特殊的图层——图片图层 MapTileLayer 之上。下面的代码在地图上增加了一个图片图层对象 layerChina：

```
<map:Map Name="map1" CredentialsProvider=…>
  <map:MapTileLayer x:Name="layerChina"/>
  <map:MapLayer x:Name="layer1" />
</map>
```

中国地图所对应的图片系统 Url 为 http://r2.tiles.ditu.live.com/tiles/r{quadkey}.png?g=41。切换到 MainPage 的 C#代码，在构造函数中基于此地址和指定矩形区域创建一个 LocationRectTileSource（图片源）对象，并将其加入 layerChina 的 TileSources 集合中：

```
public MainPage()
{
  InitializeComponent();
  Uri uri = new Uri("http://r2.tiles.ditu.live.com/tiles/r{quadkey}.png?g=41");
  LocationRectrect = new LocationRect(new Location(60, 60), new Location(13, 140));
  LocationRectTileSource ts1 = new LocationRectTileSource(uri.ToString(), rect, new Range<double>(1, 16));
  layer1.TileSources.Add(ts1);
}
```

重新编译运行程序，在网页中就可以看到如图 16-14 所示的中文地图内容。

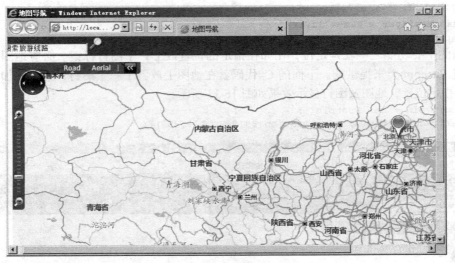

图 16-14　基于图片系统的中文地图内容

16.5.4　旅游景点地图导航

掌握了必应地图控件的基本用法之后，下面进一步规划和完善基于地图的景点导航功能。网页的初始界面是整个中文地图，其中针对每个省的省会显示一个图钉标记；单击标记将进入放大后的各省地图，并为各个景点显示图钉标记。省会和景点标记分别放置在两个不同的图层 layer1 和 layer2 上：

```
<map:Map Name="map1" CredentialsProvider=…>
  <map:MapTileLayer x:Name="layerChina"/>
  <map:MapLayer x:Name="layer1" />
  <map:MapLayer x:Name="layer2" Visibility="Collapsed" />
</map>
```

我们向 TravelSL 项目中加入了一个文本文件 prov.txt，其中记录了各个省份的省会坐标位置：

```
1 北京   39.91 116.40
2 天津   39.09 117.11
3 河北   38.02 114.28
4 山西   37.52 112.34
5 内蒙古 40.49 111.48
6 辽宁   41.50 123.24
...
```

该文件会被打包进 xap 文件并发送到客户端，以便程序能够使用离线数据。Silverlight 控件 MainPage 在启动时将调用下面的方法来逐行读入文本文件，并为每个省会创建一个 Pushpin 对象：

```
voidLoadProvinceData()
{
    StreamResourceInfo res = Application.GetResourceStream(new Uri("prov.txt", UriKind.Relative));
    StreamReader reader = new StreamReader(res.Stream);
    while (!reader.EndOfStream)
    {
        string[] ss = reader.ReadLine().Split(' ');
        Pushpin pin = new Pushpin() { Tag = ss[0] };
        ToolTipService.SetToolTip(pin, ss[1]);
        pin.Location = new Location(double.Parse(ss[2]), double.Parse(ss[3]));
        layer1.AddChild(pin, pin.Location);
        pin.MouseLeftButtonDown += new MouseButtonEventHandler(pin_MouseLeftButtonDown);
    }
    reader.Close();
}
```

在图钉上单击鼠标左键，地图将显示以图钉坐标为中心的区域并自动放大；如果单击的是直辖市，放大倍数为 7；单击其他省份的省会则放大倍数为 10。那么 Pushpin 对象的鼠标左键单击事件处理方法如下：

```
void pin_MouseLeftButtonDown(object sender, MouseButtonEventArgs e)
{
    Pushpin pin = ((Pushpin)sender);
    string id = pin.Tag.ToString();
    if (id == "1" || id == "2" || id == "9" || id == "22")
        map1.ZoomLevel = 10;
    else
        map1.ZoomLevel = 7;
    map1.Center = pin.Location;
}
```

无论是程序自动缩放还是用户手动缩放，当地图放大倍数小于 6 时，包含省会标记的图层 layer1 可见，包含景点标记的图层 layer2 不可见；倍数大于 6 时则正好相反：

```
private void map1_ViewChangeEnd(object sender, MapEventArgs e)
{
    if (map1.ZoomLevel < 6)
    {
        layer1.Visibility = Visibility.Visible;
        layer2.Visibility = Visibility.Collapsed;
    }
    else
    {
        layer1.Visibility = Visibility.Collapsed;
        layer2.Visibility = Visibility.Visible;
    }
}
```

景点坐标位置也是存放在一个文本文件 scenes.txt 中。但这里希望针对每个景点不仅仅是显示一个简单图钉，而是能够链接到景点信息网页 SceneInfo.aspx 上。为此，我们在 Silverlight 程序项目中创建了一个自定义用户控件 ScenePinControl。此控件基于 StackPanel 进行布局，其上方是一个标准的 Pushpin 控件，下方则在一个边框元素中包含了一个 HyperlinkButton 控件：

```
<StackPanel x:Name="LayoutRoot" Background="Transparent">
    <map:Pushpin Name="pin" Background="Blue" HorizontalAlignment="Left" />
    <Border Name="tPanel" BorderBrush="Red" BorderThickness="1.0" Background="White" CornerRadius="6,6,6,6" Visibility="Collapsed">
        <HyperlinkButton Name="linkBtn" Height="30" TargetName="_blank" />
    </Border>
</StackPanel>
```

定义 ScenePinControl 的 C#代码如下，其中 Text 属性表示超链接按钮上的文本，Url 属性表示超链接按钮指向的地址；默认情况下边框和其中的按钮都不可见，只有当用户将鼠标移至图钉对象上时才可见：

```
public partial class ScenePinControl : UserControl
{
    public string Text
    {
        set{linkBtn.Content = value; }
        get { return linkBtn.Content.ToString(); }
    }

    public string Url
    {
        set { linkBtn.NavigateUri = new Uri(value); }
        get { return linkBtn.NavigateUri.ToString(); }
    }

    publicScenePinControl()
    {
        InitializeComponent();
        pin.MouseEnter += (s, e) =>{ tPanel.Visibility = Visibility.Visible; };
        this.MouseLeave += (s, e) =>{ tPanel.Visibility = Visibility.Collapsed; };
    }
}
```

回到 MainPage.xaml.cs 代码文件，下面的方法 LoadSceneData 用于从文本文件读入景点坐标数据，并为每个景点创建一个 ScenePinControl 对象：

```
voidLoadSceneData()
{
    StreamResourceInfo res = Application.GetResourceStream(new Uri("scenes.txt", UriKind.Relative));
    StreamReader reader = new StreamReader(res.Stream);
    stringurl = HtmlPage.Document.DocumentUri.ToString();
    inti = url.LastIndexOf('/');
    url = url.Substring(0, i + 1);
    while (!reader.EndOfStream)
    {
        string[] ss = reader.ReadLine().Split(' ');
        ScenePinControl spc = new ScenePinControl() { Tag = ss[0] };
        spc.Text = ss[1];
        spc.Url = string.Format("{0}SceneInfo.aspx?id={1}", url, ss[0]);
        spc.pin.Location = new Location(double.Parse(ss[2]), double.Parse(ss[3]));
        layer2.AddChild(spc, spc.pin.Location);
    }
    reader.Close();
}
```

重新编译运行程序，打开"地图导航页面"，可以看到如图 16-15 所示的内容；单击省份图钉或放大地图显示倍数超过 6，则可以看到景点的分布信息，如图 16-16 所示。

图 16-15　在地图上查看各省位置

图 16-16　在地图上查看各景点位置

最后，我们希望从各景点的信息网页上也能够进入电子地图。那么可以在 SceneInfo.aspx 页面中加入一个超链接控件，其链接地址为 ViewMap.aspx?id=[id]，其中[id]为景点的编号。而在 Silverlight 程序的 MainPage 构造函数中，载入了坐标数据之后，可添加如下代码，在查询字符串 id 不为空时，突出显示对应的景点图钉：

```
this.LoadProvinceData();
this.LoadSceneData();
if (HtmlPage.Document.QueryString.ContainsKey("id"))
{
    string id = HtmlPage.Document.QueryString["id"];
    map1.ZoomLevel = 10;
    foreach (ScenePinControl spc in layer2.Children)
    {
```

```
            if (spc.Tag.ToString() == id)
            {
                map1.Center = spc.pin.Location;
                spc.pin.Background = new SolidColorBrush(Colors.Red);
                spc.pin.Focus();
                break;
            }
        }
    }
```

16.6 小　　结

Silverlight 是基于.NET Framework 的富客户端 Web 应用程序开发技术。Silverlight 程序以控件方式运行在浏览器中；它和 WPF 程序一样支持各种键盘和鼠标交互事件，也可以使用资源、样式和模板来提高代码的可重用性。使用 Microsoft 提供的必应地图 Silverlight 控件，还能够在 Web 程序中方便地集成地图应用。

16.7 习　　题

1. 简述 ASP.NET 应用程序和 Silverlight 应用程序的相同点和不同点。
2. 简述 WPF 应用程序和 Silverlight 应用程序的相同点和不同点。
3. 编写一个 Silverlight 应用程序，在页面上显示一个彩灯；当用户将鼠标移到彩灯上时，彩灯将不断地旋转和变色。
4. 编写一个 Silverlight 应用程序，在页面上绘制一些基本图形，让用户能够使用鼠标来拖曳这些图形。
5. Silverlight 应用程序运行在客户端，而其重载的 Web 程序运行在服务器端。如果要从一个程序调用另一个程序的功能，这称为跨域访问。试探讨 Silverlight 应用程序有哪些跨域访问的手段。
6. 扩展程序 P16_3，为其增加如下功能：
（1）飞船变速运动，且高度可以发生变化。
（2）用户赢得足够的分数后，游戏进入下一关，且游戏难度增大（如飞船速度和羊的下落速度加快）。
（3）用户连续通过若干关卡之后游戏结束，并播放一个庆祝视频。
7. 继续扩展旅行社网站功能，使用户能够在地图上查看旅游线路所经过的景点位置，并通过线条将这些景点连接起来。